Translation in Mitochondria
and Other Organelles

Anne-Marie Duchêne

Editor

Translation in Mitochondria and Other Organelles

Springer

Editor
Anne-Marie Duchêne
Institut de Biologie Moleculaire
  des Plantes
Strasbourg Cedex
France

QH
450.5
-T7
2013

ISBN 978-3-642-39425-6   ISBN 978-3-642-39426-3   (eBook)
DOI 10.1007/978-3-642-39426-3
Springer Heidelberg New York Dordrecht London

Library of Congress Control Number: 2013947348

Printed on acid-free paper

Springer is part of Springer Science+Business Media (www.springer.com)

Contents

Chapter 1
Insights into Structural Basis of Mammalian Mitochondrial Translation

Manjuli R. Sharma, Prem S. Kaushal, Mona Gupta, Nilesh K. Banavali and Rajendra K. Agrawal

Abstract Mitochondrial ribosomes are known to be quite divergent from cytoplasmic ribosomes in both composition and structure even as their main functional cores, such as the mRNA decoding and peptidyl transferase sites, are highly conserved. The translational factors that interact with these ribosomes to facilitate the process of protein synthesis in mitochondria have also likewise acquired unique structural features, apparently to complement the structure and function of the mitochondrial ribosome. In this chapter, we describe the current state of structural knowledge of the mammalian mitochondrial ribosome, some of its component proteins, and key translational factors.

1.1 Introduction

Mitochondrial ribosomes (mitoribosomes) are responsible for synthesizing a limited number of polypeptide chains, which form essential components of the complexes involved in oxidative phosphorylation (OXPHOS). The OXPHOS complexes reside in the mitochondrial inner membrane (mtIM) and are responsible for generating about 90 % of the energy (ATP) required by the cell. All proteins required for mammalian mitochondrial translation, including the mitochondrial

M. R. Sharma · P. S. Kaushal · M. Gupta · R. K. Agrawal (✉)
Division of Translational Medicine, Wadsworth Center,
New York State Department of Health, Albany, NY 12201-0509, USA
e-mail: agrawal@wadsworth.org

N. K. Banavali
Division of Genetics, Wadsworth Center, New York State Department of Health,
Albany, NY 12201-0509, USA

N. K. Banavali · R. K. Agrawal
Department of Biomedical Sciences, School of Public Health,
State University of New York at Albany, Albany, NY, USA

A.-M. Duchêne (ed.), *Translation in Mitochondria and Other Organelles*,
DOI: 10.1007/978-3-642-39426-3_1, © Springer-Verlag Berlin Heidelberg 2013

ribosomal proteins (MRPs), are encoded by the nuclear genome, translated in the cytoplasm and then imported into the mitochondria. The mammalian mitochondrial genome is relatively small (16.8 kb) and encodes for 37 genes, including two ribosomal RNAs (12S and 16S rRNAs), 22 mitochondrial transfer RNAs (tRNA$_{mt}$), and 13 polypeptides of the OXPHOS complexes. Unlike cytoplasmic ribosomes, whose X-ray crystallographic structures are known for several bacterial (e.g., Schuwirth et al. 2005; Selmer et al. 2006), archaeal (Ban et al. 2000) and eukaryotic (e.g., Ben-Shem et al. 2011; Klinge et al. 2011) species, the three-dimensional (3D) structures of organellar ribosomes have been studied primarily using the single-particle cryo-electron microscopy (cryo-EM) and molecular modeling (Sharma et al. 2003, 2007, 2009; Mears et al. 2006; Manuell et al. 2007). This chapter concerns the mammalian mitoribosome, whose 3D cryo-EM structure is known for *Bos taurus* (Sharma et al. 2003). Several components of the mammalian mitochondrial translational machinery, including mito-rRNAs, MRPs, tRNA$_{mt}$s, tRNA$_{mt}$ synthetases and translational factors, have been associated with a number of human genetic diseases (see O'Brien 2002; O'Brien et al. 2005; Pearce et al. 2013; Watanabe 2010). Multiple review articles have been published recently on the structure of the mammalian mitoribosome (Agrawal and Sharma 2012; Agrawal et al. 2011; Christian and Spremulli 2012; Koc et al. 2010). In the present article, we first elucidate the unique characteristic of the cryo-EM structure of the mammalian mitoribosome in comparison to those of the cytoplasmic ribosomes. We then summarize some of the mammalian mitochondrial translation factors that interact with the mitoribosome during translation and discuss their singular features. Description of the mitochondrial translation machinery is mainly presented with reference to bacterial translation since the overall mechanism for translation in mitochondria is more similar to that in bacteria rather than to that in eukaryotic cytoplasm.

1.2 The Mammalian 55S Mitoribosome

Mitochondria are thought to originate through an early endosymbiotic event (~1.8 billion years ago) between an α-protobacteria and a primitive host cell (Gray et al. 2001), and therefore, mitoribosomes were proposed to be structurally similar to their bacterial counterparts. The mammalian mitochondrial rRNAs are smaller than those in bacteria, and the MRPs are generally both larger (when they are homologous to bacteria) and greater in number. It was believed that the overall structural organization of the mitoribosome would still be very similar to its bacterial counterparts, except that the additional MRPs would structurally replace the missing bacterial rRNA segments that were deleted in the mito-rRNAs during the course of evolution. The first cryo-EM structure determined for the 55S mammalian mitoribosome revealed that this paradigm about the structural replacement of deleted bacterial rRNA segments by MRPs was only partially valid, as about 80 % of the missing bacterial rRNA segments were not found to be replaced by the MRPs (Sharma et al. 2003). In general, the loss of rRNA segments correspond to the loss of interacting bacterial proteins, suggesting a complementary evolution of the binding ribonucleoprotein (RNP) partners (Mears et al. 2006). Furthermore, the overall 3D structure of the mammalian mitoribosome was found to be significantly altered and highly porous as compared to its bacterial counterpart (Sharma et al. 2003; for recent reviews also see Agrawal et al. 2011; Agrawal and Sharma 2012), primarily due to the occupation of new spatial positions by mito-specific MRPs and extensions and insertions within several of MRPs that are homologous to their bacterial counterparts. However, the conserved mito-rRNA segments that form the functional core of a ribosome are intact in their spatial positions in the mitoribosome structure, except for few conserved rRNA segments that are present in the peripheral regions of the structure (Mears et al. 2006; Agrawal et al. 2011). Even though the molecular mass of the mammalian mitoribosome (~2.71 MDa) is very similar to that of a bacterial ribosome (~2.3 MDa) its overall dimension is much larger by about 60 Å in diameter. Its dimension is similar to that of a eukaryotic (yeast) cytoplasmic ribosome (Fig. 1.1) with a much higher molecular mass (~3.3 MDa). These observations are consistent with the observed porousness of the mammalian mitoribosome structure (Sharma et al. 2003).

Like any other known ribosome, the 55S mitoribosomes are made up of two unequally sized subunits: the small 28S subunit (SSU) composed of a 12S rRNA-molecule and about 30 nuclear genome encoded MRPs, and the large 39S subunit (LSU) composed of a 16S rRNA molecule and about 50 nuclear genome encoded MRPs. Since all MRPs are imported into the mitochondria, most possess a mitochondrial targeting sequence (MTS) at their N-terminus (Claros and Vincens 1996). Recent studies suggest that many MRPs and mitochondrial translational factors undergo acetyl-CoA-, NAD^+-, and ATP-dependent post-translational modifications in mitochondria, primarily acetylation and phosphorylation (see, Koc and Koc 2012, and Chap. 2), implying that these modifications play regulatory roles in mitochondrial translation. MRPs also participate in the formation of several mito-specific protein–protein bridges between the SSU and LSU (Sharma et al. 2003). A side-by-side comparison with the structures of the cytoplasmic ribosomes (Fig. 1.1) reveals that the exterior of the mammalian mitoribosome is predominantly shielded by MRPs, with fewer exposed rRNA regions. The inter-subunit space, which includes the mRNA decoding (Ogle et al. 2001) and peptidyl transferase (Nissen et al. 2000) sites, and tRNAs and translation factors binding sites, has a relatively conserved composition and architecture with more exposed rRNA regions (Fig. 1.1d–i). The first average structure of a mitoribosome-bound $tRNA_{mt}$ revealed by the cryo-EM study (Sharma et al. 2003) showed a canonical L-shape, but with a "caved-in" elbow region, consistent with generally reduced size of D- and/or T-stem loops in $tRNA_{mt}$s (Watanabe 2010). The corridor of the inter-subunit space, i.e., the space between the shoulder of SSU and stalk base of LSU (marked with 'sh' and 'Sb', respectively, in Fig. 1.1a–c), where most large translation factors are known to interact with the cytoplasmic ribosomes (Agrawal et al. 1998, 2000; Datta et al. 2005; Stark et al. 1997; Valle et al. 2002; Gomez-Lorenzo et al. 2000; Spahn et al. 2004; Gao et al. 2009; Schmeing et al. 2009;

Yassin et al. 2011b; Yokoyama et al. 2012), has a slightly wider opening in the mitoribosome primarily due to the deletion of bacterial rRNA segments from the shoulder region of its SSU.

The unique features of two mitoribosomal subunits and some of their MRPs are described in the next two sections. Our laboratory has been modeling 3D structures of MRPs and docking them into our most recent 7.0 Å resolution cryo-EM map to develop a pseudo-atomic model of the 55S mitoribosome. Here our focus is primarily on some of the homologous MRPs for which atomic structures and relative positions within the bacterial ribosome are known and which possess mito-specific insertion and/or extension segments.

1.2.1 The 28S Small Subunit

The overall shape of the mitoribosome small subunit (SSU) is dramatically altered compared to bacterial ribosomes (Fig. 1.1d–f) due to the deletion of a significant portion of bacterial rRNA segments (Fig. 1.2a–c). These deletions amount to almost 40 % of the SSU rRNA as there are 1,542 nucleotides (nts) in E. coli versus 950 nts in mammalian mito-SSU. The decoding-site region, which is the central functional core of the SSU (marked with asterisks in Fig. 1.1d–f) is conserved but the peripheral architecture of the mito-SSU is significantly altered. The intactness of all three characteristic domains of a SSU, the body (b), head (h) and platform (pt), suggests that the basic architecture of SSU is important for its function, despite the fact that the diameter of the mito-SSU body is reduced and its shoulder region is significantly diminished due to the absence of the bacterial rRNA helices 16 and 17.

The bovine mitoribosome contains 31 MRPs (named S1 to S31), of which 14 are bacterial homologues and 17 MRPs are specific to mitoribosome (Koc et al. 2000; Koc et al. 2001a; Suzuki et al. 2001a; Emine Koc, personal communication). One of the homologous SSU MRPs, S18 is present in three-isoforms, referred to as MRP S18A, S18B and S18C (Koc et al. 2001a). There is no homologue for the six bacterial ribosomal SSU proteins S3, S4, S8, S13, S19 and S20 in the mitoribosome. The mito-specific MRPs significantly contribute to the unique architecture of the mito-SSU by primarily occupying its head, lower body and solvent side (Sharma et al. 2003) (Fig. 1.1a–f). Addition of significantly large mass of mito-specific MRPs to the lower body is primarily responsible for an overall elongate shape and an unusually large dimension of the mito-SSU along its long axis.

The homologous SSU MRPs possess a varying size range of amino acid extension segments (8–215 amino acid) at their N- and/or C-terminus, except for MRPs S6 and S12, which are shorter in the mitoribosome (Table 1.1; Fig. 1.2d). For, example, The MRP S9, located in the head of SSU, carries the longest N-terminus extension (NTE) of 215 amino acid residues, which can be accounted by the large extra mass of mito-specific protein cryo-EM density present in the immediate vicinity of the protein's homologous segment (Sharma et al. 2003). Among all homologous SSU MRPs, S7 is the only MRP that has a 20 amino acid insertion within its

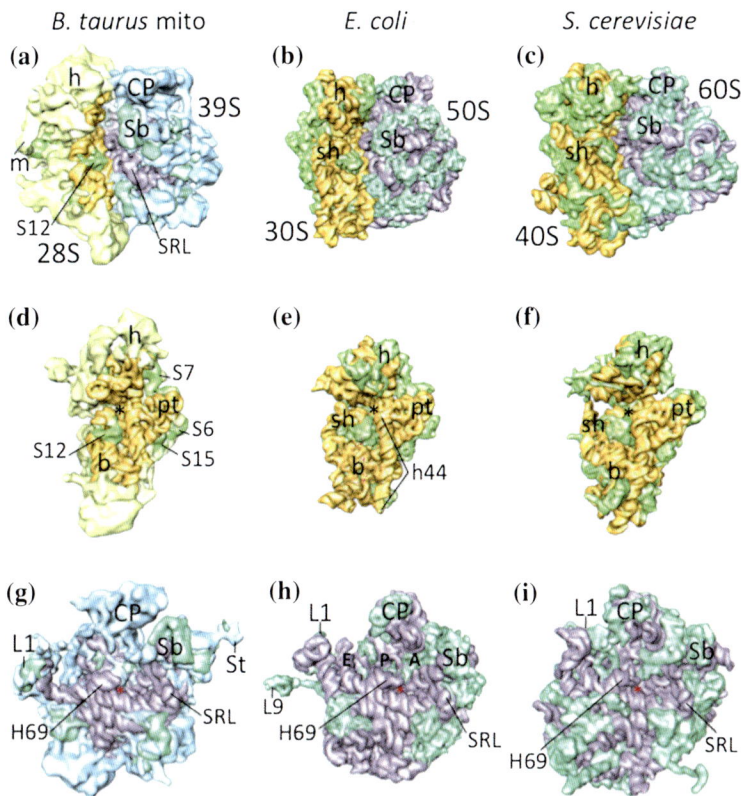

Fig. 1.1 A side-by-side comparison of structures of the mammalian mitochondrial ribosome with cytoplasmic (bacterial and eukaryotic) ribosomes. RNA–protein segmented structures of ribosomes from **a** mammalian mitochondria (55S, *Bos taurus*), **b** bacteria (70S, *Escherichia coli*), and **c** yeast cytoplasm (80S, *Sachharomyces cerevisiae*) displayed with SSU on the *left* and LSU on the *right* side, as viewed from the SSU shoulder (sh) and LSU stalk base (Sb) sides. In **b** and **c**, atomic structures of the 70S (PDB ID codes 1VOX-Y) and 80S (PDB ID codes 3U5B-E and 3IZS) ribosomes have been low-pass filtered to roughly match the resolution of the cryo-EM map of the mammalian mitoribosome shown in **a**. **d–f** Structures of SSUs from the corresponding top panel, but shown from the SSU-LSU interface side. **g–i** Structures of LSUs from corresponding top panel, but shown from the LSU-SSU interface side. Mito-specific MRPs of SSU and LSU are colored *yellow* and *blue*, respectively; conserved ribosomal proteins [here "conserved" refers to bacterial homologues present in the mitoribosome] of SSU and LSU are colored *green* and *aquamarine*, respectively; and rRNAs of SSU and LSU are colored *orange* and *purple*, respectively. Landmarks of SSU: *b* body, *h* head, *m* mRNA gate, *pt* platform, *sh* shoulder, *S12* protein S12, *Asterisks* in **d–f** point to the location of mRNA decoding site. Landmarks of LSU: *CP* central protuberance, *L1* protein L1 stalk (note the change in conformation of the L1 stalk, which is known to be a highly dynamic structure), *L9* protein L9, *Sb* Stalk base or MRP L11 region, *St* L7/L12 stalk (this region was disordered in the crystallographic structures of the 70S and 80S ribosomes, and hence absent in panels H and I), *SRL* α-sarcin-ricin stem loop, *H69* LSU rRNA helix 69, *A*, *P* and *E* in **h** indicate the positions of three canonical tRNA-binding sites (note the difference in the region analogous to E site in panel **g**), *Red asterisks* in **g–i** indicate the general location of the peptidyltransferase center. Spider (Frank et al. 2000) and Chimera (Pettersen et al. 2004) software were used respectively for surface representation and visualization of the ribosome maps

Table 1.1 Homologous MRPs of the SSU

Protein	Accession number	Length (residues)	MTS	Mito-specific segments	Identity/ similarly[a] (%)
MRP S2	P82923	293	1–49	50–75, 277–293	30/43
MRP S5	Q2KID9	430	1–88	89–209, 377–430	27/42
MRP S6	P82931	124	–	–	26/32
MRP S7	Q3T040	242	1–38	39–81, 121–141	34/56
MRP S9	Q58DQ5	396	1–52	53–267	40/53
MRP S10	P82670	201	–	1–75, 175–201	32/51
MRP S11	F1N498	197	1–24	25–68,	43/59
MRP S12	Q29RU1	139	1–30	–	52/63
MRP S14	Q6B860	128	–	1–27	37/55
MRP S15	E1BBB4	256	1–53	54–94, 188–256	28/56
MRP S16	P82915	135	1–15	97–135	45/68
MRP S17	E1BF33	130	–	90–130	29/51
MRP S18A	F1MJC2	196	1–38	39–60, 137–196	40/58
MRP S18B	F1N059	258	1–21	22–95, 170–258	28/54
MRP S18C	P82917	143	1–39	40–51, 126–143	55/71
MRP S21	P82920	87	–	1–30, 80–87	32/51

MTS mitochondrial targeting sequence
[a] With homologous proteins from *E. coli* SSU

conserved domain (Fig. 1.2d) that is tightly accommodated in an additional mass of density observed within the cryo-EM map (unpublished results from our laboratory).

One of the unique features of the mammalian mitochondrial translation is that most of its mRNAs lack a 5′ untranslated region (5′-UTR) (Temperley et al. 2010). In the cryo-EM structure of the mito-SSU the mRNA entrance has a unique gate-like feature, which is made up of mito-specific MRPs (Sharma et al. 2003; marked by "m" in Fig. 1.1a), which may be directly involved in the recruitment of mitochondrial mRNAs to initiate the translation. In the bacterial ribosome, proteins S3, S4 and S5 occupy the mRNA entrance (Yusupova et al. 2001). Of these, proteins homologous to S3 and S4 are absent in the mitoribosome, but are replaced by certain other mito-specific MRPs or the extensions of homologous MRPs. For example, MRP S5 has 120 and 52 amino acid NTE and *C*-terminal extension (CTE), respectively, that may partially compensate for the loss of bacterial S4 and S8 proteins that reside on opposite sides of S5 in the bacterial ribosome (Wimberly et al. 2000). However, these extensions would not account for the SSU's gate-like feature, which can only be explained by the presence of additional mito-specific MRPs that are yet to be identified.

1.2.2 The 39S Large Subunit

About 50 % of bacterial rRNA segments are also deleted in the mito-large subunit (LSU) (3,024 nts in *E. coli* versus 1,560 nts in mammalian mito-LSU) (Fig. 1.3a–c),

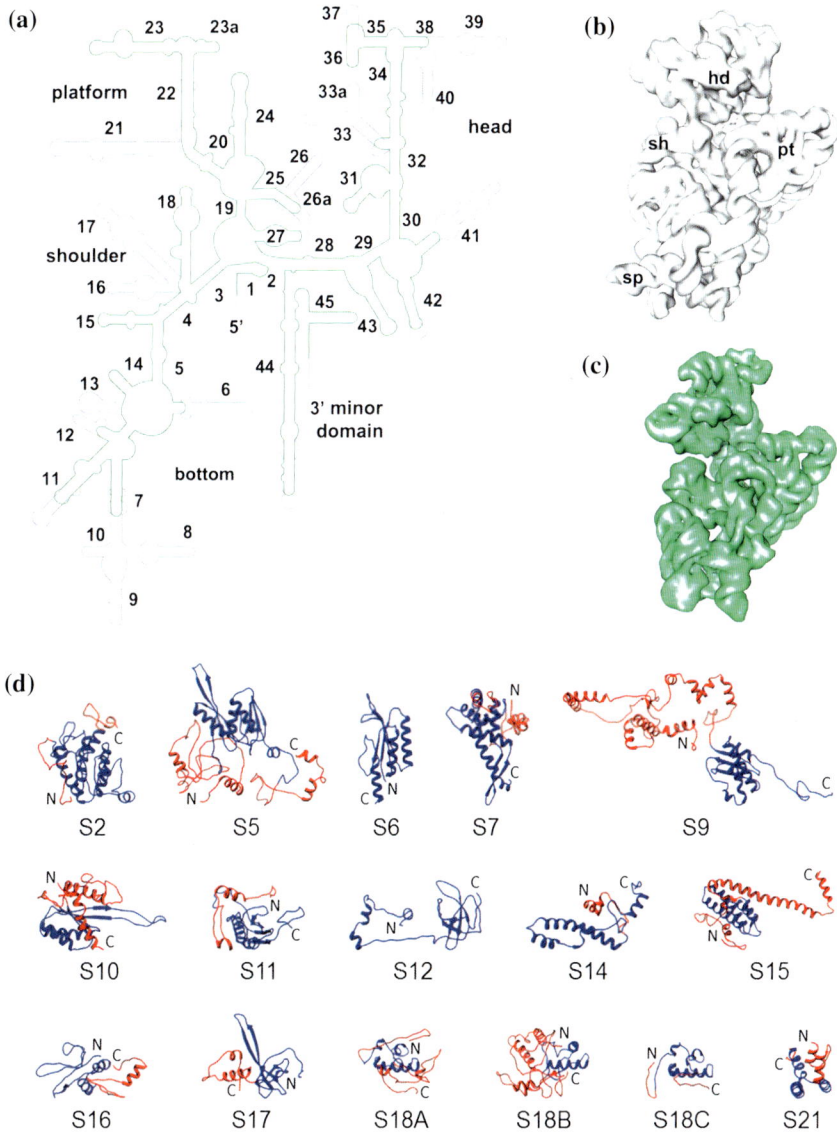

Fig. 1.2 Structural components of the mammalian mito-SSU. **a** Secondary structures of small ribosomal subunit RNAs from bacteria (16S, *grey*) and mammalian mitochondria (12S, *green*). Portions of rRNA that form components of the SSU head, body (*shoulder* and *bottom*), and platform are indicated. The rRNA helices are numbered according to bacterial 16S rRNA. **b–c** Three dimensionally folded structures of the bacterial (**b**) and mammalian mitochondrial (**c**) SSU rRNAs, as derived from X-ray crystallography (PDB ID code 1VOX) and cryo-EM (based on Sharma et al. 2003, and Sharma et al., manuscript in preparation) studies, respectively. The head (hd), platform (pt), shoulder (sh), and spur (sp) regions are labeled. Note the diminished sh region in the mito-rRNA fold due to the absence of bacterial rRNA helices 16 and 17, and a channel created in the main body due to absence of bacterial rRNA helices 12 and 21. **d** *Ab initio* models of the homologous SSU MRPs as generated using I-TASSER (Roy et al. 2010), but not fitted into the cryo-EM of the mito-SSU. The MRP segments homologous to bacterial ribosomal proteins are depicted in *blue* and the mito-specific insertions/extensions are shown in *red*. N- and C-termini are identified in each case

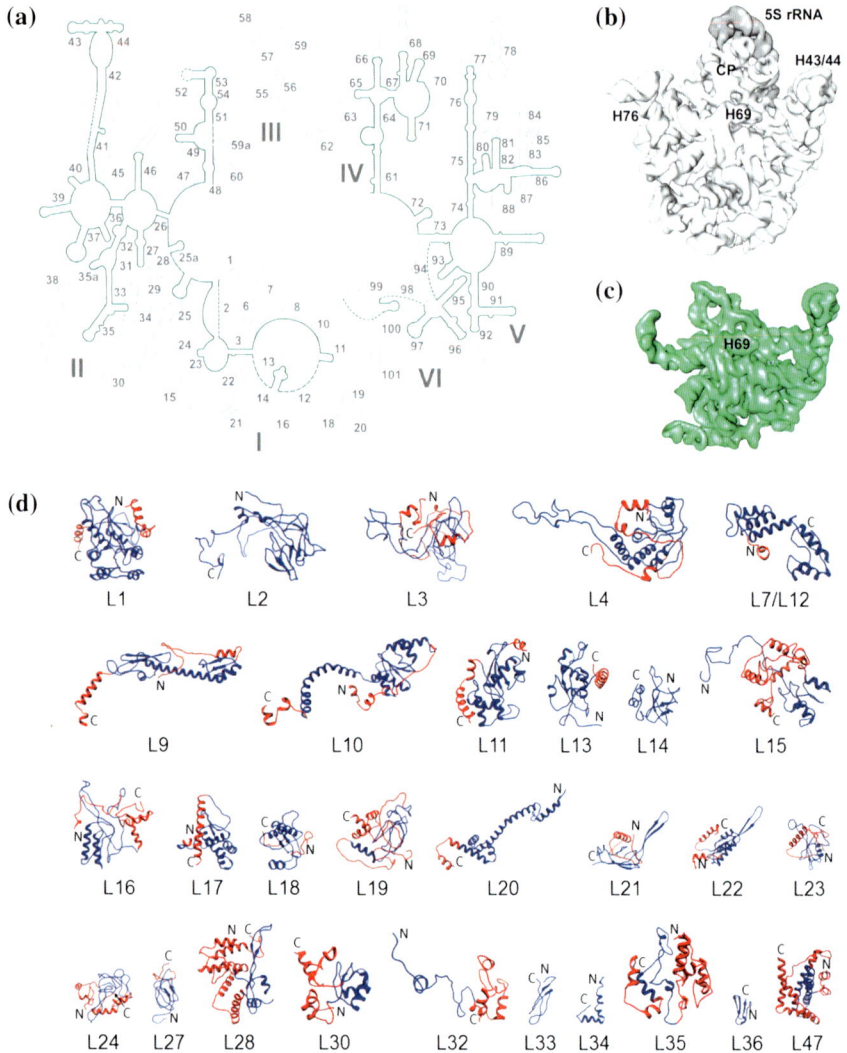

Fig. 1.3 Structural components of the mammalian mito-LSU. **a** Secondary structures of large ribosomal subunit RNAs from bacteria (23S, *grey*) and mammalian mitochondria (16S, *green*), *Dashed lines* indicate unassigned segments of the mito-rRNA. The rRNA helices are numbered according to the bacterial 23S rRNA. **b–c** Three dimensionally folded structures of the bacterial (**b**) and mammalian mitochondrial (**c**) LSU rRNAs, as derived from X-ray crystallography (PDB ID code 1VOY) and cryo-EM (based on Sharma et al. 2003; Mears et al. 2006; and Sharma et al., manuscript in preparation) studies, respectively. **d** Unfitted *ab initio* models of the homologous LSU MRPs. The MRP segments homologous to bacterial ribosomal proteins are depicted in *blue* and the mito-specific insertions/extensions are shown in *red*. MRP L47 is tentatively included in this list, as the major portion of its NTD domain appears to be a structural homolog of the bacterial L29 (see text). *N*- and *C*-termini are identified in each case

but the overall shape of the mito-LSU is not altered as dramatically as that of the mito-SSU (Fig. 1.1). The presence of all three characteristic features of a bacterial LSU, such as the L1 protuberance (L1), the central protuberance (CP) and the L7/L12 stalk (St), within the cryo-EM structure of the mito-LSU suggests that the basic architectures of both ribosomal subunits are important for their protein synthesis function. Like in mito-SSU, rRNA deletions have not affected the structure of the central functional core (the peptidyltransferase center region) of the mito-LSU, but they have significantly altered the composition of its tRNA exit site (E site), nascent polypeptide exit site, and several other peripheral regions of LSU (Fig. 1.1g–h). For example, 11 of the 12 interaction sites of tRNA involving rRNA segments at the bacterial ribosomal E site are absent in the mito-LSU (Mears et al. 2006), strongly suggesting that binding of $tRNA_{mt}$ at the putative mitoribosomal E site is either very weak or such a site does not exist on the mammalian mitoribosome. Similarly, the lower two-thirds of the nascent polypeptide exit tunnel (Nissen et al. 2000) is almost completely remodeled in the mito-LSU and is occupied primarily by the mito-specific MRPs (Sharma et al. 2003; also see, Agrawal et al. 2011).

The mito-LSU contains 50 MRPs, out of which 29 are homologous to bacterial ribosomal proteins (Table 1.2), and 21 MRPs are mito-specific (Koc et al. 2001a, b, 2010; Suzuki et al. 2001b). Homologues of the bacterial ribosomal proteins L5, L6, L25, L26 and L31 are absent in the mito-LSU. Similar to SSU MRPs, most homologous MRPs of LSU possess NTE and/or CTE, ranging in size from 8 to 120 amino acid segments, except for MRPs L2, L14, L33, L34 and L36 (Fig. 1.3d). L5, which is present within the central protuberance (CP) of the bacterial ribosome and forms the B1 group bridges with the SSU is replaced by MRPs specific to the mito-LSU (Sharma et al. 2003). The immediate binding partners of L5, the 5S rRNA and the LSU rRNA helix 84 are also absent in the mito-LSU. However, a recent report suggests that 5S rRNA is transported into mitochondria along with MRP L18 and it could become part of the mitoribosome LSU (Smirnov et al. 2011). It is possible that the 5S rRNA associates with only a small fraction of the mitoribosome, and therefore has gone undetected in the averaged 3D cryo-EM map (Sharma et al. 2003).

In mitochondria, all synthesized polypeptides are inserted into mtIM, and therefore polypeptide conducting exit tunnel in their ribosomes are tailor-made (Sharma et al. 2003, 2009; Agrawal et al. 2011), apparently to facilitate the co-translational release and incorporation of nascent hydrophobic polypeptides into the mtIM (Gruschke and Ott 2010). Unlike the situation in the structures of cytoplasmic ribosomes (e.g. Nissen et al. 2000), the polypeptide exit tunnel in mito-LSU has two solvent-accessible openings, referred to as the conventional polypeptide exit site (PES), which is located ~90 Å away from the peptidyltransferase center, and a mito-specific polypeptide-accessible site (PAS), which prematurely opens up in the middle, at only ~65 Å away from the peptidyltransferase center. In the mammalian mitoribosome, the openings at both PES and PAS are primarily occupied by MRPs. Based on overall topology of the polypeptide-exit tunnel in the mammalian mitoribosome, it is conceivable that the nascent polypeptide chain could use either the conventional PES or the mito-specific PAS

Table 1.2 Homologous MRPs of the LSU

Protein	Accession number	Length (residues)	MTS	Mito-specific segments	Identity/similarity[a] (%)
MRP L1	A6QPQ5	325	1–50	51–77, 311–325	26/48
MRP L2	Q2TA12	306	1–60	–	39/55
MRP L3	Q3ZBX6	348	1–40	41–95, 306–348	32/51
MRP L4	Q32PI6	294	1–22	23–63, 275–294	35/52
MRP L7/L12	Q7YR75	198	1–44	45–58,	32/50
MRP L9	Q2TBK2	268	1–52	53–93, 242–268	28/45
MRP L10	Q3MHY7	262	1–29	30–71, 243–262	22/44
MRP L11	Q2YDI0	192	–	1–10, 159–192	45/61
MRP L13	Q3SYS1	178	1–40	152–178	33/36
MRP L14	Q1JQ99	145	1–32	–	33/46
MRP L15	Q0VC21	297	1–49	178–297	47/57
MRP L16	Q3T0J3	251	1–29	30–57, 195–251	29/52
MRP L17	Q3T0L3	172	–	1–8, 139–172	35/56
MRP L18	Q3ZBR7	180	1–30	31–47, 173–180	39/59
MRP L19	F1MMM8	292	1–44	45–88, 206–292	33/50
MRP L20	Q2TBR2	149	1–9	126–149	42/60
MRP L21	F1MSV8	209	1–50	51–98	24/50
MRP L22	Q3SZX5	204	1–30	31–50, 168–204	33/53
MRP L23	DAA25904	153	–	1–10, 113–153	32/52
MRP L24	Q3SYS0	216	1–9	10–52, 156–216	34/53
MRP L27	Q32PC3	148	1–30	120–148	37/59
MRP L28	Q2HJJ1	256	1–25	26–74, 154–256	23/30
MRP L30	Q58DV5	161	1–35	36–63, 122–161	29/57
MRP L32	Q2TBI6	188	1–78	124–188	16/22
MRP L33	Q3SZ47	65	1–7	–	53/63
MRP L34	A8NN94	96	1–50	–	48/70
MRP L35	Q3SZA9	188	–	1–101, 159–188	32/40
MRP L36	XP_580819	148	1–112	–	47/63
MRP L47	Q08DT6	252	1–60	61–86, 155–252	25/31

MTS mitochondrial targeting sequences
[a] With homologous proteins from *E. coli* LSU

route, depending upon its folding requirements. In the bacterial ribosome, PES is encircled by four proteins L22, L23, L24 and L29 and inner lining of the lower two-thirds of the polypeptide exit tunnel is primarily made by domains I and III of the 23S rRNA. In case of mitoribosome the analogous section of the tunnel is either surrounded by MRPs or is left unoccupied, as the size of domains I and III are dramatically reduced in the mito-LSU 16S rRNA (Fig. 1.3a), giving rise to PAS (see Sharma et al. 2003; also see Agrawal et al. 2011). The conventional PES is also primarily surrounded with mito-specific MRPs. The homologous MRPs L22, L23, and L24, each with mito-specific extensions, are present in the mito-LSU (Fig. 1.3d). These extensions may account for the part of the large extra mass of mito-specific MRP density present at the mito-PES. However, there is still an ambiguity to whether or not the mito-specific MRP L47 is a homologue of bacterial L29. Bioinformatics analysis such as profile–profile search for distinct homologue has shown that MRP L47 is related to bacterial L29, whereas BLAST and Pfam analysis do not reveal any similarity among these two proteins (Smits et al. 2007). The *ab initio* model of MRP L47 shows a conserved structural domain similar to that in the bacterial L29, in addition to the large NTE and CTE (Fig. 1.3d). Based on distinct similarity between the central domain of MRP L47 3D model and structure of the bacterial L29, it is tempting to categorize MRP L47 as a bacterial L29 homologue.

In mitochondria the process of polypeptide integration into the mtIM is facilitated by the Oxa1 proteins, homologues of bacterial YidC. Oxa1L is the human homologue of the yeast Oxa1p, which is known to be involved in the biogenesis of membrane proteins and which assists in the insertion of proteins from mitochondrial matrix to the mtIM (Luirink et al. 2001; Ott and Herrmann 2010). The yeast homologue of MRP L23 (mrp20 in yeast) is known to interact with the *C*-terminus tail (CTT) of Oxa1p in order to recruit the mitoribosome for co-translational insertion of the mitochondrially-encoded polypeptides into mtIM (Jia et al. 2003; Szyrach et al. 2003; Keil et al. 2012). Interestingly, the cross-linking studies of Oxa1L-CTT with the bovine mito-LSU did not produce any crosslinks to the conventional PES proteins, such as the MRPs L22, L23, L24 and L29. However, the Oxa1L-CTT does crosslink to some other MRPs, including MRPs L13, L20, L28, L48, L49, and L51 (Haque et al. 2010). Of these, L13, L20, and L28 are known to be situated at distant locations from the polypeptide exit tunnel in the bacterial ribosome (Schuwirth et al. 2005). Of these, MRPs L13 and L20 possess only small CTEs, and therefore, are less likely to reach the polypeptide exit tunnel to interact with the mitoribosome-bound Oxa1L-CTT. However, MRP L28 has relatively large extensions on both its *N*-and C-termini (Fig. 1.3d), which could potentially reach the PAS but less likely to reach all the way to PES. Among remaining MRPs that are cross-linked to Oxa1L-CTT are mito-specific MRPs L48, L49 and L51. The exact locations of these MRPs on the mito-LSU are not presently known. As described in the previous paragraph, large mass of unidentified mito-specific MRP densities are present at both PAS and PES that could be accounted for by all three cross-linked mito-specific MRPs. Since no crosslink to MRP L47, the putative bacterial L29 homologue, was detected with the Oxa1L-CTT, it is possible that MRP L47 is shielded by mito-specific MRPs L48, L49 or L51 on the mitoribosome.

1.3 Mitochondrial Translation Factors

Like cytoplasmic ribosomes, the mitoribosome requires well-coordinated interactions with a number of mitochondrial translational factors (henceforth referred to as mito-translational factor) to conduct protein synthesis. Mito-translational factors bear greater similarity to their bacterial rather than to their eukaryotic counterparts. However, most mammalian mito-translational factors have acquired special structural features (Fig. 1.4) that may be required for effective and accurate translation on the significantly modified mitoribosomes. Like MRPs, all mito-translational factors are also encoded in the nucleus, synthesized in the cytoplasm with N-terminus MTSs (Table 1.3), and then imported into the mitochondrial matrix. In the following sections, we describe the structural organization of various mammalian mito-translational factors that directly interact with the mitoribosome and possess mito-specific amino-acid sequence insertions and/or extensions, based on sequence alignment of the mature (i.e., after the removal of MTSs) mito-translational factors (Fig. 1.4a) with those of their bacterial counterparts, and their 3D *ab initio* models (Fig. 1.4b). We describe them in a general order of their involvement in the process of protein synthesis, using bacterial translation system as the model (e.g., see, Schmeing and Ramakrishnan 2009). Accordingly, the mitochondrial translation can be divided into four stages of protein synthesis, i.e., initiation, elongation, termination, and recycling.

1.3.1 Translation Initiation

The initiation stage concludes with the docking of the fMet-tRNA$_{mt}^{Met}$ at the start codon of the mRNA onto the ribosomal peptidyl site (P site) on the 55S mitoribosome. Bacterial translation involves three initiation factors, IF1, IF2 and IF3. However, homologues of only two bacterial factors, IF2$_{mt}$ and IF3$_{mt}$, were identified in the mammalian mitochondria. It was proposed that IF2$_{mt}$ performs the task of both IF1 and IF2 in mitochondria (Spencer and Spremulli 2005; Gaur et al. 2008). Like bacterial IF3, IF3$_{mt}$ functions to actively dissociate the 55S mitoribosome into its two subunits, while it remains bound to the mito-SSU to prevent the re-association of the latter with the mito-LSU (Spencer and Spremulli 2005); whereas IF2$_{mt}$ forms a ternary complex with the initiator fMet-tRNA$_I^{Met}$ and GTP to deliver the fMet-tRNA$_I^{Met}$ into the initiator P site (e.g., see Allen et al. 2005 for IF2; and Yassin et al. 2011b for IF2$_{mt}$) of the mito-SSU. During this process, the CCA end of the initiator tRNA interacts with the conserved IF2$_{mt}$'s domain VI-C2 (Spencer and Spremulli 2004; Yassin et al. 2011b). Once the mito-SSU initiation complex is formed, IF2$_{mt}$ promotes its association with the mito-LSU via the latter's interaction with IF2$_{mt}$ domains IV (the GTPase domain) and VI. The activation of IF2$_{mt}$'s GTPase leads to the dissociation of IF2$_{mt}$ and of IF3$_{mt}$, resulting in the formation of a 55S initiation complex, which then enters the elongation phase of protein synthesis.

1.3.1.1 Mitochondrial Initiation Factor 2

The mitochondrial initiation factor 2 (IF2$_{mt}$) comprises of six structural domains (Fig. 1.4), which are homologous to domains III–VI of bacterial IF2 (Spencer and

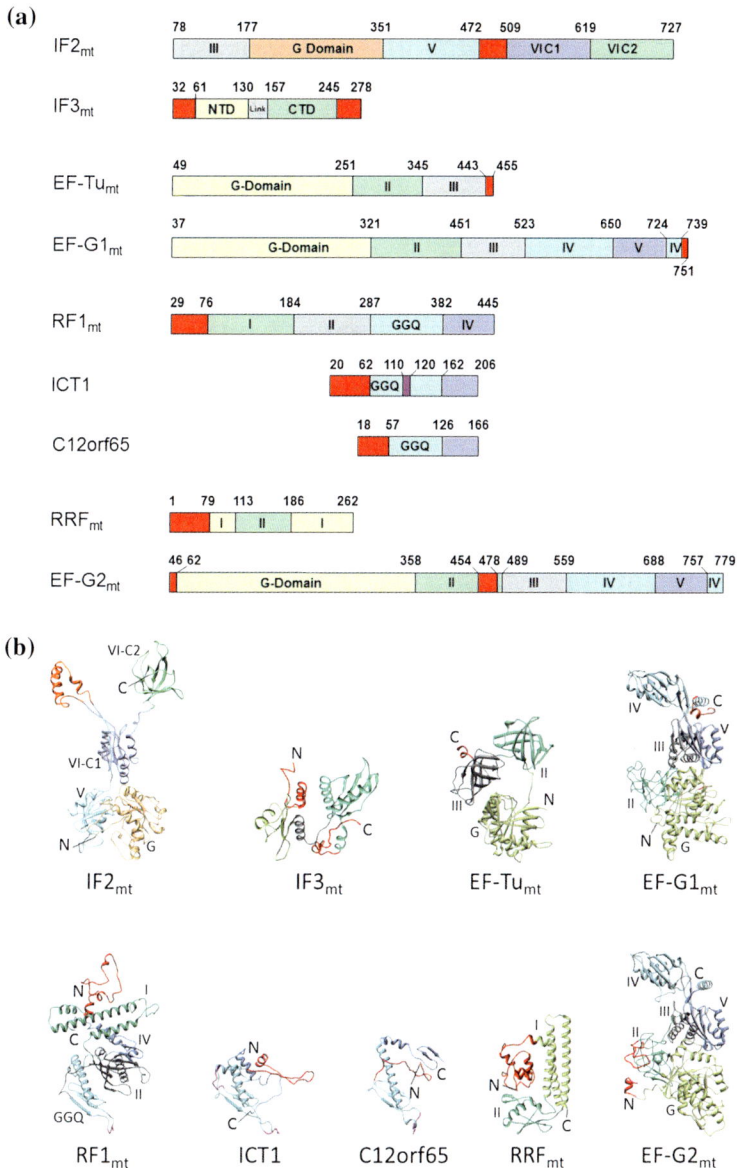

Fig. 1.4 Mammalian mitochondrial translation factors. **a** Bar diagrams showing mammalian mito-specifc segments (*red*) as compared to their bacterial counterparts. Numbers on *top* of each bar diagram refer to amino acids. Missing numbers at the *start* of each bar diagram would correspond to MTSs, except for RRF$_{mt}$, where MTS has been proposed to be a functional component of the factor. RF1a$_{mt}$ is not shown, as it does not contain a contiguous mito-specific segment. The GGQ domain of ICT1 posseses a unique 10 amino acid insertion segment (depicted in *pink*). **b** *Ab initio* models of the mitochondrial translational factors as generated using I-TASSER, except for IF2$_{mt}$, which is based on cryo-EM study (Yassin et al. 2011b), and EF-Tu$_{mt}$, which is known from X-ray-crystallography (Jeppesen et al. 2005). The nomenclature of structural domains of various factors is based on general consensus in the field, and domains are identified mostly by roman numerals, except for the GTP-binding domains, which are labeled as G-Domain. NTD and CTD refer to IF3$_{mt}$'s *N*- and *C*-terminal domains, respectively; in **b**, *N*- and *C*-termini are also identified in each case. The color codes used for various domains of corresponding factors are matched between **a** and **b**

Table 1.3 Mitochondrial translational factors with mito-specific segments

Protein	Accession number	Length (residues)	MTS	Mito-specific segments	Identity/similarity[a] (%)
IF2$_{mt}$	NP_001005369	727	1–77	472–508	40/60
IF3$_{mt}$	NP_001159734	278	1–31	32–60, 245–278	26/48
EF-Tu$_{mt}$	NP_003312	455	1–48	443–455	55/75
EF-G1$_{mt}$	NP_079272	751	1–36	739–751	45/63
RF1$_{mt}$	NP_004285	445	1–28	29–75	40/60
ICT1	NP_001536	206	1–19	20–61	28/52
C12orf65	NP_001137377	166	1–17	18–56	28/43
RRF$_{mt}$	NP_620132	262	1–26[b]	1–78	30/52
EF-G2$_{mt}$	NP_115756	779	1–45	46–61, 454–477	38/55

MTS, mitochondrial targeting sequences

[a] With homologous E. coli protein

[b] MTS has been proposed to be a functional component of RRF$_{mt}$

Spremulli 2005; Gaur et al. 2008; Yassin et al. 2011b), except for a mito-specific 37 amino-acid residues insertion domain. The insertion domain is present in the inter-domain linker between domains V and VI-C1. Mutation of several conserved basic residues of the insertion domain reduces the ability of the *Bos taurus* IF2$_{mt}$ to bind mito-SSU, implying that the insertion domain makes direct contacts with the SSU (Spencer and Spremulli 2005). Among all mitochondrial translation factors, the ribosome-bound structure is known only for the IF2$_{mt}$, which was obtained by cryo-EM. The cryo-EM study of the IF2$_{mt}$•GDPNP•fMet-tRNA$_i^{Met}$ in complex with the bacterial 70S ribosome (Yassin et al. 2011b) strongly supported the previous biochemical (Spencer and Spremulli 2005) and genetic (Gaur et al. 2008) studies. It showed that the insertion domain protrudes from rest of the IF2$_{mt}$ mass onto the ribosomal-SSU's aminoacyl-tRNA binding site (A site), where it interacts with conserved ribosomal elements (the SSU rRNA helices 18 and 44, and protein S12) that are also known to interact with the bacterial IF1 (Fig. 1.5) as well as the A-site tRNA (Carter et al. 2001). Thus, the study suggested that the mito-specific insertion domain mimics the function of the bacterial IF1 by sterically precluding the binding of initiator tRNA to the ribosomal A site (Yassin et al. 2011b). However, some questions remain unresolved as (i) the insertion domain had to be linked to the rest of IF2$_{mt}$ in the homology model through very long unstructured regions on both its ends (Yassin et al. 2011a), and (ii) there is neither a sequence homology nor any structural similarity between the insertion domain and the bacterial IF1 (Fig. 1.5c, d). However, both structures bear similar surface charge distributions, which might play an important role in recognition of the same binding pocket on the SSU. Owing to the long unstructured linker region, it is conceivable that the insertion domain is a highly dynamic structure that remains folded onto the rest of the IF2$_{mt}$ molecule in its unbound state and the extended conformation is attained only upon its interaction with the ribosome, as has been observed for bacterial class I release factors (e.g., Rawat et al. 2003). Moreover, the ribosome-bound IF2$_{mt}$ structure presents a

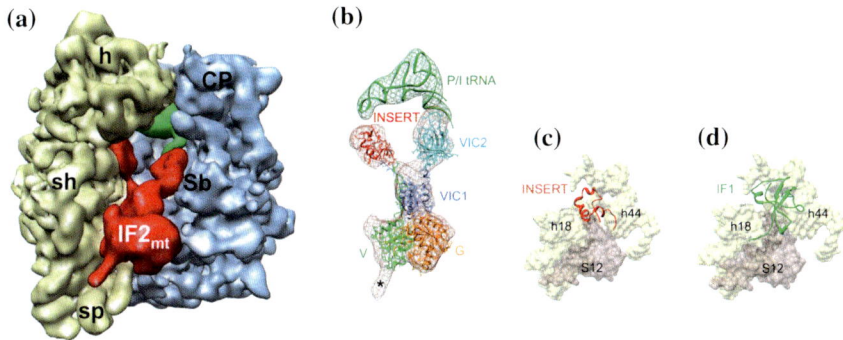

Fig. 1.5 Structure of the ribosome-bound mammalian IF2$_{mt}$, as derived by cryo-EM. **a** The cryo-EM map of the *E. coli* 70S ribosome (30S subunit, *yellow*; 50S subunit, *blue*) in complex with IF2$_{mt}$ (*red*), initiator tRNA (*green*) at the P/I position. **b** Fitting of atomic models of the IF2$_{mt}$ and initiator tRNA (P/I stands for P-site tRNA at initiator position) into the corresponding cryo-EM densities (meshwork) extracted from the map of the 70S•IF2$_{mt}$•GMPPNP•fMet-tRNA complex shown in **a**. The color codes used for various domains of IF2$_{mt}$ are the same as in Fig. 1.4. *Asterisk* (*) point to the region that would correspond to domain III of IF2$_{mt}$. Binding positions of the insertion domain (*red*, **c**) and IF1 (*green*, **d**) onto a common binding pocket of SSU of the ribosome. Landmarks of the ribosome are as in Figs. 1.1 and 1.2 (adopted from Yassin et al. 2011b)

compelling evidence of the integration of function of bacterial IF1 onto IF2$_{mt}$ that might have occurred during the course of evolution so that one less protein had to be transported into the mitochondrial matrix. Integration of relatively small IF1 feature to IF2$_{mt}$ would also ensure its efficient transport to the protein synthesis site within highly dense mitochondrial matrix.

1.3.1.2 Mitochondrial Initiation Factor 3

Mitochondrial initiation factor 3 (IF3$_{mt}$) shares the basic domain organization of the eubacterial IF3, with an *N*-terminal domain (NTD) and a *C*-terminal domain (CTD) connected by a flexible helical linker (Biou et al. 1995; Christian and Spremulli 2009; Petrelli et al. 2001). In addition, both its conserved domains harbor mito-specific NTE and CTE of ~30 amino acid residues each (Fig. 1.4). IF3$_{mt}$ stimulates initiation complex formation on the mito-specific leaderless mRNAs (Christian and Spremulli 2010). Both its mito-specific NTE and CTE have been implicated in optimizing the binding of IF3$_{mt}$ to enable its own dissociation from the 55S mitoribosome (Bhargava and Spremulli 2005; Christian and Spremulli 2009; Haque and Spremulli 2008). The CTE and the linker region of IF3$_{mt}$ have also been implicated in dissociation of the fMet-tRNA•IF2$_{mt}$ from the mito-SSU in the absence of mRNA, suggesting their role in preventing the premature occupation of the P site.

The binding position of bacterial IF3 on the bacterial SSU is known from cryo-EM (McCutcheon et al. 1999; Julian et al. 2011) and by hydroxyl radical probing

(Dallas and Noller 2001) studies. Interestingly, each of these studies places IF3 on the rim of the platform of the ribosomal SSU, but with different assignments of orientation for the two IF3 domains. While the structure of a mito-SSU•IF3$_{mt}$ is not yet published, our preliminary cryo-EM reconstructions suggest that the over-all configuration of IF3$_{mt}$ binding on the mito-SSU would be similar to what was proposed earlier for the bacterial ribosome. The cross-linking studies of IF3$_{mt}$ with the mito-SSU (Haque et al. 2011) identified several MRPs that may be present in immediate vicinity of the platform rim of the mito-SSU. This includes homologous MRPs S5, S9, S10 and S18 and several mito-specific MRPs, S29, S32, S36, and PTCD3. Among the homologous MRPs, S18 is situated on the platform and therefore, its direct interaction with IF3$_{mt}$ can be readily explained. The presence of the globular portions of S5, S9, and S10 on the solvent side of the bacterial SSU (Wimberly et al. 2000) suggests that the interaction of these MRPs with IF3$_{mt}$, which sits on the rim of the SSU platform, would be unlikely. However, C-terminus of S9 is exposed on the SSU-LSU interface side and MRPs S5 and S10 both possess long mito-specific extensions (Fig. 1.2d), which have potential to reach to the inter-subunit face to produce crosslinks with mito-SSU bound IF3$_{mt}$. In addition, there is a large mass of unidentified cryo-EM density within the mito-SSU head region that could account for the cross-linked mito-specific MRPs. Of these, docking of the homology model of PTCD3 into the cryo-EM map nicely explains the bulk of unassigned density in the SSU head at the interface of the mRNA channel (unpublished results in our laboratory), from where it could potentially crosslink to IF3$_{mt}$.

1.3.2 Translation Elongation

Translation elongation is a cyclic process (see for example, Agrawal et al. 2000; Schmeing and Ramakrishnan 2009) where an amino acid residue, as specified by the mRNA codon, is added to the growing nascent peptide, followed by pro-gression of the ribosome along the mRNA by a codon step. The elongation stage alternates between aminoacyl-tRNA (aa-tRNA) delivery and translocation of the mRNA•(tRNA)$_2$ complex on the ribosome, and ends when the ribosome encoun-ters a stop codon. Like in bacterial translation system, mitochondrial transla-tion also involves two canonical translation elongation factors (EFs) that interact directly with the mitoribosome. (i) EF-Tu$_{mt}$, which promotes the accurate bind-ing of aa-tRNA$_{mt}$, in form of a ternary complex aa-tRNA$_{mt}$•EF-Tu$_{mt}$•GTP, to the vacant aa-tRNA binding site (A site) of the ribosome; and (ii) EF-G1$_{mt}$, which, after the peptide-bond formation, binds to the ribosome as EF-G1$_{mt}$•GTP and promotes translocation of the mRNA•peptidyl-tRNA complex to free up the ribo-somal A site, or the mRNA decoding site, for the next round of elongation. Two forms of EF-G$_{mt}$ are present in most organisms (Hammarsund et al. 2001). The second isoform, EF-G2$_{mt}$, is exclusively involved at the recycling stage, and is described later under the Sect. 1.3.4.2.

1.3.2.1 Mitochondrial Elongation Factor Tu

Mitochondrial elongation factor Tu (EF-Tu$_{mt}$) is the most highly conserved among all translation factors that associate with the mitoribosome. EF-Tu$_{mt}$ has an overall 55 and 75 % sequence identity and similarity, respectively, with bacterial EF-Tu. It is also the only mitochondrial translational factor for which an atomic structure is currently available, in complex with its GTP exchange factor, EF-Ts$_{mt}$ (Jeppesen et al. 2005), and as expected, it shows an overall structural similarity to its bacterial counterpart (Kjeldgaard et al. 1993; Nissen et al. 1995). It comprises of three domains, the G domain (the GTP-binding GTPase domain), and domains II and III. The 3′ end of the aa-tRNA resides in the crevice between the G domain and domain II as its CCA arm interacts with domain III (Schmeing et al. 2009; Akama et al. 2010). Although the factor-binding region on the mitoribosome is significantly open due to deletion of the bacterial SSU rRNA helices h16 and h17 in the mito-SSU rRNA (Sharma et al. 2003) (Figs. 1.1a, d, and 1.2a–c), most of the ribosomal components that are known to interact with the functional sites of bacterial EF-Tu (G domain and domain II) and bound tRNAs (anticodon stem-loop ASL region) as part of the ternary complex (Valle et al. 2002; Schmeing et al. 2009; Schuette et al. 2009; Agirrezabala et al. 2011) are conserved in the mitoribosome. These include components such as α-sarcin-ricin stem loop (SRL), GTPase-associated center (comprised of LSU rRNA helices 43, 44 and protein L11), SSU protein S12, and the decoding site (comprised of portions of SSU rRNA helices 18 and 44). Thus, all essential structural elements that are involved in the proofreading step in bacterial ribosome are conserved in the mitoribosome. However, EF-Tu$_{mt}$ possesses a CTE of 11 amino acid residues (Fig. 1.4), whose functional significance is unknown.

The *E. coli* ternary complex can deliver the aa-tRNA to the mitoribosome when constituted with canonical aa-tRNAs, but not when in complex with aa-tRNA$_{mt}$s (Bullard et al. 1999), suggesting that the nature of the interaction of the aa-tRNA with EF-Tu$_{mt}$ must also account for the differences in the shape and stability of tRNA$_{mt}$s, several of which are smaller in size and lack their D- and/or T-arms (Hanada et al. 2000; Watanabe 2010). Furthermore, the tRNA itself participates in the signal transduction process of decoding. On cognate codon-anticodon interaction, the anticodon stem-loop is pulled into the A site and gets distorted. Part of that signal is transmitted via the tRNA scaffold to the G domain triggering the GTPase activity of EF-Tu$_{mt}$ (Valle et al. 2002; Schmeing et al. 2009; Schuette et al. 2009). The coupling of tRNA$_{mt}$ structure to that of the EF-Tu$_{mt}$ suggests that the mito-specific CTE may play a role in positioning the shorter aa-tRNA$_{mt}$s effectively in the decoding site for proofreading. It may compensate for the non-canonical shape and size of mammalian tRNA$_{mt}$s, in a mechanism that may be similar to that of the much longer CTE of *C. elegans* EF-Tu1$_{mt}$. The CTE in *C. elegans* EF-Tu1$_{mt}$ has been proposed to compensate for the lack of the T-arm in the *C. elegans* tRNA$_{mt}$ (Ohtsuki and Watanabe 2007). In the bacterial ribosome•aa-tRNA•EF-Tu complex, domain III interacts with the T stem-loop of the tRNA such that its C-terminus points towards the shoulder of the SSU (Schmeing et al. 2009). Accordingly, the CTE in EF-Tu$_{mt}$ may also interact with the structurally diminished shoulder of the mito-SSU. Such an interaction would help stabilizing the aa-tRNA$_{mt}$•EF-Tu$_{mt}$•GTP ternary complex interaction with the mitoribosome. A structural characterization of the mitoribosome•aa-tRNA$_{mt}$•EF-Tu$_{mt}$•GTP complex could help in delineating the function of the CTE in EF-Tu$_{mt}$.

1.3.2.2 Mitochondrial Elongation Factor G1

Mitochondrial elongation factor G1 (EF-G1$_{mt}$) catalyzes the translocation of the mRNA•(tRNA$_{mt}$)$_2$ complex on the mitoribosome. It carries a mito-specific CTE besides the well-defined five structurally conserved domains, which are homologous to the bacterial EF-G (Ævarsson et al. 1994; Czworkowski et al. 1994) (Fig. 1.4). Unlike its bacterial counterpart, EF-G1$_{mt}$ is inactive in ribosome-recycling (Tsuboi et al. 2009) and is highly resistant to fusidic acid (Chung and Spremulli 1990; Bhargava et al. 2004). Bacterial ribosomal elements that are known to interact with EF-G domains are mostly conserved in the mitoribosome (Bhargava et al. 2004), except for certain rRNA components in the mito-SSU shoulder (Sharma et al. 2003). Phylogenetic analysis indicates that the EF-G1$_{mt}$ like proteins have evolved to facilitate the translocation process on the ribosome (Atkinson and Baldauf 2011). In bacterial ribosomes, EF-G binds and stabilizes the ratcheted state of ribosome to catalyze the translocation step (Agrawal et al. 1999; Frank and Agrawal 2001; Agirrezabala and Frank 2009). Cryo-EM data from our laboratory suggests that the inter-subunit ratcheting is less pronounced in the mitoribosome (Sharma et al., manuscript in preparation), as compared to that observed for the bacterial ribosome (Agrawal et al. 1999; Frank and Agrawal 2000). It is possible that the remodeling of EF-G1$_{mt}$ enables it to function on the mitoribosome without the requirement of substantial ratchet-like reorganization. In addition to the mito-specifc CTE, which appears to be directly involved in facilitating translocation on the mitoribosome (Sharma et al., manuscript in preparation), the G domain of EF-G1$_{mt}$ has also been extensively remodeled, including an insertion of GEV in the highly conserved switch I loop (Atkinson and Baldauf 2011). Such a remodeling of the G domain has been implicated in conferring fusidic acid resistance to EF-G1$_{mt}$. In addition to the mito-specific features in EF-G1$_{mt}$, the central protuberance of the mito-LSU bears a unique and dynamic structural feature that may also contribute to the tRNA$_{mt}$ translocation process. This feature was identified as the P-site finger (Sharma et al. 2003), as it interacts with the T-stem loop side of the P-site tRNA$_{mt}$ on the mitoribosome.

1.3.3 Translation Termination

A stop codon (UAA/UAG) at the A site marks the end of the open-reading frame (ORF) and it is recognized by the class I release factors (RFs) that catalyse the hydrolysis of the ester bond of peptidyl-tRNA, leading to release of the nascent polypeptide chain form the ribosome and translation termination. In mitochondria, at least four proteins possessing the peptide-hydrolyzing domain of a typical RF, the GGQ domain, have been identified, which include $RF1_{mt}$, $RF1a_{mt}$, immature colon carcinoma transcript-1 (ICT1), and C12orf65 (Fig. 1.4, note that the figure includes only those factors that possess mito-specific insertions or extensions, suggesting that $RF1a_{mt}$ does not carry a mito-specific segment). Of these, so far only $RF1a_{mt}$ has been characterized as a canonical RF in mammalian mitochondria (see Richter et al. 2010). However, in the following sections, we briefly describe each of these four factors.

1.3.3.1 Mitochondrial Release Factor 1a

Mitochondrial release factor 1a ($RF1a_{mt}$) is a mitochondrial class 1 release factor that recognizes both UAA/UAG stop codons and terminates translation of all 13 mitochondrially-encoded polypeptides (Chrzanowska-Lightowlers et al. 2011). The domain organization of $RF1a_{mt}$ is similar to that of bacterial class I factors (Vestergaard et al. 2001). Domain II, the codon-recognition domain of the factor, contains structural elements that identify the stop codon; the codon recognition loop with the conserved tri-peptide motif PXT and the tip of helix $\alpha5$. Domain III bears the universal GGQ motif, which interacts with the acceptor end of the P-site tRNA. The glutamine residue of the GGQ motif catalyzes the hydrolysis of the peptidyl-tRNA by coordinating a water molecule, as shown by structural studies of analogous bacterial complexes (Rawat et al. 2003; Laurberg et al. 2008; Weixlbaumer et al. 2008). Correct stop codon recognition by domain II is required for proper placement of the GGQ motif of domain III into the peptidyltransferase center. The codon recognition signal is transmitted via LSU rRNA helix 69 and the inter-domain switch loop within class I RF (Laurberg et al. 2008; Korostelev 2011). Since there is no expected difference between the structural organizations of the bacterial RF1/2 and $RF1a_{mt}$, it is likely that $RF1a_{mt}$ interacts with the mitoribosome in a similar fashion as its bacterial counterpart (Laurberg et al. 2008; Weixlbaumer et al. 2008).

1.3.3.2 Mitochondrial Release Factor 1

Mitochondrial release factor 1 ($RF1_{mt}$) has been shown to be inactive as a peptidyl hydrolyase on the bacterial ribosome (Soleimanpour-Lichaei et al. 2007), apparently because the codon recognition elements of $RF1_{mt}$ are significantly different from those of canonical RFs. The conserved tripeptide of the codon recognition loop of RF1 is PXV instead of the conserved PXT, which is followed by a GXS insertion. Another codon recognition element on the tip of α-helix 5 has an RT insertion in its preceding loop (Soleimanpour-Lichaei et al. 2007; Chrzanowska-Lightowlers et al. 2011; Huynen et al. 2012). Besides the conserved domain organization of a class I RF, $RF1_{mt}$ has a 48 amino acid long mito-specific NTE. The function of this RF, including its mito-specific extension (Fig. 1.4a), is yet unknown. An analysis of a ribosome-bound homology model of $RF1_{mt}$ alludes to RF1$_{mt}$ as having a role analogous to that of the tmRNA in bacterial ribosome (Huynen et al. 2012). The study suggests that bulkier RF1$_{mt}$ codon recognition motif is unlikely to be accommodated in an mRNA occupied A site and that it probably recognizes the empty A site of a stalled mitoribosome.

1.3.3.3 Immature Colon Carcinoma Transcript-1

Immature colon carcinoma transcript-1 (ICT1) is a component of the mito-LSU and its activity is essential for cell viability. Having the universal GGQ motif but lacking the codon recognition domain, it is active as a ribosome-dependent pep-tidyl-tRNA hydrolyase in a non-codon dependent manner. Its activity has been implicated in the hydrolysis of prematurely terminated peptidyl-tRNA in the stalled mitoribosome (Richter et al. 2010). This functionality is similar to that of the bacterial YaeJ (Antonicka et al. 2010; Gagnon et al. 2012). However, its position on the mitoribosome is yet to be mapped, which is required to understand how this factor, being a component of the mitoribosome LSU, disengages itself from the nascent polypeptide hydrolysis site when not required.

1.3.3.4 C12orf65

C12orf65 is another essential mitochondrial RF lacking the stop codon recognition domain. However, despite the presence of the universal GGQ domain, its peptidyl-hydrolyase activity has not been demonstrated *in vitro*. The fact that the C12orf65 deficiency can be suppressed by over expression of ICT1 suggests that the protein is catalytically active *in vivo* (Antonicka et al. 2010). Since it is present in the mitochondrial matrix, C12orf65 has been suggested to play a role in recycling the abortive peptidyl-tRNA species, released from the mitoribosome during the elongation phase of translation. Phylogenetic analysis suggests that though C12orf65 is of eukaryotic origins, it shares a *C*-terminal helix rich in basic residues with YaeJ and ICT1 (Duarte et al. 2012). In YaeJ this helix is responsible for sensing the empty mRNA channel on a stalled ribosome (Gagnon et al. 2012).

1.3.4 Ribosome Recycling

After the translation termination, the post-termination ribosome complex (PoTC) remains occupied by the translated mRNA and a deacylated tRNA at the P/E position. As in bacteria (see Yokoyama et al. 2012), two mito-translational factors, RRF$_{mt}$ and EF-G2$_{mt}$, work in conjunction to facilitate the recycling step in mitochondria (Tsuboi et al. 2009) to release the deacylated tRNA$_{mt}$ and mRNA from the PoTC, and perhaps, to dissociate the 55S mitoribosome into its two subunits with the involvement of the third factor, IF3$_{mt}$. The cryo-EM studies of the bacterial PoTC•RRF and PoTC•RRF•EF-G complexes suggest that alternate inter-ribosomal subunit ratcheting (upon RRF binding; Barat et al. 2007; Yokoyama et al. 2012) and unratcheting (upon subsequent EF-G binding; Yokoyama et al. 2012) in conjunction with a steric clash between domain II of RRF_{mt} and SSU (Barat et al. 2007) facilitates disassembly of the PoTC.

1.3.4.1 Mitochondrial Ribosome Recyling Factor

The conserved fold of the mitochondrial ribosome recyling factor (RRF_{mt}) is similar to that of a bacterial RRF that contains two well-defined structural domains (Selmer et al. 1999). Domain I is a three helix bundle connected by flexible elbow linkers to domain II, a βαβ sandwich. In addition, RRF_{mt} has a mito-specific 78 amino acid residues long NTE (Rorbach et al. 2008). The homology model (Fig. 1.4b) suggests that its NTE is mainly α-helical. Most components of the ribosome (LSU rRNA helices, 69, 71, 80 and 93, and SSU protein S12) that are known to interact with the bacterial RRF (Agrawal et al. 2004; Gao et al. 2005; Barat et al. 2007; Weixlbaumer et al. 2007; Pai et al. 2008; Dunkle et al. 2011; Yokoyama et al. 2012) or the chloroplast ribosome (Sharma et al. 2007, 2010) are conserved in the mitoribosome. Therefore, it is likely that the conserved domains I and II of RRF_{mt} make similar contacts on the mitoribosome.

In bacteria, RRF stabilizes the ribosome in its ratcheted state, which causes the destabilization of several inter-subunit bridges (B1 group bridges, and bridges B2a and B3; see Gabashvili et al. 2000; Yusupov et al. 2001; Sharma et al. 2003, for the bridge positions and nomenclature), and apparently primes the ribosome for the subsequent EF-G binding (Barat et al. 2007; Yokoyama et al. 2012), which catalyzes the final disassembly step. As indicated earlier (Sect. 1.3.2.2), the mitoribosome does not undergo ratcheting to the same degree as the bacterial ribosome does. It is possible that the long mito-specific NTE of RRF_{mt} is involved in disruption of the mitoribosome's B1 group bridges, which are also made of mito-specific MRPs, to offset the requirement of a pronounced ratcheted state.

1.3.4.2 Mitochondrial Elongation Factor G2

Mitochondrial elongation factor G2 ($EF\text{-}G2_{mt}$) is a mito-specific paralog of EF-G that interacts with the 55S mitoribosome•RRF_{mt} complex to catalyze the disassembly of the PoTC, therefore the factor has been also referred to as $RRF2_{mt}$ (Tsuboi et al. 2009). Unlike the canonical EF-G, $EF\text{-}G2_{mt}$ is unable to catalyze translocation and does not require hydrolysis of GTP to accomplish the ribosomal subunit splitting. The ability of $EF\text{-}G2_{mt}$ to functionally interact with RRF_{mt} and the ribosome, to bring about ribosome splitting, mainly lies within its domains III and IV (Tsuboi et al. 2009). There is a mito-specific 25 amino acid insertion within $EF\text{-}G2_{mt}$'s domain II. The cryo-EM study of the bacterial PoTC•RRF•EF-G complex shows that domains III, IV and V (but not domain II) of the structurally homologous bacterial EF-G interact with the domain II of RRF (Yokoyama et al. 2012). Thus, the function of the mito-specific insertion in domain II of $EF\text{-}G2_{mt}$

remains elusive. However, based on its location on the 3D model of EF-G2$_{mt}$ (Fig. 1.4b) it is conceivable and as suggested earlier for the mito-specific extension in EF-Tu$_{mt}$ (Sect. 1.3.2.1), that the insertion may be involved in facilitating the factor's interaction with the structurally depleted shoulder of the mito-SSU, rather than directly interacting with the mito-specific NTE of RRF$_{mt}$. Further structural studies in context of the mitoribosome would be needed to resolve this intriguing interplay between the two recycling factors of the mitochondrial translation.

1.4 Concluding Remarks

An overall comparison of the previously determined cryo-EM structure of the mammalian mitoribosome with atomic structures of the cytoplasmic ribosomes is presented in this article, highlighting some of the unique features of the mitoribosome. The retention of key architectural elements in the mitoribosome underpins a notably conserved basic functioning despite compositional changes during its long structural evolution. The cryo-EM structure suggests that the mammalian mitoribosome has acquired several novel features related to mitochondrial protein synthesis. We have come a long way in improving the resolution of the cryo-EM structures of the mitoribosome, which is currently at 7 Å resolution in our laboratory. In the absence of any atomic structures of MRPs or mitochondrial translational factors, with the exception of EF-Tu$_{mt}$, molecular interpretation of the cryo-EM structures of the mitoribosome and its functional complexes is currently based on experimentally supported docking of homology models into the cryo-EM maps. However, identification and modeling of 38 non-homologous mito-specific MRPs in the cryo-EM map, especially those with undefined secondary structure motifs, continues to be a challenging task. The main barrier in achieving a high resolution structure appears to be an inherently heterogeneous composition of the mitoribosome, primarily due to a dramatic reduction in size of mito-rRNAs and significant increase in the number of MRPs. Many of these MRPs may be loosely attached to the rest of the mitoribosome in the absence of a direct interaction with the main rRNA scaffolds. This situation is dramatically different from the cytoplasmic ribosomes, which possess large rRNA scaffolds for interaction with their mostly basic ribosomal proteins, to produce relatively stable complex amenable to X-ray crystallographic structure determination. While the compositional fragility of the mitoribosome poses a challenge for X-ray crystallographic technique, it is highly suited to the single-particle cryo-EM method. This technique can provide structures for the mitoribosome and its functional complexes at ever increasing resolution to understand the functions of various insertion and extension sequences in both MRPs and mitochondrial translation factors, and to unravel mechanistic and molecular details of mitochondrial protein synthesis.

Acknowledgments This work was supported by the National Institutes of Health grant R01 GM61576 (to R.K.A.).

References

Ævarsson A, Brazhnikov E, Garber M, Zheltonosova J, Chirgadze Y, al-Karadaghi S, Svensson LA, Liljas A (1994) Three-dimensional structure of the ribosomal translocase: elongation factor G from *Thermus thermophilus*. EMBO J 13:3669–3677

Agirrezabala X, Frank J (2009) Elongation in translation as a dynamic interaction among the ribosome, tRNA, and elongation factors EF-G and EF-Tu. Q Rev Biophys 42:159–200

Agirrezabala X, Schreiner E, Trabuco LG, Lei J, Ortiz-Meoz RF, Schulten K, Green R, Frank J (2011) Structural insights into cognate versus near-cognate discrimination during decoding. EMBO J 30:1497–1507

Agrawal RK, Sharma MR (2012) Structural aspects of mitochondrial translational apparatus. Curr Opin Struct Biol 22:797–803

Agrawal RK, Penczek P, Grassucci RA, Frank J (1998) Visualization of elongation factor G on the *Escherichia coli* 70S ribosome: the mechanism of translocation. Proc Natl Acad Sci U S A 95:6134–6138

Agrawal RK, Heagle AB, Penczek P, Grassucci RA, Frank J (1999) EF-G-dependent GTP hydrolysis induces translocation accompanied by large conformational changes in the 70S ribosome. Nat Struct Biol 6:643–647

Agrawal RK, Spahn CM, Penczek P, Grassucci RA, Nierhaus KH, Frank J (2000) Visualization of tRNA movements on the *Escherichia coli* 70S ribosome during the elongation cycle. J Cell Biol 150:447–460

Agrawal RK, Sharma MR, Kiel MC, Hirokawa G, Booth TM, Spahn CM, Grassucci RA, Kaji A, Frank J (2004) Visualization of ribosome-recycling factor on the *Escherichia coli* 70S ribosome: functional implications. Proc Natl Acad Sci U S A 101:8900–8905

Agrawal RK, Sharma MR, Yassin AS, Lahiri I, Spremulli L (2011) Structure and function of organellar ribosomes as revealed by cryo-EM. In: Rodnina M, Wintermeyer W, Green R (eds) Ribosomes: structure, function, and dynamics. SpringerWien, New York, pp 83–96

Akama K, Christian BE, Jones CN, Ueda T, Takeuchi N, Spremulli LL (2010) Analysis of the functional consequences of lethal mutations in mitochondrial translational elongation factors. Biochim Biophys Acta 1802:692–698

Allen GS, Zavialov A, Gursky R, Ehrenberg M, Frank J (2005) The cryo-EM structure of a translation initiation complex from *Escherichia coli*. Cell 121:703–712

Antonicka H, Ostergaard E, Sasarman F, Weraarpachai W, Wibrand F, Pedersen AM, Rodenburg RJ, van der Knaap MS, Smeitink JA, Chrzanowska-Lightowlers ZM et al (2010) Mutations in C12orf65 in patients with encephalomyopathy and a mitochondrial translation defect. Am J Hum Genet 87:115–122

Atkinson GC, Baldauf SL (2011) Evolution of elongation factor G and the origins of mitochondrial and chloroplast forms. Mol Biol Evol 28:1281–1292

Ban N, Nissen P, Hansen J, Moore PB, Steitz TA (2000) The complete atomic structure of the large ribosomal subunit at 2.4 Å resolution. Science 289:905–920

Barat C, Datta PP, Raj VS, Sharma MR, Kaji H, Kaji A, Agrawal RK (2007) Progression of the ribosome recycling factor through the ribosome dissociates the two ribosomal subunits. Mol Cell 27:250–261

Ben-Shem A, Garreau de Loubresse N, Melnikov S, Jenner L, Yusupova G, Yusupov M (2011) The structure of the eukaryotic ribosome at 3.0 Å resolution. Science 334:1524–1529

Bhargava K, Spremulli LL (2005) Role of the N- and C-terminal extensions on the activity of mammalian mitochondrial translational initiation factor 3. Nucleic Acids Res 33:7011–7018

Bhargava K, Templeton P, Spremulli LL (2004) Expression and characterization of isoform 1 of human mitochondrial elongation factor G. Protein Expr Purif 37:368–376

Biou V, Shu F, Ramakrishnan V (1995) X-ray crystallography shows that translational initiation factor IF3 consists of two compact alpha/beta domains linked by an alpha-helix. EMBO J 14:4056–4064

Bullard JM, Cai YC, Zhang Y, Spremulli LL (1999) Effects of domain exchanges between *Escherichia coli* and mammalian mitochondrial EF-Tu on interactions with guanine nucleotides, aminoacyl-tRNA and ribosomes. Biochim Biophys Acta 1446:102–114

Carter AP, Clemons WM Jr, Brodersen DE, Morgan-Warren RJ, Hartsch T, Wimberly BT, Ramakrishnan V (2001) Crystal structure of an initiation factor bound to the 30S ribosomal subunit. Science 291:498–501

Christian BE, Spremulli LL (2009) Evidence for an active role of IF3mt in the initiation of translation in mammalian mitochondria. Biochemistry 48:3269–3278

Christian BE, Spremulli LL (2010) Preferential selection of the 5'-terminal start codon on leaderless mRNAs by mammalian mitochondrial ribosomes. J Biol Chem 285:28379–28386

Christian BE, Spremulli LL (2012) Mechanism of protein biosynthesis in mammalian mitochondria. Biochim Biophys Acta 1819:1035–1054

Chrzanowska-Lightowlers ZM, Pajak A, Lightowlers RN (2011) Termination of protein synthesis in mammalian mitochondria. J Biol Chem 286:34479–34485

Chung HK, Spremulli LL (1990) Purification and characterization of elongation factor G from bovine liver mitochondria. J Biol Chem 265:21000–21004

Claros MG, Vincens P (1996) Computational method to predict mitochondrially imported proteins and their targeting sequences. Eur J Biochem 241:779–786

Czworkowski J, Wang J, Steitz TA, Moore PB (1994) The crystal structure of elongation factor G complexed with GDP, at 2.7 Å resolution. EMBO J 13:3661–3668

Dallas A, Noller HF (2001) Interaction of translation initiation factor 3 with the 30S ribosomal subunit. Mol Cell 8:855–864

Datta PP, Sharma MR, Qi L, Frank J, Agrawal RK (2005) Interaction of the G' domain of elongation factor G and the C-terminal domain of ribosomal protein L7/L12 during translocation as revealed by cryo-EM. Mol Cell 20:723–731

Duarte I, Nabuurs SB, Magno R, Huynen M (2012) Evolution and diversification of the organellar release factor family. Mol Biol Evol 29:3497–3512

Dunkle JA, Wang L, Feldman MB, Pulk A, Chen VB, Kapral GJ, Noeske J, Richardson JS, Blanchard SC, Cate JH (2011) Structures of the bacterial ribosome in classical and hybrid states of tRNA binding. Science 332:981–984

Frank J, Agrawal RK (2000) A ratchet-like inter-subunit reorganization of the ribosome during translocation. Nature 406:318–322

Frank J, Agrawal RK (2001) Ratchet-like movements between the two ribosomal subunits: their implications in elongation factor recognition and tRNA translocation. Cold Spring Harb Symp Quant Biol 66:67–75

Frank J, Penczek P, Agrawal RK, Grassucci RA, Heagle AB (2000) Three-dimensional cryoelectron microscopy of ribosomes. Methods Enzymol 317:276–291

Gabashvili IS, Agrawal RK, Spahn CM, Grassucci R, Svergun D, Frank J, Penczek P (2000) Solution structure of the *E. coli* 70S ribosome at 11.5 Å resolution. Cell 100:537–549

Gagnon MG, Seetharaman SV, Bulkley D, Steitz TA (2012) Structural basis for the rescue of stalled ribosomes: structure of YaeJ bound to the ribosome. Science 335:1370–1372

Gao N, Zavialov AV, Li W, Sengupta J, Valle M, Gursky RP, Ehrenberg M, Frank J (2005) Mechanism for the disassembly of the posttermination complex inferred from cryo-EM studies. Mol Cell 18:663–674

Gao YG, Selmer M, Dunham CM, Weixlbaumer A, Kelley AC, Ramakrishnan V (2009) The structure of the ribosome with elongation factor G trapped in the posttranslocational state. Science 326:694–699

Gaur R, Grasso D, Datta PP, Krishna PD, Das G, Spencer A, Agrawal RK, Spremulli L, Varshney U (2008) A single mammalian mitochondrial translation initiation factor functionally replaces two bacterial factors. Mol Cell 29:180–190

Gomez-Lorenzo MG, Spahn CM, Agrawal RK, Grassucci RA, Penczek P, Chakraburtty K, Ballesta JP, Lavandera JL, Garcia-Bustos JF, Frank J (2000) Three-dimensional cryo-electron microscopy localization of EF2 in the *Saccharomyces cerevisiae* 80S ribosome at 17.5 Å resolution. EMBO J 19:2710–2718

Gray, M.W., Burger, G., and Lang, B.F. (2001). The origin and early evolution of mitochondria. Genome Biol 2, REVIEWS1018

Gruschke S, Ott M (2010) The polypeptide tunnel exit of the mitochondrial ribosome is tailored to meet the specific requirements of the organelle. BioEssays 32:1050–1057

Hammarsund M, Wilson W, Corcoran M, Merup M, Einhorn S, Grander D, Sangfelt O (2001) Identification and characterization of two novel human mitochondrial elongation factor genes, hEFG2 and hEFG1, phylogenetically conserved through evolution. Hum Genet 109:542–550

Hanada T, Suzuki T, Watanabe K (2000) Translation activity of mitochondrial tRNA with unusual secondary structure. Nucleic Acids Symp Ser 44:249–250

Haque ME, Spremulli LL (2008) Roles of the N- and C-terminal domains of mammalian mitochondrial initiation factor 3 in protein biosynthesis. J Mol Biol 384:929–940

Haque ME, Spremulli LL, Fecko CJ (2010) Identification of protein–protein and protein-ribosome interacting regions of the C-terminal tail of human mitochondrial inner membrane protein Oxa1L. J Biol Chem 285:34991–34998

Haque ME, Koc H, Cimen H, Koc EC, Spremulli LL (2011) Contacts between mammalian mitochondrial translational initiation factor 3 and ribosomal proteins in the small subunit. Biochim Biophys Acta 1814:1779–1784

Huynen MA, Duarte I, Chrzanowska-Lightowlers ZM, Nabuurs SB (2012) Structure based hypothesis of a mitochondrial ribosome rescue mechanism. Biol Direct 7:14

Jeppesen MG, Navratil T, Spremulli LL, Nyborg J (2005) Crystal structure of the bovine mitochondrial elongation factor Tu.Ts complex. J Biol Chem 280:5071–5081

Jia L, Dienhart M, Schramp M, McCauley M, Hell K, Stuart RA (2003) Yeast Oxa1 interacts with mitochondrial ribosomes: the importance of the C-terminal region of Oxa1. EMBO J 22:6438–6447

Julian P, Milon P, Agirrezabala X, Lasso G, Gil D, Rodnina MV, Valle M (2011) The Cryo-EM structure of a complete 30S translation initiation complex from Escherichia coli. PLoS Biol 9:e1001095

Keil M, Bareth B, Woellhaf MW, Peleh V, Prestele M, Rehling P, Herrmann JM (2012) Oxa1-ribosome complexes coordinate the assembly of cytochrome C oxidase in mitochondria. J Biol Chem 287:34484–34493

Kjeldgaard M, Nissen P, Thirup S, Nyborg J (1993) The crystal structure of elongation factor EF-Tu from Thermus aquaticus in the GTP conformation. Structure 1:35–50

Klinge S, Voigts-Hoffmann F, Leibundgut M, Arpagaus S, Ban N (2011) Crystal structure of the eukaryotic 60S ribosomal subunit in complex with initiation factor 6. Science 334:941–948

Koc EC, Koc H (2012) Regulation of mammalian mitochondrial translation by post-translational modifications. Biochim Biophys Acta 1819:1055–1066

Koc EC, Burkhart W, Blackburn K, Moseley A, Koc H, Spremulli LL (2000) A proteomics approach to the identification of mammalian mitochondrial small subunit ribosomal proteins. J Biol Chem 275:32585–32591

Koc EC, Burkhart W, Blackburn K, Moseley A, Spremulli LL (2001a) The small subunit of the mammalian mitochondrial ribosome. Identification of the full complement of ribosomal proteins present. J Biol Chem 276:19363–19374

Koc EC, Burkhart W, Blackburn K, Moyer MB, Schlatzer DM, Moseley A, Spremulli LL (2001b) The large subunit of the mammalian mitochondrial ribosome. Analysis of the complement of ribosomal proteins present. J Biol Chem 276:43958–43969

Koc EC, Haque ME, Spremulli LL (2010) Current views of the structure of the mammalian mitochondrial ribosome. Isr J Chem 50:45–59

Korostelev AA (2011) Structural aspects of translation termination on the ribosome. RNA 17:1409–1421

Laurberg M, Asahara H, Korostelev A, Zhu J, Trakhanov S, Noller HF (2008) Structural basis for translation termination on the 70S ribosome. Nature 454:852–857

Luirink J, Samuelsson T, de Gier JW (2001) YidC/Oxa1p/Alb3: evolutionarily conserved mediators of membrane protein assembly. FEBS Lett 501:1–5

Manuell AL, Quispe J, Mayfield SP (2007) Structure of the chloroplast ribosome: novel domains for translation regulation. PLoS Biol 5:e209

McCutcheon JP, Agrawal RK, Philips SM, Grassucci RA, Gerchman SE, Clemons WM Jr, Ramakrishnan V, Frank J (1999) Location of translational initiation factor IF3 on the small ribosomal subunit. Proc Natl Acad Sci U S A 96:4301–4306

Mears JA, Sharma MR, Gutell RR, McCook AS, Richardson PE, Caulfield TR, Agrawal RK, Harvey SC (2006) A structural model for the large subunit of the mammalian mitochondrial ribosome. J Mol Biol 358:193–212

Nissen P, Kjeldgaard M, Thirup S, Polekhina G, Reshetnikova L, Clark BF, Nyborg J (1995) Crystal structure of the ternary complex of Phe-tRNAPhe, EF-Tu, and a GTP analog. Science 270:1464–1472

Nissen P, Hansen J, Ban N, Moore PB, Steitz TA (2000) The structural basis of ribosome activity in peptide bond synthesis. Science 289:920–930

O'Brien TW (2002) Evolution of a protein-rich mitochondrial ribosome: implications for human genetic disease. Gene 286:73–79

O'Brien TW, O'Brien BJ, Norman RA (2005) Nuclear MRP genes and mitochondrial diseases. Gene 354:147–151

Ogle JM, Brodersen DE, Clemons WM Jr, Tarry MJ, Carter AP, Ramakrishnan V (2001) Recognition of cognate transfer RNA by the 30S ribosomal subunit. Science 292:897–902

Ohtsuki T, Watanabe Y (2007) T-armless tRNAs and elongated elongation factor Tu. IUBMB Life 59:68–75

Ott M, Herrmann JM (2010) Co-translational membrane insertion of mitochondrially encoded proteins. Biochim Biophys Acta 1803:767–775

Pai RD, Zhang W, Schuwirth BS, Hirokawa G, Kaji H, Kaji A, Cate JH (2008) Structural Insights into ribosome recycling factor interactions with the 70S ribosome. J Mol Biol 376:1334–1347

Pearce S, Nezich CL, Spinazzola A (2013) Mitochondrial diseases: translation matters. Mol Cell Neurosci 55:1–12

Petrelli D, LaTeana A, Garofalo C, Spurio R, Pon CL, Gualerzi CO (2001) Translation initiation factor IF3: two domains, five functions, one mechanism? EMBO J 20:4560–4569

Pettersen EF, Goddard TD, Huang CC, Couch GS, Greenblatt DM, Meng EC, Ferrin TE (2004) UCSF Chimera–a visualization system for exploratory research and analysis. J Comput Chem 25:1605–1612

Rawat UB, Zavialov AV, Sengupta J, Valle M, Grassucci RA, Linde J, Vestergaard B, Ehrenberg M, Frank J (2003) A cryo-electron microscopic study of ribosome-bound termination factor RF2. Nature 421:87–90

Richter R, Rorbach J, Pajak A, Smith PM, Wessels HJ, Huynen MA, Smeitink JA, Lightowlers RN, Chrzanowska-Lightowlers ZM (2010) A functional peptidyl-tRNA hydrolase, ICT1, has been recruited into the human mitochondrial ribosome. EMBO J 29:1116–1125

Rorbach J, Richter R, Wessels HJ, Wydro M, Pekalski M, Farhoud M, Kuhl I, Gaisne M, Bonnefoy N, Smeitink JA et al (2008) The human mitochondrial ribosome recycling factor is essential for cell viability. Nucleic Acids Res 36:5787–5799

Roy A, Kucukural A, Zhang Y (2010) I-TASSER: a unified platform for automated protein structure and function prediction. Nat Protoc 5:725–738

Schmeing TM, Ramakrishnan V (2009) What recent ribosome structures have revealed about the mechanism of translation. Nature 461:1234–1242

Schmeing TM, Voorhees RM, Kelley AC, Gao YG, Murphy FVt, Weir JR JR, Ramakrishnan V (2009) The crystal structure of the ribosome bound to EF-Tu and aminoacyl-tRNA. Science 326:688–694

Schuette JC, Murphy FVt, Kelley AC, Weir JR, Giesebrecht J, Connell SR, Loerke J, Mielke T, Zhang W, Penczek PA et al (2009) GTPase activation of elongation factor EF-Tu by the ribosome during decoding. EMBO J 28:755–765

Schuwirth BS, Borovinskaya MA, Hau CW, Zhang W, Vila-Sanjurjo A, Holton JM, Cate JH (2005) Structures of the bacterial ribosome at 3.5 Å resolution. Science 310:827–834

Selmer M, Al-Karadaghi S, Hirokawa G, Kaji A, Liljas A (1999) Crystal structure of *Thermotoga maritima* ribosome recycling factor: a tRNA mimic. Science 286:2349–2352

Selmer M, Dunham CM, Murphy FVt, Weixlbaumer A, Petry S, Kelley AC, Weir JR, Ramakrishnan V (2006) Structure of the 70S ribosome complexed with mRNA and tRNA. Science 313:1935–1942

Sharma MR, Koc EC, Datta PP, Booth TM, Spremulli LL, Agrawal RK (2003) Structure of the mammalian mitochondrial ribosome reveals an expanded functional role for its component proteins. Cell 115:97–108

Sharma MR, Wilson DN, Datta PP, Barat C, Schluenzen F, Fucini P, Agrawal RK (2007) Cryo-EM study of the spinach chloroplast ribosome reveals the structural and functional roles of plastid-specific ribosomal proteins. Proc Natl Acad Sci U S A 104:19315–19320

Sharma MR, Booth TM, Simpson L, Maslov DA, Agrawal RK (2009) Structure of a mitochondrial ribosome with minimal RNA. Proc Natl Acad Sci U S A 106:9637–9642

Sharma MR, Dönhöfer A, Barat C, Marquez V, Datta PP, Fucini P, Wilson DN, Agrawal RK (2010) PSRP1 is not a ribosomal protein, but a ribosome-binding factor that is recycled by the ribosome-recycling factor (RRF) and elongation factor G (EF-G). J Biol Chem 285:4006–4014

Smirnov A, Entelis N, Martin RP, Tarassov I (2011) Biological significance of 5S rRNA import into human mitochondria: role of ribosomal protein MRP-L18. Genes Dev 25:1289–1305

Smits P, Smeitink JA, van den Heuvel LP, Huynen MA, Ettema TJ (2007) Reconstructing the evolution of the mitochondrial ribosomal proteome. Nucleic Acids Res 35:4686–4703

Soleimanpour-Lichaei HR, Kuhl I, Gaisne M, Passos JF, Wydro M, Rorbach J, Temperley R, Bonnefoy N, Tate W, Lightowlers R et al (2007) mtRF1a is a human mitochondrial translation release factor decoding the major termination codons UAA and UAG. Mol Cell 27:745–757

Spahn CM, Gomez-Lorenzo MG, Grassucci RA, Jorgensen R, Andersen GR, Beckmann R, Penczek PA, Ballesta JP, Frank J (2004) Domain movements of elongation factor eEF2 and the eukaryotic 80S ribosome facilitate tRNA translocation. EMBO J 23:1008–1019

Spencer AC, Spremulli LL (2004) Interaction of mitochondrial initiation factor 2 with mitochondrial fMet-tRNA. Nucleic Acids Res 32:5464–5470

Spencer AC, Spremulli LL (2005) The interaction of mitochondrial translational initiation factor 2 with the small ribosomal subunit. Biochim Biophys Acta 1750:69–81

Stark H, Rodnina MV, Rinke-Appel J, Brimacombe R, Wintermeyer W, van Heel M (1997) Visualization of elongation factor Tu on the *Escherichia coli* ribosome. Nature 389:403–406

Suzuki T, Terasaki M, Takemoto-Hori C, Hanada T, Ueda T, Wada A, Watanabe K (2001a) Proteomic analysis of the mammalian mitochondrial ribosome. Identification of protein components in the 28S small subunit. J Biol Chem 276:33181–33195

Suzuki T, Terasaki M, Takemoto-Hori C, Hanada T, Ueda T, Wada A, Watanabe K (2001b) Structural compensation for the deficit of rRNA with proteins in the mammalian mitochondrial ribosome. Systematic analysis of protein components of the large ribosomal subunit from mammalian mitochondria. J Biol Chem 276:21724–21736

Szyrach G, Ott M, Bonnefoy N, Neupert W, Herrmann JM (2003) Ribosome binding to the Oxa1 complex facilitates co-translational protein insertion in mitochondria. EMBO J 22:6448–6457

Temperley RJ, Wydro M, Lightowlers RN, Chrzanowska-Lightowlers ZM (2010) Human mitochondrial mRNAs-like members of all families, similar but different. Biochim Biophys Acta 1797:1081–1085

Tsuboi M, Morita H, Nozaki Y, Akama K, Ueda T, Ito K, Nierhaus KH, Takeuchi N (2009) EF-G2mt is an exclusive recycling factor in mammalian mitochondrial protein synthesis. Mol Cell 35:502–510

Valle M, Sengupta J, Swami NK, Grassucci RA, Burkhardt N, Nierhaus KH, Agrawal RK, Frank J (2002) Cryo-EM reveals an active role for aminoacyl-tRNA in the accommodation process. EMBO J 21:3557–3567

Vestergaard B, Van LB, Andersen GR, Nyborg J, Buckingham RH, Kjeldgaard M (2001) Bacterial polypeptide release factor RF2 is structurally distinct from eukaryotic eRF1. Mol Cell 8:1375–1382

Watanabe K (2010) Unique features of animal mitochondrial translation systems. The non-universal genetic code, unusual features of the translational apparatus and their relevance to human mitochondrial diseases. Proc Jpn Acad Ser B Phys Biol Sci 86:11–39

Weixlbaumer A, Petry S, Dunham CM, Selmer M, Kelley AC, Ramakrishnan V (2007) Crystal structure of the ribosome recycling factor bound to the ribosome. Nat Struct Mol Biol 14:733–737

Weixlbaumer A, Jin H, Neubauer C, Voorhees RM, Petry S, Kelley AC, Ramakrishnan V (2008) Insights into translational termination from the structure of RF2 bound to the ribosome. Science 322:953–956

Wimberly BT, Brodersen DE, Clemons WM Jr, Morgan-Warren RJ, Carter AP, Vonrhein C, Hartsch T, Ramakrishnan V (2000) Structure of the 30S ribosomal subunit. Nature 407:327–339

Yassin AS, Agrawal RK, Banavali NK (2011a) Computational exploration of structural hypotheses for an additional sequence in a mammalian mitochondrial protein. PLoS ONE 6:e21871

Yassin AS, Haque ME, Datta PP, Elmore K, Banavali NK, Spremulli LL, Agrawal RK (2011b) Insertion domain within mammalian mitochondrial translation initiation factor 2 serves the role of eubacterial initiation factor 1. Proc Natl Acad Sci U S A 108:3918–3923

Yokoyama T, Shaikh TR, Iwakura N, Kaji H, Kaji A, Agrawal RK (2012) Structural insights into initial and intermediate steps of the ribosome-recycling process. EMBO J 31:1836–1846

Yusupov MM, Yusupova GZ, Baucom A, Lieberman K, Earnest TN, Cate JH, Noller HF (2001) Crystal structure of the ribosome at 5.5 Å resolution. Science 292:883–896

Yusupova GZ, Yusupov MM, Cate JH, Noller HF (2001) The path of messenger RNA through the ribosome. Cell 106:233–241

Chapter 2
Mechanism and Regulation of Protein Synthesis in Mammalian Mitochondria

Emine C. Koc and Hasan Koc

Abstract The mammalian mitochondrial translation machinery is responsible for the synthesis of 13 mitochondrially encoded proteins that are essential for energy production. These proteins are subunits of the oxidative phosphorylation complexes embedded in the inner membrane of mitochondria. Mitochondrial protein synthesis is highly similar to that of bacterial systems; however, there are subtle differences between these systems in terms of their mechanisms and components. In this review, we will discuss the elements of mitochondrial translation, including the stages of protein synthesis and the factors involved in these processes. Although much still waits to be learned about the regulation of this system, a summary of what is currently known about the regulation of its protein components by post-translational modifications, specifically concerning energy metabolism, will also be included in this chapter.

2.1 Background

2.1.1 The Role of Mitochondrial Translation in Energy Metabolism

Mitochondria provide more than 90 % of the energy used by mammalian cells through the process of oxidative phosphorylation (OXPHOS). The 13 mitochondrially encoded proteins that are synthesized by the mitochondrial translation machinery are integral

E. C. Koc (✉)
Department of Biochemistry and Microbiology, Joan C. Edwards School of Medicine, Marshall University, Huntington, USA
e-mail: koce@marshall.edu

H. Koc
Department of Pharmacological Science and Research, School of Pharmacy, Marshall University, Huntington, USA
e-mail: kocha@marshall.edu

A.-M. Duchêne (ed.), *Translation in Mitochondria and Other Organelles*, DOI: 10.1007/978-3-642-39426-3_2, © Springer-Verlag Berlin Heidelberg 2013 components of the electron transfer and ATP synthase complexes and are essential for energy metabolism. They are localized in the inner membrane (IM) of mitochondria and include seven subunits of complex I (NADH:ubiquinone oxidoreductase), one subunit of complex III (ubiquinone:cytochrome c oxidoreductase), three subunits of complex IV (cytochrome c:oxygen oxidoreductase), and two subunits of complex V (ATP synthase) (Fig. 2.1). The remaining subunits of the OXPHOS complexes, in addition to the ~1,500 proteins that support energy metabolism, are the products of nuclear genes. These proteins are synthesized by cytoplasmic ribosomes and imported into the mitochondria.

Mitochondrial energy metabolism, including the production of mitochondrial translation components, requires coordination of mitochondrial transcription and cytoplasmic translation. As described below, all the proteins supporting mitochondrial protein synthesis are synthesized by the cytoplasmic ribosomes. However, little is known about the retrograde regulation of mitochondrial energy metabolism and translation. The availability of the high energy molecules, Acetyl-CoA, NADH/NAD$^+$, and ATP, could be the major regulator of energy metabolism and protein synthesis in mammalian mitochondria via post-translational modifications by reversible acetylation and phosphorylation (Fig. 2.1). A brief analysis of this hypothesis will be discussed at the end of this chapter.

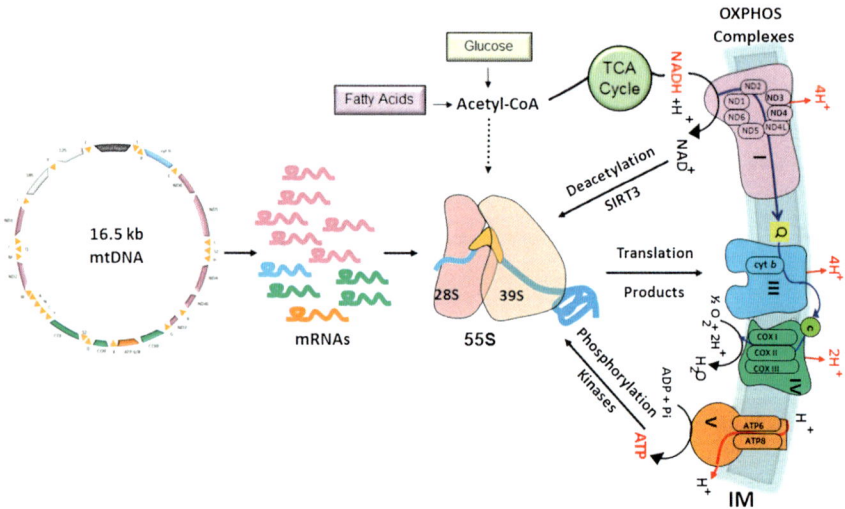

Fig. 2.1 Oxidative phosphorylation and regulation of the mitochondrial translation machinery. Mammalian mitochondria contain a 16.5 kb circular genome (mtDNA) which encodes for 22 tRNAs, two rRNAs, and nine monocistronic and two dicistronic mRNAs. Mammalian mitochondrial ribosomes (28S and 39S subunits) are responsible for the synthesis of 13 mitochondrially encoded proteins that are subunits of complex I (*blue*), III (*pink*), IV (*cyan*), and ATP synthase, also known as complex V (*orange*). Metabolic levels of acetyl-CoA, NADH/NAD$^+$, and ATP are important for the regulation of the mitochondrial translation machinery through reversible acetylation and phosphorylation

2.1.2 The Mitochondrial Genome

In mammals, mitochondrial DNA (mtDNA) contains about 16.5 kilobase pairs. It is a circular genome that encodes for 13 of the proteins of the OXPHOS complexes, as well as 22 tRNAs and two rRNAs required for protein synthesis (Fig. 2.1). Nuclear DNA encodes for the majority of mitochondrial proteins and most subunits of the OXPHOS complexes. The 13 polypeptides encoded in the mammalian mitochondrial genome are synthesized from nine monocistronic mRNAs and two dicistronic mRNAs, which have overlapping reading frames (Jackson 1991; Anderson et al. 1982; Wolstenholme 1992). These mRNAs are quite unusual, almost entirely lacking 5' and 3' untranslated nucleotides. The translational start codon is generally located within three nucleotides of the 5' end of the mRNA; therefore, mammalian mitochondrial mRNAs are defined as leaderless mRNAs (Anderson et al. 1982; Montoya et al. 1981). Analysis of the 5' ends of human mitochondrial mRNAs reveals post-transcriptional processing of the 5' ends in 8 of the 11 mRNAs (Montoya et al. 1981). Moreover, they have minimal secondary structures at their 5' ends (Jones et al. 2008).

In animal mitochondria, the 22 tRNAs are encoded by the mitochondrial genome. They are shorter than other tRNAs and lack the conserved nucleotides that play important roles in tRNA-folding (Watanabe 2010). These tRNAs fold into the basic cloverleaf structure of canonical tRNAs. However, they lack a number of the tertiary interactions that are highly conserved in prokaryotic and eukaryotic cytoplasmic tRNA (Watanabe 2010; Helm et al. 2000). Although no crystal structure information is available for mitochondrial tRNAs, an L-shaped tRNA with a caved-in elbow region was found to be tightly bound to the mitochondrial ribosome at the P-site in cryo-EM reconstitution studies (Sharma et al. 2003).

Another interesting feature of the mitochondrial genome is the presence of only two rRNA species, 12S and 16S, for the small and large subunit rRNAs, respectively (Pietromonaco et al. 1991). A recent report suggests that a 5S rRNA is also imported into mitochondria; however, it has not been confirmed whether the imported 5S rRNA is incorporated into the mitochondrial ribosome (Smirnov et al. 2011). In addition to leaderless and short tRNAs, mitochondria contain rRNAs that are considerably smaller than their counterparts in bacterial and eukaryotic cytoplasmic ribosomes (Koc et al. 2010). The compactness of the mammalian mitochondrial genome suggests that this phenomenon is not coincidental and has possibly evolved to minimize damage from the oxidative environment of mammalian mitochondria.

2.1.3 Mitochondrial Translation Machinery

The mammalian mitochondrial translation machinery is composed of ribosomes, tRNAs, mRNAs, recycling factors, and the factors of translation initiation, elongation, and termination. Although emerging studies suggest RNA import into the mitochondria, all of the RNA components that are involved directly in the translation machinery, including rRNAs, tRNAs, and mRNAs, are encoded by the mitochondrial genome (Attardi 1985; Attardi et al. 1982). All the other components (translation factors and ribosomal proteins) are products of nuclear-encoded genes and are translated into proteins by cytoplasmic ribosomes before they are imported into the mitochondria.

The largest component of the translation machinery is the mitochondrial 55S ribosome, which is comprised of 28S and 39S subunits, called the small and large subunits, respectively (O'Brien 1971). Mitochondrial ribosomes have a molecular mass of about 2.7×10^6 Da, approximately the size of bacterial ribosomes. As mentioned above, mitochondrial rRNAs are shorter than their bacterial counterparts. They are truncated in specific locations, either in certain regions of their secondary structure or in entire domains (Koc et al. 2010). The missing regions in mammalian mitochondrial rRNAs are, for the most part, located on the periphery of bacterial rRNAs (Zweib et al. 1981; Glotz et al. 1981) and appear to be replaced by ribosomal proteins (Koc et al. 1999, 2000, 2001a, b, c). In fact, cryo-EM reconstitution studies have revealed that the truncated rRNA regions and domains are substituted with mitochondrial ribosomal proteins (Sharma et al. 2003). The combination of shorter rRNAs and a greater quantity of proteins has led us to describe mammalian mitochondrial ribosomes as "protein-rich" (Sharma et al. 2003; Koc et al. 2001b, c). For a more detailed and comprehensive analysis of the structural characteristics of mammalian mitochondrial ribosomes, readers should refer to Chap. 1.

The small subunit of the bovine mitochondrial ribosome has about 29 proteins, of which 14 have homologs in prokaryotic ribosomes. The remaining 15 proteins are specific to mitochondrial ribosomes. In contrast, the large subunit of bovine mitochondrial ribosomes contains about 48 proteins. Of these, 28 are homologs of bacterial ribosomal proteins, while the remaining 20 are unique to mitochondrial ribosomes. Only 15 of the mitochondria-specific proteins have homologs in the yeast mitochondrial ribosome (Koc et al. 2001b, c; Smits et al. 2007). This observation indicates that there is a significant divergence between the protein composition of mitochondrial ribosomes in higher and lower eukaryotes. Again, this high protein and shortened rRNA arrangement was probably favored during the evolution of mitochondria from endosymbiotic bacteria in order to protect rRNA, which is more prone to oxidative damage than proteins.

Mammalian mitochondrial ribosomes resemble bacterial ribosomes more closely than eukaryotic cytoplasmic ribosomes, as shown in the homology of bacterial and mammalian mitochondrial protein components. About half of the mitochondrial ribosomal proteins (MRPs) have homologs in bacterial ribosomes, while the remaining proteins represent a new class of ribosomal proteins that is specific to mitochondria (Koc et al. 2001b, c, 2010). Conversely, bacterial ribosomes contain certain proteins not preserved in mitochondrial ribosomes. The distribution of mitochondria-specific proteins on the exterior surface of the ribosome is visible in cryo-EM studies; however, additional high-resolution structural information is still needed to determine the exact function of these proteins in translation (Sharma et al. 2003). Our studies suggest that mitochondria-specific proteins are replacements of the bacterial 70S ribosomal proteins that do not have clear homologs in mitochondrial ribosomes (Koc et al. 2001b, c; Suzuki et al. 2001). In addition to their primary functions in protein synthesis, several mitochondria-specific MRPs, including MRPS29 (also known as DAP3) and MRPS30, are reported to be involved in apoptosis and various diseases (Henning 1993; Kissil et al. 1995; Mariani et al. 2001; Takeda et al. 2007; Bhatti et al. 2010; Stacey et al. 2008; Woolcott et al. 2009). The disease-causing mutants and defects of these MRPs and other mitochondrial translation components has been elegantly summarized in several recent reviews (Rotig 2011; Smits et al. 2010; Christian and Spremulli 2011).

2.1.4 Interactions of Mammalian Mitochondrial Ribosomes with the Inner Membrane

Along with mtDNA and transcription machinery, mitochondrial ribosomes are associated with the IM, and mitochondrially encoded subunits are co-translationally inserted into the OXPHOS complexes (Bogenhagen 2009; Wang and Bogenhagen 2006; Liu and Spremulli 2000; Mick et al. 2012; Ott and Herrmann 2010). These 13 proteins are located at the hydrophobic cores of the OXPHOS complexes, possibly to prevent proton leakage and provide tightly coupled mitochondria. Studies in yeast provided most of our knowledge of the interactions of mitochondrial ribosomes with the IM and the assembly of mitochondrially encoded subunits of OXPHOS (Ott and Herrmann 2010; Bonnefoy et al. 2001; Naithani et al. 2003; Hell et al. 1998). The majority of the proteins involved in these processes are either absent or not highly conserved in mammals. In mammals, only three IM proteins, Oxa1L, Cox18, and LetM, have been found to interact with mitochondrial ribosomes, which are homologous to yeast IM proteins Oxa1p (a homolog of bacterial Yidc, Cox18p or Oxa2, and Mdm38p, respectively (Bonnefoy et al. 2009; Frazier et al. 2006; Gaisne and Bonnefoy 2006). It was recently proposed that the components conserved in both systems regulate mitochondrial translation in response to the assembly state of the OXPHOS complexes (Mick et al. 2012). It is possible that this response mechanism, which depends on the availability of its components, is the most efficient way to regulate mitochondrial biogenesis at multiple stages. Interaction of mitochondrial ribosomes with the IM and co-translational insertion of mitochondrially encoded components can be found in several recent reviews (Christian and Spremulli 2011; Ott and Herrmann 2010; Fox 2012).

2.2 Protein Synthesis in Mammalian Mitochondria

Protein synthesis occurs in four stages, designated as initiation, elongation, termination, and ribosome recycling. Here, we will briefly describe the protein factors and mechanisms involved in these processes. The majority of factors involved in protein synthesis in mammalian mitochondria were initially discovered and characterized in the Spremulli laboratory (see recent reviews by Spremulli et al. (2010, 2011, 2011) for the detailed structural and mechanistic aspects of these proteins).

2.2.1 Translation Initiation

Translation initiation starts with the dissociation of ribosomes into their subunits. The formylated Met-tRNA (fMet-tRNA) is base-paired to the start codon on the mRNA in the P-site of the small subunit. This process is stimulated by three translation initiation factors in bacteria: IF1, IF2, and IF3. Although IF1 has been viewed as a universal translational initiation factor in bacteria, only two mammalian mitochondrial initiation factors, mitochondrial initiation factor 2 (IF2mt) and mitochondrial initiation factor 3 (IF3mt), have been identified to date (Koc and Spremulli 2002; Liao and Spremulli 1990, 1991). We have shown that these two factors are sufficient to assemble an initiation complex on 55S ribosomes with fMet-tRNA in vitro (Koc and Spremulli 2002; Grasso et al. 2007).

In the current model (Fig. 2.2), the first step is the dissociation of the 55S ribosome into its subunits by IF3mt and formation of the 28S-IF3mt complex. It is proposed that mRNA could enter via a protein-rich mRNA entrance and bind to the 28S subunit first (Sharma et al. 2003); however, the exact mechanism of mRNA binding to the ribosome is not known. Currently, the order of mRNA,

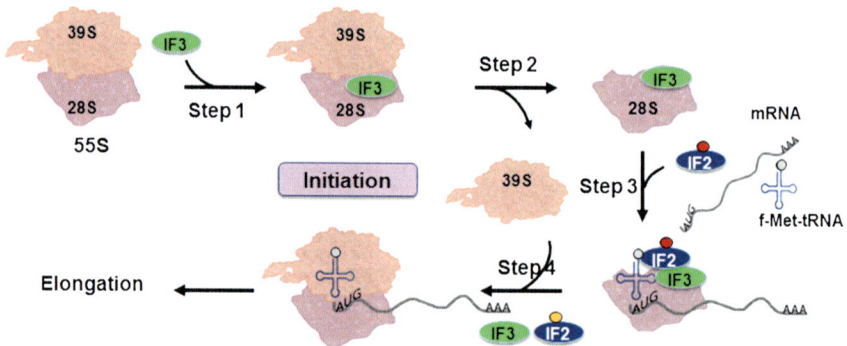

Fig. 2.2 Model for the initiation stage mitochondrial protein synthesis. The process of initiation begins with the 39S and 28S subunits associated a 55S ribosome. Mitochondrial initiation factor 3 (IF3) binds and dissociates the 55S ribosomes into its subunits (Step 1). The IF3mt remains bound to the 28S subunit (Step 2). Thereafter, fMet-tRNA, mRNA, and mitochondrial initiation factor 2 (IF2), which is bound to GTP (*red circle*), bind to the 28S subunit (Step 3). The presence of the proper start codon (AUG) allows the mRNA to be locked into place by codon–anticodon interactions, and this setup forms the 28S initiation complex. The 39S subunit joins the initiation complex, and GTP is hydrolyzed to GDP (*orange circle*) to allow IF2mt and IF3mt to dissociate (Step 4)

fMet-tRNA, and IF2mt:GTP binding to the 28S-IF3mt complex is not clear (Christian and Spremulli 2011; Grasso et al. 2007). For simplicity, these processes are shown together in Step 3. Toeprinting analysis suggests that mRNA movement is paused on the 28S subunit to inspect the codon at the 5′ end of the mRNA (Christian and Spremulli 2010). During this pause, IF2mt:GTP can promote the binding of fMet-tRNA to the 5′ AUG start codon of leaderless mitochondrial mRNAs on the 28S the ribosome (Step 3). Next, the large subunit joins the 28S initiation complex, IF2mt hydrolyzes GTP to GDP, and the initiation factors are released, resulting in a complete 55S initiation complex (Step 4).

As mentioned above, mammalian mitochondrial mRNAs lack the 5′ methylguanylate cap structure and the canonical Shine-Dalgarno helix found in eukaryotic and bacterial mRNAs, respectively, to position the start codon at the P-site of the ribosome. However, the location of the start codon within one or two nucleotides of the 5′-end of leaderless mRNAs has been shown to be critical in initiation complex formation *in vitro* (Christian and Spremulli 2010). These unusual features of leaderless mRNAs suggest the presence of a novel mRNA recognition and binding mechanism to ribosomes, possibly provided by mitochondria-specific ribosomal proteins or additional protein factors in mammalian mitochondria. It is possible that this is one of the most highly regulated steps in mitochondrial translation, but this remains to be investigated.

2.2.1.1 Mitochondrial Translation Initiation Factor 2

Mammalian mitochondrial IF2mt, which is homologous to bacterial IF2, was initially characterized and purified from bovine liver (Schwartzbach et al. 1996). The full-length human, bovine, and mouse IF2mt are all 727 amino acids (aa) long. The predicted mature form, which includes residues 78-727, can be stably expressed in *E. coli* (Claros and Vincens 1996). This recombinant protein is capable of promoting the binding of the initiator tRNA (fMet-tRNA) to mitochondrial 28S subunits or 55S ribosomes in the presence of GTP and synthetic mRNAs. Mammalian IF2mt can also stimulate the binding of fMet-tRNA to bacterial ribosomes; however, bacterial IF2 cannot stimulate formation of the initiation complex on mitochondrial ribosomes (Ma and Spremulli 1995).

When compared to the six-domain model of *E. coli* IF2, mammalian IF2mt covers domains III–VI, with an additional small insertion domain that is only found in animal mitochondria (Christian and Spremulli 2011; Atkinson et al. 2012; Yassin et al. 2011; Spremulli et al. 2004). The electron density map of the initiation complex formed with bovine IF2mt and 70S ribosomes at 10.8 Å reveals a three-dimensional model for the structure of this factor (Fig. 2.3a) (Yassin et al. 2011). Although domain III could not be modeled in this structure, the deletion mutations of IF2mt suggest that this domain interacts with the ribosome and makes important contacts with the 28S subunit (Spencer and Spremulli 2005). Domain IV, which is called the G-domain and is the most highly conserved domain, is also responsible for effective binding to 28S subunits. The presence of

the G-domain is required for the proper arrangement of IF2mt domains, specifically domain V and its interactions with the large subunit near the ribosomal L7/L12
stalk and the sarcin-ricin loop (Allen et al. 2005; La Teana et al. 2001). The small
insertion domain is located between domains V and VIC1, and its structural
aspects are discussed in Chap. 1 by Agrawal et al. The region of IF2mt responsible for interacting with fMet-tRNA has been mapped to the VIC2 subdomain
(Fig. 2.3a) (Spencer and Spremulli 2004). The mammalian mitochondrial genome
encodes only a single tRNA^{Met} gene that is partitioned between initiation and
elongation after aminoacylation. This is also a unique feature that is only found
in the animal mitochondrial translation system. Translation initiation starts with
a formylated Met-tRNA (fMet-tRNA); therefore, only a fraction of Met-tRNA is
formylated by a mitochondrial Met-tRNA formylase (Takeuchi et al. 1998). The
formylated form interacts with IF2mt and is used in translation initiation, while
the unformylated Met-tRNA interacts with EF-Tumt and is channeled into elongation. It is postulated that this partitioning begins with the competition between
transformylase and EF-Tumt (Spencer and Spremulli 2004).

2.2.1.2 Mitochondrial Translation Initiation Factor 3

We discovered mammalian IF3mt in homology searches of human and mouse
ESTs using IF3 sequences from various species as queries (Koc and Spremulli
2002). The coding region of human IF3mt contains 278 amino acid residues.

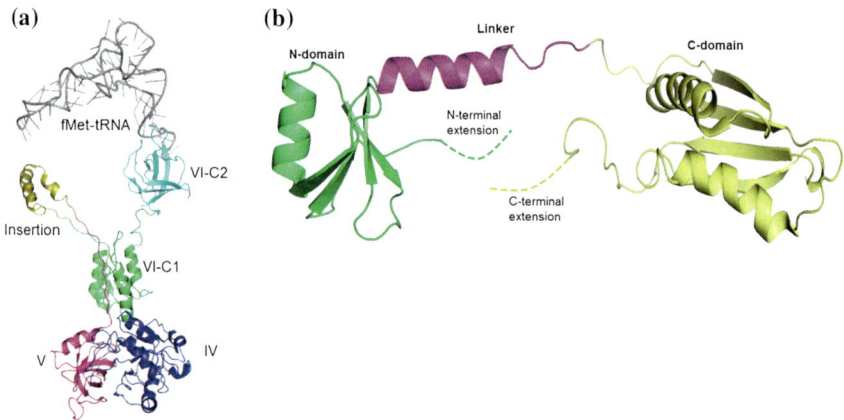

Fig. 2.3 Structural models of mammalian mitochondrial initiation factors. **a** Model for the 3-D
structure of IF2mt, based on the cryo-EM map of IF2mt:70S initiation complex (PDB# 3IZY).
Domain III has been omitted from the structure due to the resolution obtained with the cryo-
EM images. Domain IV is shown in *blue*, domain V is shown in *purple*, the insertion domain
is shown in *orange*, domain VI-C1 is shown in *green*, and domain VI-C2 is shown in *pink*.
b Structural model of IF3mt based on the crystal structure of the *N*-terminal domain of *G. stearothermophilus* IF3 and the NMR structure of the murine IF3mt *C*-terminal domain (PDB# 1TIF
and 2CRQ)

IF3mt is highly conserved in mammals, while the sequence homology is 35–40 % in vertebrates. Recently, a clear homolog in *Saccharomyces cerevisiae* was reported (Atkinson et al. 2012). Recombinant human IF3mt was generated after removing the 31-residue signal peptide predicted by MitoProtII (Koc and Spremulli 2002). Bacterial factors contain two globular domains separated by a linker region (Biou et al. 1995). The structural model of IF3mt based on the crystal structure of the *N*-terminal and linker domains of *Geobacillus stearothermophilus* IF3 and the NMR structure of the *C*-terminal domain of mouse IF3mt indicates that the mammalian factor is also organized in a dumbbell shape (Fig. 2.3b) (Moreau et al. 1997). The *N*-terminal homology domain of IF3mt contains an α-helix folded against four β-sheets (Fig. 2.3b). The *C*-terminal domain of IF3mt is folded into a similar structure, except with two α-helices (Fig. 2.3b). Although the linker region contains a helical segment in bacteria, the mitochondrial linker is predicted to be a partial α-helix. Deletion mutations in IF3mt biochemical studies suggest that this linker is highly flexible prior to binding to the ribosome (Christian and Spremulli 2009).

IF3mt promotes the dissociation of 55S ribosomes into their subunits and stimulates the binding of fMet-tRNA to mitochondrial 28S subunits in the presence of IF2mt and mRNA (Koc and Spremulli 2002). It also prevents premature binding of fMet-tRNA to ribosomes prior to mRNA binding (Bhargava and Spremulli 2005). However, the proofreading activity found in bacterial IF3 is not conserved in IF3mt (Petrelli et al. 2001). This lack of proofreading activity is postulated to be due to the recognition of both AUG and AUA as start codons and to the partitioning of tRNA^{Met} between translation initiation and elongation (Christian and Spremulli 2011). The deletion and point mutations of IF3mt and their roles in translation initiation have been studied extensively (Christian and Spremulli 2009; Bhargava and Spremulli 2005; Haque and Spremulli 2008). The removal of the *N*-terminal domain and the linker region only slightly affects IF3mt activity, whereas truncation of the *C*-terminal domain completely inactivates IF3mt function and its binding to the 28S subunit. It has also been shown that the *C*-terminal domain of IF3mt is sufficient by itself to promote initiation complex formation with the 55S ribosome (Christian and Spremulli 2009). In fact, Ala mutations of several critical residues at the *C*-terminal domain of human IF3mt, located at positions 170–171 and 175, result in loss of its dissociation function, and therefore, initiation complex formation (Christian and Spremulli 2009). This observation clearly supports the IF3mt mechanism of action proposed in Fig. 2.2. The *N*- and *C*-terminal extensions of IF3mt also play roles in monitoring the sequence of initiation complex formation by reducing the affinity of the factor for the 39S subunit and preventing the premature binding of fMet-tRNA, respectively (Fig. 2.4) (Christian and Spremulli 2009, 2011).

In bacteria, IF3 binds to the platform region of the small subunit, and this region is one of the most highly conserved regions of the mitochondrial ribosome (Sharma et al. 2003; Koc et al. 2001b; McCutcheon et al. 1999; Pioletti et al. 2001). We have identified the contacts between IF3mt and the 28S subunit using cross-linking assays in combination with identification of cross-linked ribosomal

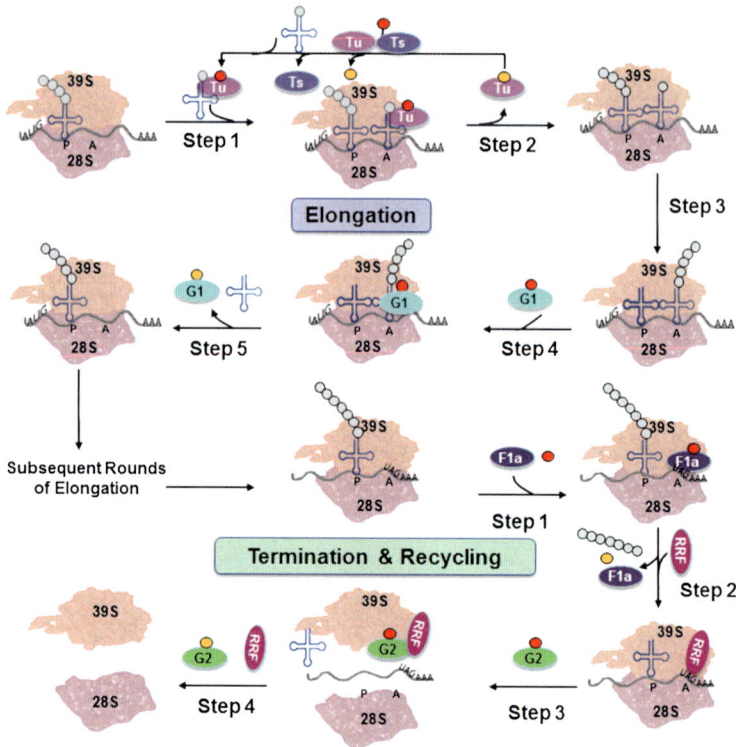

Fig. 2.4 Model for the elongation and termination steps of mitochondrial protein synthesis. During elongation, the tRNA with the growing polypeptide chain or fMet-tRNA is located in the P-site of the ribosome. A second aa-tRNA is brought to the A-site of the ribosome by EF-Tumt (Tu), which is bound to GTP (shown as a *red circle*) (Step 1). As GTP is hydrolyzed to GDP (depicted as an *orange circle*), EF-Tumt is released from the ribosome, and at this step, EF-Tsmt (Ts) is required for GDP-GTP exchange (Step 2). In the next step, peptide bond synthesis is catalyzed and the growing polypeptide chain is transferred to the aa-tRNA in the A-site (Step 3). Next, EF-G1mt:GTP (G1 with a *red circle*) binds to the ribosome's A-site and catalyzes its translocation (Step 4). This moves the deacetylated tRNA out of the P-site and shifts the newly acetylated tRNA to the P-site (Step 5). Elongation proceeds until the ribosome recognizes a termination codon. The beginning of termination is signaled by the termination codon (shown here as UAG), which enters the ribosome's A-site (Step 1). Both mtRF1a (F1a) and GTP bind to the A-site, and the polypeptide chain is released with hydrolysis of GTP (Step 2). The mechanism for the release of mtRF1a and GDP from the ribosome is still unknown. RRF1mt (RRF) then binds to the A-site, followed by EF-G2mt (G2, also known as RRF2mt) (Step 3). These two release factors promote the dissociation of the ribosomal subunits as well as the release of the mRNA and deacetylated tRNA (Step 4). The ribosome is free to perform further cycles of translation after RRF1mt and EF-G2mt are released

proteins by mass spectrometry (Haque et al. 2011). In this analysis, MRPs with bacterial homologs (such as MRPS5, MRPS9, MRPS10, and MRPS18-2) and proteins with no homologs in bacterial ribosomes (such as MRPS29, MRPS32,

MRPS36, and PTCD3) are identified as the 28S subunit proteins that interact with IF3mt. The interaction between IF3mt and MRPS29 (DAP3) is unexpected, because IF3mt was found to be located at the lower part of the 28S subunit by immunoelectron microscopy (O'Brien et al. 2005). The locations of MRPS32, MRPS36, and PTCD3 in the small subunit are not known; however, we have shown that PTCD3 is one of the proteins that interacts with mRNA on the 28S subunit (Koc and Spremulli 2003). The group of bacterial homologs of MRPs cross-linked to IF3mt is different from the ribosomal proteins that interact with the bacterial IF3 in the platform of the 30S subunit. This difference is possibly due to the mitochondria-specific ribosomal proteins surrounding the platform region in the 28S subunit (Koc and Spremulli 2003).

2.2.2 Translation Elongation in Mammalian Mitochondria

During elongation, mRNA is decoded sequentially as amino acids are incorporated into the growing polypeptide chain (Fig. 2.4). In contrast to the differences observed between the mitochondrial and bacterial translation initiation processes, elongation is highly conserved in these systems (Spremulli et al. 2004). The basic mechanism of mitochondrial translation elongation is summarized in Fig. 2.4. In the first step, a ternary complex formed with GTP, mitochondrial elongation factor Tu (EF-Tumt), and aminoacyl-tRNA (aatRNA) enters the A-site of the 55S ribosome, which is occupied with either fMet-tRNA or aa-tRNA at the P-site (Step 1). Formation of the correct cognate ternary complex hydrolyzes GTP and causes the release of EF-Tu:GDP in Step 2. Elongation Factor Ts (EF-Tsmt) supports the GDP/GTP exchange by forming an EF-Tumt:EF-Tsmt complex. The peptidyl-transferase activity of the ribosome catalyzes peptide bond formation and extends the peptide chain by one residue on the peptidyl-tRNA at the A-site, leaving a deacylated tRNA in the P-site (Step 3). In the next step, deacetylated tRNA is removed (Step 4) and peptidyl-tRNA at the A-site is translocated into the P-site (Step 5) by mitochondrial Elongation Factor G1 (EF-G1mt). In bacteria, the deacylated tRNA is progressively released from the E-site; however, a corresponding E-site tRNA-binding site was not found in the cryo-EM studies of the mitochondrial ribosome (Sharma et al. 2003). At the final stage of elongation, ribosome with peptidyl-tRNA at the P-site is ready for either a subsequent elongation cycle or for termination. Mammalian mitochondrial elongation factors, specifically EF-Tumt and EF-Tsmt, have been extensively studied, either using their native forms from bovine mitochondria or recombinant proteins from *E. coli* (Schwartzbach and Spremulli 1989; Woriax et al. 1995, 1996; Xin et al. 1995). Here, we will briefly discuss the properties of the mammalian mitochondrial elongation factors. For more comprehensive information on mitochondrial elongation factors, readers are referred to previous reviews (Christian and Spremulli 2011; Spremulli et al. 2004).

2.2.2.1 Mitochondrial Elongation Factors Tu and Ts

EF-Tumt and EF-Tsmt were the first mitochondrial translation factors identified and characterized from bovine mitochondria (Schwartzbach and Spremulli
1989). EF-Tumt is a 452 aa long protein in mammalian mitochondria. The
mature form of the bovine protein is 409 residues long and is 55–60 % identical
to bacterial EF-Tu. Crystal structures of the bacterial EF-Tu ternary complexes
from several different species have been solved (Nissen et al. 1995, 1999). The
crystal structure of bovine EF-Tumt has similarities to bacterial EF-Tu and is
folded into three domains (Jeppesen et al. 2005) (Fig. 2.5). Domain I binds
guanine nucleotides and forms the 3′-end of the aa-tRNA-binding site along
with domain II. These domains also interact with the small subunit during the
tRNA delivery to the peptidyl-transferase active center. The extended acceptor-
TΨC helix of the aa-tRNA interacts with domain III. A binding pocket for the
5′-end and the acceptor stem of the aa-tRNA is provided by all three domains
(Fig. 2.5).

Studies performed with chimeric bacterial EF-Tu and EF-Tumt proteins reveal
that domains I and II of the mitochondrial factor are primarily responsible for
aa-tRNA delivery to the ribosome (Bullard et al. 1999; Hunter and Spremulli
2004a). It is also suggested that the formation of codon–anticodon interactions
possibly causes conformational changes in the body of the tRNA and sends a
signal to domain I, triggering GTP hydrolysis. This observation is extremely

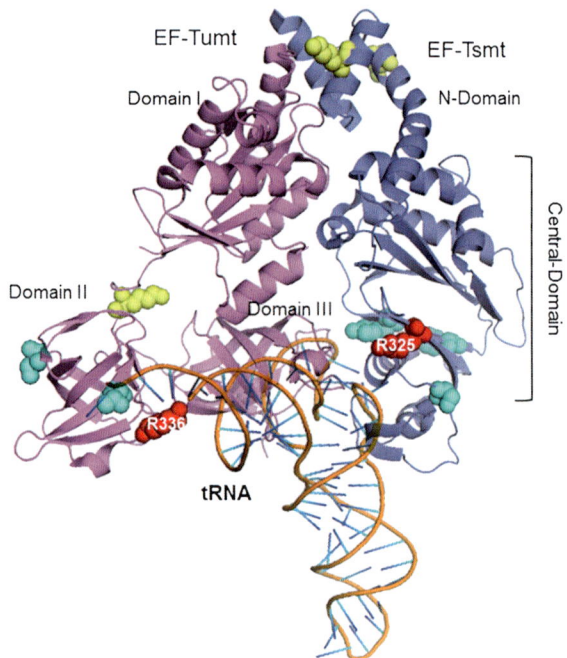

Fig. 2.5 Model of the bovine EF-Tumt:EF-Tsmt complex with tRNA. In this model of the bovine EF-Tumt:EF-Tsmt complex (PDB# 1XB2) with *E. coli* Cys-tRNA^{Cys} (PDB# IB23), EF-Tumt, EF-Tsmt, and tRNA are shown in *violet*, *blue*, and *orange*, respectively. Individual domains observed in the structure are indicated. The positions of two disease-causing mutations found in EF-Tumt (R336) and EF-Tsmt (R325) are labeled in *red*. EF-Tumt is phosphorylated at highly conserved Thr273 and Ser312 residues (shown in *cyan*) near the tRNA-binding site and acetylated at Lys88 and Lys 256 (shown in *yellow*)

valuable in explaining mitochondrial myopathies caused by mutations in mito-chondrial tRNAs, specifically those causing conformational changes, rendering them inefficient in triggering the GTPase activity of EF-Tu (Hunter and Spremulli 2004b, c, d; Shi et al. 2012; Hao and Moraes 1997; Kelley et al. 2000; Ling et al. 2007) (Fig. 2.5).

In mammals, the most common form of EF-Tsmt is 325 amino acid resi-dues in length, and the mature protein is about 31 kDa (Xin et al. 1995). Although bacterial EF-Ts is found as a free protein, EF-Tsmt has been identi-fied to be complexed with EF-Tumt (Schwartzbach and Spremulli 1989). EF-Tsmt is only 25–30 % identical to its bacterial counterpart. In agreement with this low sequence homology, the structure of EF-Tsmt is also less conserved in comparison to the structural homology between mitochondrial and bacte-rial EF-Tus. One of the most remarkable differences is the complete loss of the coiled-coil domain structure of the C-terminal domain, in contrast to the similar folds found in the N-terminal domain (Fig. 2.5). Likewise, the β-strand folds located in the central domain of EF-Tsmt are organized differently in the bac-terial protein. The central domain of EF-Tsmt interacts with domains I and II in EF-Tumt, and this region undergoes substantial conformational changes dur-ing the nucleotide exchange process between EF-Tumt and EF-Tsmt (Fig. 2.5) (Jeppesen et al. 2005).

The mechanistic and structural aspects of mammalian mitochondrial translation elongation factors described above have been extensively studied and reviewed by the Spremulli laboratory (Christian and Spremulli 2011; Spremulli et al. 2004). The *in vitro* studies performed by this laboratory have shed light on the role of these factors in metabolic diseases and cancers, as reported recently (Shi et al. 2012; Smeitink et al. 2006; Skrtic et al. 2011; Akama et al. 2010). For a com-prehensive representation of the roles of these factors in human diseases, readers should refer to the previous reviews (Rotig 2011; Akama et al. 2010).

2.2.2.2 Mitochondrial Elongation Factor G1

There are two forms of EF-Gmt: EF-G1mt and EF-G2mt (Tsuboi et al. 2009; Hammarsund et al. 2001). EF-G1mt catalyzes the translocation of peptidyl-tRNA from the A-site to the P-site while dislocating P-site tRNA from the ribosome (Fig. 2.4). EF-G2mt is involved in termination. Full-length EF-G1mt is 751 aa res-idues, and both native and recombinant forms of mammalian EF-G1mt are active in catalyzing translocation in both mitochondrial and bacterial translation elonga-tion *in vitro* (Tsuboi et al. 2009; Chung and Spremulli 1990; Bhargava et al. 2004). On the other hand, bacterial EF-G does not support mitochondrial translocation (Chung and Spremulli 1990). Another unusual feature of mammalian EF-G1mt is the resistance of this factor to fusidic acid (Bhargava et al. 2004). The specific-ity of EF-G1mt can be attributed to the structural differences between the EF-G sequence and the L7/L12 stalk region of mitochondrial and bacterial ribosomes (Bhargava et al. 2004; Terasaki et al. 2004).

2.2.3 Termination of Mitochondrial Translation and Ribosome Recycling

The final stage of protein synthesis is termination and the dissociation of the ribosome complex for recycling. In the proposed model, UAA and UAG serve as stop codons in mammalian mitochondria (Fig. 2.4a). In the first step of termination, these stop codons move into the A-site and are recognized by mtRF1a (referred to as F1a in Fig. 2.4). In humans, two arginine codons, AGA and AGG, also function as terminal codons; however, these codons promote -1 frameshifting and reposition the standard UAG codon at the A-site for termination (Temperley et al. 2010). The binding of mtRF1a to the ribosome triggers GTP hydrolysis and the release of the completed peptide during Step 2. In the following step, ribosomes are ready to be recycled and to release mtRF1a. This step requires RF3 in bacteria. However, no factor homologous to RF3 has been found in mitochondria; thus, the mechanism of mtRF1a dissociation is not known. In Step 3, mitochondrial ribosome recycling factor (RRF1mt, abbreviated as RRF in Fig. 2.4) and EFG2mt (also known as RRF2mt and abbreviated as G2 in Fig. 2.4) act together to release mRNA and P-site tRNA and to dissociate 55S ribosomes into the subunits. The release of these factors from the ribosomes, accompanied by GTP hydrolysis, prepares them for the next round of protein synthesis in Step 4 (Fig. 2.4) (Chrzanowska-Lightowlers et al. 2011).

2.2.3.1 Mitochondrial Release Factors

There are three mitochondrial proteins with significant homology to bacterial RF1: mtRF1a, c12orf65, and ICT1. All of these factors contain the GGQ motif, which is critical for the termination of protein synthesis because they all hydrolyze the peptidyl-tRNA bond (Chrzanowska-Lightowlers et al. 2011). Full-length human mtRF1a contains 445 amino acid residues and has been shown to participate in the termination of 11 human mitochondrial mRNAs at UAA and UAG codons (Zhang and Spremulli 1998; Soleimanpour-Lichaei et al. 2007). Therefore, it is described as the mitochondrial release factor used in the decoding process (Christian and Spremulli 2011; Soleimanpour-Lichaei et al. 2007). Currently, the function of c12orf65 is not known; however, ICT1 functions as a ribosome-dependent peptidyl-tRNA hydrolase because it is tightly associated with the large subunit of the mitochondrial ribosome (Richter et al. 2010). It is also proposed to have a role in the recycling of stalled ribosomes on damaged mRNAs or mRNAs lacking stop codons (Chrzanowska-Lightowlers et al. 2011). This is a possible replacement for the tmRNA mechanism found in bacteria (Christian and Spremulli 2011; Haque and Spremulli 2010).

2.2.3.2 Ribosome Recycling in Mammalian Mitochondria

The coding region of the mitochondrial ribosome recycling factor (mtRRF) is a 262 aa residue-long protein in mammals and is 25–30 % identical to bacterial RRF. In human cell lines, mtRRF has been shown to be essential (Rorbach et al. 2008). There are several alternatively spliced isoforms of mtRRF listed in protein databases such as NCBI and UniProt; however, it is not known whether these shorter isoforms function in the termination of protein synthesis. The bacterial form of mtRRF works with EF-G, the only translocation factor found in bacteria. In mammals, however, mtRRF and EF-G2mt cooperate to disassemble the ribosome:mRNA:tRNA complex in the termination steps (Steps 3 and 4) of mitochondrial translation (Fig. 2.4).

As described above, two proteins homologous to bacterial EF-G (EF-G1mt and EF-G2mt) are present in mammalian mitochondria. While EF-G1mt is responsible for translocation during chain elongation, EF-G2mt, also known as RRF2mt, functions in combination with mtRRF at the termination step (Tsuboi et al. 2009). In many organisms, there is only a single EF-G involved in both elongation and termination, as occurs in bacteria. However, mammalian EF-G2mt only serves in the termination process, and GTP hydrolysis is required for its release, along with mtRRF, from the ribosome (Tsuboi et al. 2009). In bacterial termination, GTP hydrolysis is necessary for ribosome recycling.

2.3 Regulation of Mitochondrial Protein Synthesis

Most of our knowledge of mammalian mitochondrial protein synthesis comes from either *in vitro* studies or from studies of disease-causing or lethal mutations in humans. Yet, the regulation of the mitochondrial translation machinery, which is essential for cellular homeostasis and survival, has not been explored in great detail. The regulation of mitochondrial protein synthesis requires coordinated expression of nuclear and mitochondrial genes at the transcriptional and translational levels. This retrograde signaling is proposed to be regulated by molecules derived from mitochondrial metabolism, such as reactive oxygen species (ROS), NAD^+, Acetyl-coA, and ATP (Wallace 2012; Finley and Haigis 2009; Schieke and Finkel 2006). In addition, it has been suggested that organelle biogenesis in plants and animals is regulated by dual or multiple localizations of proteins and small RNA species, such as 5S rRNA and microRNAs, into the nucleus, cytosol, and mitochondria (Smirnov et al. 2011; Duchene and Giege 2012; Vedrenne et al. 2012; Ernoult-Lange et al. 2012). Regulation of protein synthesis by these small RNAs in yeast and mammalian mitochondria will be discussed in Chap. 4.

Recent technological progress in system biological approaches has demonstrated that these approaches have the potential to solve metabolite-regulated signaling processes in mammalian mitochondria. As indicated above, Acetyl-coA, NAD^+, and ATP are the most essential metabolites of oxidative energy metabolism and are used for a variety of cellular functions in health and disease (Fig. 2.1). More importantly, these metabolites are involved in post-translational modifications and the modulation of protein activities by reversible acetylation and phosphorylation in mammalian mitochondria (O'Rourke et al. 2011; Hebert et al. 2013; Zhao et al. 2010). High-throughput proteomic surveys of mitochondrial proteins employed by many different laboratories have confirmed that these post-translational modifications are very common and are regulated by the activities of mitochondrially localized kinases and NAD^+-dependent deacetylases (sirtuins), including SIRT3, SIRT4, and SIRT5 (Hebert et al. 2013; Zhao et al. 2010; Choudhary et al. 2009a; Kim et al. 2006; Michishita et al. 2005; Schwer et al. 2006; Onyango et al. 2002; Bell and Guarente 2011).

Protein synthesis in eukaryotic cytoplasm by 80S ribosomes is predominantly regulated by the phosphorylation of translation factors, specifically eIF2a (Kimball et al. 1998; Hershey 1989). Phosphorylation of eIF2a at Ser 51 results in the differential expression of mRNAs and directs many critical cellular events, such as cell growth and apoptosis. However, the regulatory role of phosphorylation on bacterial and mammalian mitochondrial translation systems has been overlooked until recently. Our recent studies have revealed that the regulation of mitochondrial translation by post-translational modifications, specifically by phosphorylation and acetylation, is far more widespread than originally thought (Miller et al. 2008, 2009; Yang et al. 2010). Additionally, the phosphorylation and acetylation of mammalian mitochondrial ribosomal proteins and translation factors have been mapped in high-throughput proteomics approaches using tandem mass spectrometry (Zhao et al. 2010; Choudhary et al. 2009a; He et al. 2001; Zhang et al. 2009; Sajid et al. 2011; Koc and Koc 2012). However, the regulation mechanism of mitochondrial translation factors by reversible acetylation and phosphorylation remains largely to be discovered. A comprehensive list of post-translationally modified bacterial and mammalian mitochondrial ribosomal proteins and translation factors can be found in a recent review by our group (Koc and Koc 2012).

Surprisingly, we observe that post-translationally modified aa residues are found in the functionally critical regions of translation factors and MRPs, supporting the hypothesis that these modifications play a role in the regulation of mitochondrial translation and, consequently, OXPHOS. For example, inhibition of bacterial protein synthesis by phosphorylation of EF-Tu has long been known; however, phosphorylation of Thr 118 has been recently demonstrated to inhibit tRNA binding in *Mycoplasma tuberculosis* EF-Tu (Sajid et al. 2011; Alexander et al. 1995). In two recent large-scale mouse phosphoproteome analyses, phosphorylation at Thr273 and Ser312 residues was mapped by mass spectrometry, and, interestingly, these two residues are located in the tRNA-binding region (Fig. 2.5) (Choudhary et al. 2009b; Huttlin et al. 2010). Moreover, most post-translationally modified regions of MRPs that are exposed to solvent are typically rich in Lys, Ser, Thr, and Tyr residues to support acetylation and phosphorylation. One of the unexpected observations included in our review is the conservation of modifications in both bacterial and mitochondrial ribosomes (Miller et al. 2009; Yang et al. 2010; Koc and Koc 2012; Soung et al. 2009). It is clear that these reversible modifications in the exposed parts of ribosomal proteins are likely to be involved in the recruitment of translation factors, rRNA and mRNA interactions, and subunit association.

One of the most active regions of mitochondrial and bacterial ribosomes during translation is the *L7-L12 stalk*. It is involved in the recruitment of translation factors IF2, EF-Tu, and EF-G in both bacteria and mammalian mitochondrial ribosomes (Uchiumi et al. 1999, 2002). This flexible region of the large subunit is highly conserved and is composed of L11, L10, and multiple copies of L7/L12, the exact number depending on the species. There is ample evidence that phosphorylation and acetylation in this region are involved in regulatory roles in both mitochondrial and bacterial ribosomes (Ilag et al. 2005; Traugh and Traut 1972; Mikulik et al. 2011; Gordiyenko et al. 2008). We have demonstrated that MRPL10 is the major acetylated protein in the mammalian mitochondrial ribosome and that an NAD^+-dependent deacetylase, SIRT3, is involved in the deacetylation of MRPL10 (Yang et al. 2010). In SIRT3 knock-out mice mitochondria, we have shown that the acetylation of MRPL10 results in the enhancement of mitochondrial translation. MRPL12 was also found to be acetylated in unpublished studies from our laboratory and in a recent high-throughput mapping of acetylated proteins in SIRT3 knock-out mice mitochondria (Hebert et al. 2013; Yang et al. 2010; Han et al. 2013). This observation suggests that the role of L12 acetylation in bacteria is also conserved in mammalian mitochondrial ribosomes (Gordiyenko et al. 2008; Ramagopal and Subramanian 1974). Due to the fact that the stalk itself is very flexible and that mitochondria contain free pools of MRPL12 copies, it is likely that the components of the stalk transiently associate with the ribosome to regulate protein synthesis in mitochondria. In fact, a free pool of MRPL12 has been reported to interact with mitochondrial RNA polymerase (Wang et al. 2007; Surovtseva et al. 2011).

In addition to the MRPs with bacterial homologs, we observed extensive modifications of mitochondria-specific proteins found in the small subunit of the 55S ribosome (Miller et al. 2009). One of the most extensively studied pro-apoptotic proteins of the 28S subunit, MRPS29 (DAP3), is both acetylated and phosphorylated. Its phosphorylation near the GTP-binding domain has been shown to be crucial to its pro-apoptotic function in various cell lines (Takeda et al. 2007; Miller et al. 2008; Miyazaki et al. 2004). Aberrant expression of DAP3 has been detected in thyroid oncocytic, brain, and breast cancer tissues (Mariani et al. 2001; Jacques et al. 2009; Wazir et al. 2012). Clearly, further studies need to be done with mitochondrial translation factors and ribosomal proteins to shed light on their roles and the mechanisms by which they are regulated, including and starting with PTMs, to support cellular homeostasis.

2.4 Future Directions

Major milestones have been reached toward the complete characterization and identification of the components of mammalian mitochondrial translation. However, some mechanistic and structural aspects of mitochondrial translation have yet to be understood. One of the major unanswered questions that remains concerning the mechanism of mitochondrial protein synthesis is the recognition of leaderless mRNAs. The evidence so far points to a protein-dominated mechanism for mRNA recognition and binding to the mitochondrial ribosome. In addition, the components that facilitate co-translational insertion of mitochondrially encoded subunits of OXPHOS complexes into the IM need to be discovered and studied in greater detail in mammals. Retrograde signaling pathway(s) providing communication links between the mitochondrial and cytoplasmic translation machineries remain to be uncovered. Finally, the modulation of the translation machinery by reversible acetylation and phosphorylation and the enzymes involved in these reversible modifications are urgently waiting to be investigated.

Acknowledgments We thank Ms. Tamara B. Trout for critical reading of this chapter and for technical assistance in the generation of figures.

References

Agrawal RK, Sharma MR, Yassin A, Lahiri I, Spremulli LL (2011) Structure and function of organaller ribosomes as revealed by cryo-EM in ribosomes: structure, functions and dynamics. Springer, New York

Akama K, Christian BE, Jones CN, Ueda T, Takeuchi N, Spremulli LL (2010) Analysis of the functional consequences of lethal mutations in mitochondrial translational elongation factors. Biochim Biophys Acta 1802:692–698

Alexander C, Bilgin R, Lindschau C, Mestes J, Kraal B, Hilgenfeld R, Erdmann V, Lippmann C (1995) Phosphorylation of elongation factor Tu prevents ternary complex formation. J Biol Chem 270:14541–14547

Allen GS, Zavialov A, Gursky R, Ehrenberg M, Frank J (2005) The cryo-EM structure of a translation initiation complex from *Escherichia coli*. Cell 121:703–712

Anderson S, de Brujin M, Coulson A, Eperon I, Sanger F, Young I (1982) Complete sequence of bovine mitochondrial DNA: conserved features of the mammalian mitochondrial genome. J Mol Biol 156:683–717

Atkinson GC, Kuzmenko A, Kamenski P, Vysokikh MY, Lakunina V, Tankov S, Smirnova E, Soosaar A, Tenson T, Hauryliuk V (2012) Evolutionary and genetic analyses of mitochondrial translation initiation factors identify the missing mitochondrial IF3 in *S. cerevisiae*. Nucleic Acids Res 40:6122–6134

Attardi G (1985) Animal mitochondrial DNA: an extreme example of genetic economy. Int Rev Cytology 93:93–145

Attardi G, Chomyn A, Montoya J, Ojala D (1982) Identifaction and mapping of human mitochondrial genes. Cytogenet Cell Genet 32:85–98

Bell EL, Guarente L (2011) The SirT3 divining rod points to oxidative stress. Mol Cell 42:561–568

Bhargava K, Spremulli LL (2005) Role of the N- and C-terminal extensions on the activity of mammalian mitochondrial translational initiation factor 3. Nucleic Acids Res 33:7011–7018

Bhargava K, Templeton P, Spremulli LL (2004) Expression and characterization of isoform 1 of human mitochondrial elongation factor G. Protein Expr Purif 37:368–376

Bhatti P, Doody MM, Rajaraman P, Alexander BH, Yeager M, Hutchinson A, Burdette L, Thomas G, Hunter DJ, Simon SL (2010) Novel breast cancer risk alleles and interaction with ionizing radiation among U.S. radiologic technologists. Radiat Res 173:214–224

Biou V, Shu F, Ramakrishnan V (1995) X-ray crystallography shows that translational initiation factor IF3 consists of two compact α/β domains linked by an α-helix. EMBO J 14:4056–4064

Bogenhagen DF (2009) Biochemical isolation of mtDNA nucleoids from animal cells. Methods Mol Biol 554:3–14

Bonnefoy N, Bsat N, Fox TD (2001) Mitochondrial translation of *Saccharomyces cerevisiae* COX2 mRNA is controlled by the nucleotide sequence specifying the pre-Cox2p leader peptide. Mol Cell Biol 21:2359–2372

Bonnefoy N, Fiumera HL, Dujardin G, Fox TD (2009) Roles of Oxa1-related inner-membrane translocases in assembly of respiratory chain complexes. Biochim Biophys Acta 1793:60–70

Bullard JM, Cai Y-C, Zhang Y, Spremulli LL (1999) Effects of domain exchanges between *Escherichia coli* and mammalian mitochondrial EF-Tu on interactions with guanine nucleotides, aminoacyl-tRNA and ribosomes. Biochim Biophys Acta 1446:102–114

Choudhary C, Kumar C, Gnad F, Nielsen ML, Rehman M, Walther TC, Olsen JV, Mann M (2009a) Lysine acetylation targets protein complexes and co-regulates major cellular functions. Science 325:834–840

Choudhary C, Olsen JV, Brandts C, Cox J, Reddy PN, Bohmer FD, Gerke V, Schmidt-Arras DE, Berdel WE, Muller-Tidow C et al (2009b) Mislocalized activation of oncogenic RTKs switches downstream signaling outcomes. Mol Cell 36:326–339

Christian BE, Spremulli LL (2009) Evidence for an active role of IF3mt in the initiation of translation in mammalian mitochondria. Biochemistry 48:3269–3278

Christian BE, Spremulli LL (2010) Preferential selection of the 5′-terminal start codon on leaderless mRNAs by mammalian mitochondrial ribosomes. J Biol Chem 285:28379–28386

Christian BE, Spremulli LL (2011) Mechanism of protein biosynthesis in mammalian mitochondria. Biochim Biophys Acta 1819:1035–1054

Chrzanowska-Lightowlers ZM, Pajak A, Lightowlers RN (2011) Termination of protein synthesis in mammalian mitochondria. J Biol Chem 286:34479–34485

Chung HK, Spremulli LL (1990) Purification and characterization of elongation factor G from bovine liver mitochondria. J Biol Chem 265:21000–21004

Claros MG, Vincens P (1996) Computational method to predict mitochondrially imported proteins and their targeting sequences. Eur J Biochem 241:770–786

Duchene AM, Giege P (2012) Dual localized mitochondrial and nuclear proteins as gene expression regulators in plants? Front Plan Sci 3:221

Ernoult-Lange M, Benard M, Kress M, Weil D (2012) P-bodies and mitochondria: which place in RNA interference? Biochimie 94:1572–1577

Finley LW, Haigis MC (2009) The coordination of nuclear and mitochondrial communication during aging and calorie restriction. Ageing Res Rev 8:173–188

Fox TD (2012) Mitochondrial protein synthesis, import, and assembly. Genetics 192:1203–1234

Frazier AE, Taylor RD, Mick DU, Warscheid B, Stoepel N, Meyer HE, Ryan MT, Guiard B, Rehling P (2006) Mdm38 interacts with ribosomes and is a component of the mitochondrial protein export machinery. J Cell Biol 172:553–564

Gaisne M, Bonnefoy N (2006) The COX18 gene, involved in mitochondrial biogenesis, is functionally conserved and tightly regulated in humans and fission yeast. FEMS Yeast Res 6:869–882

Glotz C, Zweib C, Brimacombe R (1981) Secondary structure of the large subunit ribosomal RNA from *Escherichia coli*, *Zea mays* chloroplasts and human and mouse mitochondrial ribosomes. Nuc Acids Res 9:3287–3306

Gordiyenko Y, Deroo S, Zhou M, Videler H, Robinson CV (2008) Acetylation of L12 increases interactions in the *Escherichia coli* ribosomal stalk complex. J Mol Biol 380:404–414

Grasso DG, Christian BE, Spencer A, Spremulli LL (2007) Overexpression and purification of mammalian mitochondrial translational initiation factor 2 and initiation factor 3. Methods Enzymol 430:59–78

Hammarsund M, Wilson W, Corcoran M, Merup M, Einhorn S, Grander D, Sangfelt O (2001) Identification and characterization of two novel human mitochondrial elongation factor genes, hEFG2 and hEFG1, phylogenetically conserved through evolution. Hum Genet 109:542–550

Han M-JC, H.; Koc H, Yang Y, Tong Q, Koc EC (2013) SIRT3 regulates mitochondrial translation by modulating MRPL12 binding to the ribosome (in preparation)

Hao H, Moraes CT (1997) A disease-associated G5703A mutation in human mitochondrial DNA causes a conformational change and a marked decrease in steady-state levels of mitochondrial tRNA^{ASN}. Mol Cell Biochem 17:6831–6837

Haque ME, Spremulli LL (2008) Roles of the N- and C-terminal domains of mammalian mitochondrial initiation factor 3 in protein biosynthesis. J Mol Biol 384:929–940

Haque ME, Spremulli LL (2010) ICT1 comes to the rescue of mitochondrial ribosomes. EMBO J 29:1019–1020

Haque ME, Koc H, Cimen H, Koc EC, Spremulli LL (2011) Contacts between mammalian mitochondrial translational initiation factor 3 and ribosomal proteins in the small subunit. Biochim Biophys Acta 1814:1779–1784

He H, Chen M, Scheffler NK, Gibson BW, Spremulli LL, Gottlieb RA (2001) Phosphorylation of mitochondrial elongation factor Tu in ischemic myocardium: basis for chloramphenicol-mediated cardioprotection. Circ Res 89:461–467

Hebert AS, Dittenhafer-Reed KE, Yu W, Bailey DJ, Selen ES, Boersma MD, Carson JJ, Tonelli M, Balloon AJ, Higbee AJ et al (2013) Calorie restriction and SIRT3 trigger global reprogramming of the mitochondrial protein acetylome. Mol Cell 49:186–199

Hell K, Herrmann JM, Pratje E, Neupert W, Stuart RA (1998) Oxa1p, an essential component of the N-tail protein export machinery in mitochondria. Proc Natl Acad Sci U S A 95:2250–2255

Helm M, Brule H, Friede D, Giege R, Putz D, Florentz C (2000) Search for characteristic structural features of mammalian mitochondrial tRNAs. RNA 6:1356–1379

Henning KA (1993) The molecular genetics of human diseases with defective DNA damage processing. PhD thesis, Stanford University, Stanford, USA

Hershey JW (1989) Protein phosphorylation controls translation rates. J Biol Chem 264:20823–20826

Hunter SE, Spremulli LL (2004a) Interaction of mitochondrial elongation factor Tu with aminoacyl-tRNAs. Mitochondrion 4:21–29

Hunter SE, Spremulli LL (2004b) Mutagenesis of Arg335 in bovine mitochondrial elongation factor Tu and the corresponding residue in the *Escherichia coli* factor affects interactions with mitochondrial aminoacyl-tRNAs. RNA Biol 1:95–102

Hunter SE, Spremulli LL (2004c) Mutagenesis of glutamine 290 in *Escherichia coli* and mitochondrial elongation factor Tu affects interactions with mitochondrial aminoacyl-tRNAs and GTPase activity. Biochemistry 43:6917–6927

Hunter SE, Spremulli LL (2004d) Effects of mutagenesis of residue 221 on the properties of bacterial and mitochondrial elongation factor EF-Tu. Biochim Biophys Acta 1699:173–182

Huttlin EL, Jedrychowski MP, Elias JE, Goswami T, Rad R, Beausoleil SA, Villen J, Haas W, Sowa ME, Gygi SP (2010) A tissue-specific atlas of mouse protein phosphorylation and expression. Cell 143:1174–1189

Ilag LL, Videler H, McKay AR, Sobott F, Fucini P, Nierhaus KH, Robinson CV (2005) Heptameric (L12)6/L10 rather than canonical pentameric complexes are found by tandem MS of intact ribosomes from thermophilic bacteria. Proc Natl Acad Sci U S A 102:8192–8197

Jackson RJ (1991) The ATP requirement for initiation of eukaryotic translation varies according to mRNA species. Eur J Biochem 200:285–294

Jacques C, Fontaine JF, Franc B, Mirebeau-Prunier D, Triau S, Savagner F, Malthiery Y (2009) Death-associated protein 3 is overexpressed in human thyroid oncocytic tumours. Br J Cancer 101:132–138

Jeppesen MG, Navratil T, Spremulli LL, Nyborg J (2005) Crystal structure of the bovine mitochondrial elongation factor Tu·Ts complex. J Biol Chem 280:5071–5081

Jones CN, Wilkinson KA, Hung KT, Weeks KM, Spremulli LL (2008) Lack of secondary structure characterizes the 5′ ends of mammalian mitochondrial mRNAs. RNA 14:862–871

Kelley SO, Steinberg SV, Schimmel P (2000) Functional defects of pathogenic human mitochondrial tRNAs related to structural fragility. Nat Struct Biol 7:862–865

Kim SC, Sprung R, Chen Y, Xu Y, Ball H, Pei J, Cheng T, Kho Y, Xiao H, Xiao L et al (2006) Substrate and functional diversity of lysine acetylation revealed by a proteomics survey. Mol Cell 23:607–618

Kimball S, Fabian J, Pavitt G, Hinnebusch A, Jefferson L (1998) Regulation of Guanine nucleotide exchange through phosphorylation of eukaryotic initiation factor eIF2α; Role of the α- and σ-subunits of eIF2B. J Biol Chem 273:12841–12845

Kissil JL, Deiss LP, Bayewitch M, Raveh T, Khaspekov G, Kimchi A (1995) Isolation of DAP3, a novel mediator of interferon-gamma-induced cell death. J Biol Chem 270:27932–27936

Koc EC, Koc H (2012) Regulation of mammalian mitochondrial translation by post-translational modifications. Biochim Biophys Acta 1819:1055–1066

Koc EC, Spremulli LL (2002) Identification of mammalian mitochondrial translational initiation factor 3 and examination of its role in initiation complex formation with natural mRNAs. J Biol Chem 277:35541–35549

Koc EC, Spremulli LL (2003) RNA-binding proteins of mammalian mitochondria. Mitochondrion 2:277–291

Koc EC, Blackburn K, Burkhart W, Spremulli LL (1999) Identification of a mammalian mitochondrial homolog of ribosomal protein S7. Biochem Biophys Res Comm 266:141–146

Koc EC, Burkhart W, Blackburn K, Moseley A, Koc H, Spremulli LL (2000) A proteomics approach to the identification of mammalian mitochondrial small subunit ribosomal proteins. J Biol Chem 275:32585–32591

Koc EC, Burkhart W, Blackburn K, Koc H, Moseley A, Spremulli LL (2001a) Identification of four proteins from the small subunit of the mammalian mitochondrial ribosome using a proteomics approach. Prot Sci 10:471–481

Koc EC, Burkhart W, Blackburn K, Moseley A, Spremulli LL (2001b) The small subunit of the mammalian mitochondrial ribosome: identification of the full complement of ribosomal proteins present. J Biol Chem 276:19363–19374

Koc EC, Burkhart W, Blackburn K, Moyer MB, Schlatzer DM, Moseley A, Spremulli LL (2001c) The large subunit of the mammalian mitochondrial ribosome. Analysis of the complement of ribosomal proteins present. J Biol Chem 276:43958–43969

Koc EC, Haque ME, Spremulli LL (2010) Current views of the structure of the mammalian mitochondrial ribosome. Israel J Chem 50:45–59

La Teana A, Gualerzi CO, Dahlberg AE (2001) Initiation factor IF 2 binds to the alpha-sarcin loop and helix 89 of *Escherichia coli* 23S ribosomal RNA. RNA 7:1173–1179

Liao H-X, Spremulli LL (1990) Identification and initial characterization of translational initiation factor 2 from bovine mitochondria. J Biol Chem 265:13618–13622

Liao H-X, Spremulli LL (1991) Initiation of protein synthesis in animal mitochondria: purification and characterization of translational initiation factor 2. J Biol Chem 266:20714–20719

Ling J, Roy H, Qin D, Rubio MA, Alfonzo JD, Fredrick K, Ibba M (2007) Pathogenic mechanism of a human mitochondrial tRNAPhe mutation associated with myoclonic epilepsy with ragged red fibers syndrome. Proc Natl Acad Sci U S A 104:15299–15304

Liu M, Spremulli LL (2000) Interaction of mammalian mitochondrial ribosomes with the inner membrane. J Biol Chem 275:29400–29406

Ma L, Spremulli LL (1995) Cloning and sequence analysis of the human mitochondrial translational initiation factor 2 cDNA. J Biol Chem 270:1859–1865

Mariani L, Beaudry C, McDonough WS, Hoelzinger DB, Kaczmarek E, Ponce F, Coons SW, Giese A, Seiler RW, Berens ME (2001) Death-associated protein 3 (Dap-3) is overexpressed in invasive glioblastoma cells in vivo and in glioma cell lines with induced motility phenotype in vitro. Clin Cancer Res 7:2480–2489

McCutcheon J, Agrawal R, Philips SM, Grassucci R, Gerchman S, Clemons WM, Ramakrishnan V, Frank J (1999) Location of translational initiation factor IF3 on the small ribosomal subunit. Proc Natl Acad Sci U S A 96:4301–4306

Michishita E, Park JY, Burneskis JM, Barrett JC, Horikawa I (2005) Evolutionarily conserved and nonconserved cellular localizations and functions of human SIRT proteins. Mol Biol Cell 16:4623–4635

Mick DU, Dennerlein S, Wiese H, Reinhold R, Pacheu-Grau D, Lorenzi I, Sasarman F, Weraarpachai W, Shoubridge EA, Warscheid B et al (2012) MITRAC links mitochondrial protein translocation to respiratory-chain assembly and translational regulation. Cell 151:1528–1541

Mikulik K, Bobek J, Zikova A, Smetakova M, Bezouskova S (2011) Phosphorylation of ribosomal proteins influences subunit association and translation of poly (U) in Streptomyces coelicolor. Mol Biosystems 7:817–823

Miller JL, Koc H, Koc EC (2008) Identification of phosphorylation sites in mammalian mitochondrial ribosomal protein DAP3. Protein Sci 17:251–260

Miller JL, Cimen H, Koc H, Koc EC (2009) Phosphorylated proteins of the mammalian mitochondrial ribosome: implications in protein synthesis. J Proteome Res 8:4789–4798

Miyazaki T, Shen M, Fujikura D, Tosa N, Kon S, Uede T, Reed JC (2004) Functional role of death associated protein 3 (DAP3) in anoikis. J Biol Chem 279:44667–44672

Montoya J, Ojala D, Attardi G (1981) Distinctive features of the 5′-terminal sequences of the human mitochondrial mRNAs. Nature 290:465–470

Moreau M, de Cock E, Fortier P-L, Garcia C, Albaret C, Blanquet S, Lallemand J-Y, Dardel F (1997) Heteronuclear NMR studies of E. coli translation initiation factor IF3. Evidence that the inter-domain region is disordered in solution. J Mol Biol 266:15–22

Naithani S, Saracco SA, Butler CA, Fox TD (2003) Interactions among COX1, COX2, and COX3 mRNA-specific translational activator proteins on the inner surface of the mitochondrial inner membrane of Saccharomyces cerevisiae. Mol Biol Cell 14:324–333

Nissen P, Kjeldgaard M, Thirup S, Polekhina G, Reshetnikova L, Clark B, Nyborg J (1995) Crystal structure of the ternary complex of Phe-tRNA^{phe}, EF-Tu and a GTP analog. Science 270:1464–1472

Nissen P, Thirup S, Kjeldgaard M, Nyborg J (1999) The crystal structure of Cys-tRNACys-EF-Tu-GDPNP reveals general and specific features in the ternary complex and in tRNA. Structure 7:143–156

O'Brien TW (1971) The general occurrence of 55S ribosomes in mammalian liver mitochondria. J Biol Chem 246:3409–3417

O'Brien TW, O'Brien BJ, Norman RA (2005) Nuclear MRP genes and mitochondrial disease. Gene 354:147–151

Onyango P, Celic I, McCaffery JM, Boeke JD, Feinberg AP (2002) SIRT3, a human SIR2 homologue, is an NAD-dependent deacetylase localized to mitochondria. Proc Natl Acad Sci U S A 99:13653–13658

O'Rourke B, Van Eyk JE, Foster DB (2011) Mitochondrial protein phosphorylation as a regulatory modality: implications for mitochondrial dysfunction in heart failure. Congest Heart Fail 17:269–282

Ott M, Herrmann JM (2010) Co-translational membrane insertion of mitochondrially encoded proteins. Biochim Biophys Acta 1803:767–775

Petrelli D, LaTeana A, Garofalo C, Spurio R, Pon CL, Gualerzi CO (2001) Translation initiation factor IF3: two domains, five functions, one mechanism? EMBO J 20:4560–4569

Pietromonaco SF, Denslow ND, O'Brien TW (1991) Proteins of mammalian mitochondrial ribosomes. Biochimie 73:827–835

Pioletti M, Schlunzen F, Harms J, Zarivach R, Gluhmann M, Avila H, Bashan A, Bartels H, Auerbach T, Jacobi C et al (2001) Crystal structures of complexes of the small ribosomal subunit with tetracycline, edeine and IF3. EMBO J 20:1829–1839

Ramagopal S, Subramanian AR (1974) Alteration in the acetylation level of ribosomal protein L12 during growth cycle of Escherichia coli. Proc Natl Acad Sci U S A 71:2136–2140

Richter R, Rorbach J, Pajak A, Smith PM, Wessels HJ, Huynen MA, Smeitink JA, Lightowlers RN, Chrzanowska-Lightowlers ZM (2010) A functional peptidyl-tRNA hydrolase, ICT1, has been recruited into the human mitochondrial ribosome. EMBO J 29:1116–1125

Rorbach J, Richter R, Wessels HJ, Wydro M, Pekalski M, Farhoud M, Kuhl I, Gaisne M, Bonnefoy N, Smeitink JA et al (2008) The human mitochondrial ribosome recycling factor is essential for cell viability. Nucleic Acids Res 36:5787–5799

Rotig A (2011) Human diseases with impaired mitochondrial protein synthesis. Biochim Biophys Acta 1807:1198–1205

Sajid A, Arora G, Gupta M, Singhal A, Chakraborty K, Nandicoori VK, Singh Y (2011) Interaction of Mycobacterium tuberculosis elongation factor Tu with GTP is regulated by phosphorylation. J Bacteriol 193:5347–5358

Schieke SM, Finkel T (2006) Mitochondrial signaling, TOR, and life span. Biol Chem 387:1357–1361

Schwartzbach C, Spremulli LL (1989) Bovine mitochondrial protein synthesis elongation factors: identification and initial characterization of an elongation factor Tu-elongation factor Ts complex. J Biol Chem 264:19125–19131

Schwartzbach C, Farwell M, Liao H-X, Spremulli LL (1996) Bovine mitochondrial initiation and elongation factors. Methods Enzymol 264:248–261

Schwer B, Bunkenborg J, Verdin RO, Andersen JS, Verdin E (2006) Reversible lysine acetylation controls the activity of the mitochondrial enzyme acetyl-CoA synthetase 2. Proc Natl Acad Sci U S A 103:10224–10229

Sharma MR, Koc EC, Datta PP, Booth TM, Spremulli LL, Agrawal RK (2003) Structure of the mammalian mitochondrial ribosome reveals an expanded functional role for its component proteins. Cell 115:97–108

Shi H, Hayes M, Kirana C, Miller R, Keating J, Macartney-Coxson D, Stubbs R (2012) TUFM is a potential new prognostic indicator for colorectal carcinoma. Pathology 44:506–512

Skrtic M, Sriskanthadevan S, Jhas B, Gebbia M, Wang X, Wang Z, Hurren R, Jitkova Y, Gronda M, Maclean N et al (2011) Inhibition of mitochondrial translation as a therapeutic strategy for human acute myeloid leukemia. Cancer Cell 20:674–688

Smeitink JA, Elpeleg O, Antonicka H, Diepstra H, Saada A, Smits P, Sasarman F, Vriend G, Jacob-Hirsch J, Shaag A et al (2006) Distinct clinical phenotypes associated with a mutation in the mitochondrial translation elongation factor EFTs. Am J Hum Genet 79:869–877

Smirnov A, Entelis N, Martin RP, Tarassov I (2011) Biological significance of 5S rRNA import into human mitochondria: role of ribosomal protein MRP-L18. Genes Dev 25:1289–1305

Smits P, Smeitink JA, van den Heuvel LP, Huynen MA, Ettema TJ (2007) Reconstructing the evolution of the mitochondrial ribosomal proteome. Nucleic Acids Res 35:4686–4703

Smits P, Smeitink J, van den Heuvel L (2010) Mitochondrial translation and beyond: processes implicated in combined oxidative phosphorylation deficiencies. J Biomed Biotech 2010:737385

Soleimanpour-Lichaei HR, Kuhl I, Gaisne M, Passos JF, Wydro M, Rorbach J, Temperley R, Bonnefoy N, Tate W, Lightowlers R et al (2007) mtRF1a is a human mitochondrial translation release factor decoding the major termination codons UAA and UAG. Mol Cell 27:745–757

Soung GY, Miller JL, Koc H, Koc EC (2009) Comprehensive analysis of phosphorylated proteins of Escherichia coli ribosomes. J Proteome Res 8:3390–3402

Spencer AC, Spremulli LL (2004) Interaction of mitochondrial initiation factor 2 with mitochondrial fMet-tRNA. Nucleic Acids Res 32:5464–5470

Spencer AC, Spremulli LL (2005) The interaction of mitochondrial translational initiation factor 2 with the small ribosomal subunit. Biochim Biophys Acta 1750:69–81

Spremulli LL, Coursey A, Navratil T, Hunter SE (2004) Initiation and elongation factors in mammalian mitochondrial protein biosynthesis. Prog Nucleic Acid Res Mol Biol 77:211–261

Stacey SN, Manolescu A, Sulem P, Thorlacius S, Gudjonsson SA, Jonsson GF, Jakobsdottir M, Bergthorsson JT, Gudmundsson J, Aben KK (2008) Common variants on chromosome 5p12 confer susceptibility to estrogen receptor-positive breast cancer. Nat Genet 40:703–706

Surovtseva YV, Shutt TE, Cotney J, Cimen H, Chen SY, Koc EC, Shadel GS (2011) Mitochondrial Ribosomal Protein L12 selectively associates with human mitochondrial RNA polymerase to activate transcription. Proc Natl Acad Sci U S A 108:17921–17926

Suzuki T, Terasaki M, Takemoto-Hori C, Hanada T, Ueda T, Wada A, Watanabe K (2001) Structural compensation for the deficit of rRNA with proteins in the mammalian mitochondrial ribosome. Systematic analysis of protein components of the large ribosomal subunit from mammalian mitochondria. J Biol Chem 276:21724–21736

Takeda S, Iwai A, Nakashima M, Fujikura D, Chiba S, Li HM, Uehara J, Kawaguchi S, Kaya M, Nagoya S et al (2007) LKB1 is crucial for TRAIL-mediated apoptosis induction in osteosarcoma. Anticancer Res 27:761–768

Takeuchi N, Kawakami M, Omori A, Ueda T, Spremulli LL, Watanabe K (1998) Mammalian mitochondrial methionyl-tRNA transformylase from bovine liver: purification, characterization and gene structure. J Biol Chem 273:15085–15090

Temperley R, Richter R, Dennerlein S, Lightowlers RN, Chrzanowska-Lightowlers ZM (2010) Hungry codons promote frameshifting in human mitochondrial ribosomes. Science 327:301

Terasaki M, Suzuki T, Hanada T, Watanabe K (2004) Functional compatibility of elongation factors between mammalian mitochondrial and bacterial ribosomes: characterization of GTPase activity and translation elongation by hybrid ribosomes bearing heterologous L7/12 proteins. J Mol Biol 336:331–342

Traugh JA, Traut RR (1972) Phosphorylation of ribosomal proteins of *Escherichia coli* by protein kinase from rabbit skeletal muscle. Biochemistry 11:2503–2509

Tsuboi M, Morita H, Nozaki Y, Akama K, Ueda T, Ito K, Nierhaus KH, Takeuchi N (2009) EF-G2mt is an exclusive recycling factor in mammalian mitochondrial protein synthesis. Mol Cell 35:502–510

Uchiumi T, Hori K, Nomura T, Hachimori A (1999) Replacement of L7/L12.L10 protein complex in *Escherichia coli* ribosomes with the eukaryotic counterpart changes the specificity of elongation factor binding. J Biol Chem 274:27578–27582

Uchiumi T, Honma S, Nomura T, Dabbs ER, Hachimori A (2002) Translation elongation by a hybrid ribosome in which proteins at the GTPase center of the *Escherichia coli* ribosome are replaced with rat counterparts. J Biol Chem 277:3857–3862

Vedrenne V, Gowher A, De Lonlay P, Nitschke P, Serre V, Boddaert N, Altuzarra C, Mager-Heckel AM, Chretien F, Entelis N et al (2012) Mutation in PNPT1, which encodes a polyribonucleotide nucleotidyltransferase, impairs RNA import into mitochondria and causes respiratory-chain deficiency. Am J Hum Genet e 91:912–918

Wallace DC (2012) Mitochondria and cancer. Nat Rev Cancer 12:685–698

Wang Y, Bogenhagen DF (2006) Human mitochondrial DNA nucleoids are linked to protein folding machinery and metabolic enzymes at the mitochondrial inner membrane. J Biol Chem 281:25791–25802

Wang Z, Cotney J, Shadel GS (2007) Human mitochondrial ribosomal protein MRPL12 interacts directly with mitochondrial RNA polymerase to modulate mitochondrial gene expression. J Biol Chem 282:12610–12618

Watanabe K (2010) Unique features of animal mitochondrial translation systems. The non-universal genetic code, unusual features of the translational apparatus and their relevance to human mitochondrial diseases. Proc Jpn Acad Ser B Phys Biol Sci 86:11–39

Wazir U, Jiang WG, Sharma AK, Mokbel K (2012) The mRNA expression of DAP3 in human breast cancer: correlation with clinicopathological parameters. Anticancer Res 32:671–674

Wolstenholme D (1992) Animal mitochondrial DNA. In: Wolstenholme D, Jeon K (eds) Mitochondrial genomes. Academic Press, New York, pp 173–216

Woolcott CG, Maskarinec G, Haiman CA, Verheus M, Pagano IS, Le Marchand L, Henderson BE, Kolonel LN (2009) Association between breast cancer susceptibility loci and mammographic density: the Multiethnic Cohort. Breast Cancer Res 11:R10

Woriax V, Burkhart W, Spremulli LL (1995) Cloning, sequence analysis and expression of mammalian mitochondrial protein synthesis elongation factor Tu. Biochim Biophys Acta 1264:347–356

Woriax V, Spremulli G, Spremulli LL (1996) Nucleotide and aminoacyl-tRNA specificity of the mammalian mitochondrial elongation factor EF-Tu:Ts complex. Biochim Biophys Acta 1307:66–72

Xin H, Woriax VL, Burkhart W, Spremulli LL (1995) Cloning and expression of mitochondrial translational elongation factor Ts from bovine and human liver. J Biol Chem 270:17243–17249

Yang Y, Cimen H, Han MJ, Shi T, Deng JH, Koc H, Palacios OM, Montier L, Bai Y, Tong Q et al (2010) NAD$^+$-dependent deacetylase SIRT3 regulates mitochondrial protein synthesis by deacetylation of the ribosomal protein MRPL10. J Biol Chem 285:7417–7429

Yassin AS, Haque ME, Datta PP, Elmore K, Banavali NK, Spremulli LL, Agrawal RK (2011) Insertion domain within mammalian mitochondrial translation initiation factor 2 serves the role of eubacterial initiation factor 1. Proc Natl Acad Sci U S A 108:3918–3923

Zhang Y, Spremulli LL (1998) Identification and cloning of human mitochondrial translational release factor 1 and the ribosome recycling factor. Biochim Biophys Acta 1443:245–250

Zhang J, Sprung R, Pei J, Tan X, Kim S, Zhu H, Liu CF, Grishin NV, Zhao Y (2009) Lysine acetylation is a highly abundant and evolutionarily conserved modification in *Escherichia coli*. Mol Cell Prot 8:215–225

Zhao S, Xu W, Jiang W, Yu W, Lin Y, Zhang T, Yao J, Zhou L, Zeng Y, Li H et al (2010) Regulation of cellular metabolism by protein lysine acetylation. Science 327:1000–1004

Zweib C, Glotz C, Brimacombe R (1981) Secondary structure comparisons between small subunit ribosomal RNA molecules from six different species. Nuc Acids Res 9:3621–3640

Chapter 3
Translation in Mammalian Mitochondria: Order and Disorder Linked to tRNAs and Aminoacyl-tRNA Synthetases

Catherine Florentz, Joern Pütz, Frank Jühling, Hagen Schwenzer, Peter F. Stadler, Bernard Lorber, Claude Sauter and Marie Sissler

Abstract Transfer RNAs (tRNAs) and aminoacyl-tRNA synthetases (aaRSs) are key actors in all translation machineries. AaRSs aminoacylate cognate tRNAs with a specific amino acid that is transferred to the growing protein chain on the ribosome. Mammalian mitochondria possess their own translation machinery for the synthesis of 13 proteins only, all subunits of the respiratory chain complexes involved in the synthesis of ATP. While 22 tRNAs and two ribosomal RNAs are also coded by the mitochondrial genome, aaRSs are nuclear encoded and become imported. The fact that the two cellular genomes, nuclear and mitochondrial,

C. Florentz (✉) · J. Pütz · F. Jühling · H. Schwenzer · B. Lorber · C. Sauter · M. Sissler
Institut de Biologie Moléculaire et Cellulaire du CNRS, Architecture et Réactivité de l'ARN,
Université de Strasbourg, 15 rue René Descartes, 67084 Strasbourg, Cedex, France
e-mail: C.Florentz@ibmc-cnrs.unistra.fr

J. Pütz
e-mail: J.Puetz@ibmc-cnrs.unistra.fr

F. Jühling
e-mail: frank@bioinf.uni-leipzig.de

H. Schwenzer
e-mail: H.Schwenzer@ibmc-cnrs.unistra.fr

B. Lorber
e-mail: B.Lorber@ibmc-cnrs.unistra.fr

C. Sauter
e-mail: C.Sauter@ibmc-cnrs.unistra.fr

M. Sissler
e-mail: M.Sissler@ibmc-cnrs.unistra.fr

F. Jühling · P. F. Stadler
Bioinformatics Group, Department of computer Science and Interdisciplinary Center
for Bioinformatics, University of Leipzig, Härtelstrasse 16-18, 04107 Leipzig, Germany
e-mail: studla@bioinf.uni-leipzig.de

A.-M. Duchêne (ed.), *Translation in Mitochondria and Other Organelles*,
DOI: 10.1007/978-3-642-39426-3_3, © Springer-Verlag Berlin Heidelberg 2013

evolve at different rates raises numerous questions as to the co-evolution of partner macromolecules. Herein we review the present state-of-the-art on structural,
biophysical, and functional peculiarities of mammalian mitochondrial tRNAs and
aaRSs, and of their partnership in their wild-type state. Then, we oppose this mitochondrial "order" to the "disorder" generated by the presence of a variety of mutations occurring in the corresponding human genes that have been correlated to an
increasing number of diseases. So far, more than 230 mutations in mitochondrial
tRNA genes and a rapidly growing number of mutations in mitochondrial aaRS
genes have been reported as causative of a large variety of pathologies. The molecular incidence of mutations on structural, biophysical and functional properties
of the related macromolecules will be summarized. Mutations in mitochondrial
tRNA genes lead to complex mosaic effects with a major impact on tRNA structure. Some mutations affecting mitochondrial aaRS genes do not interfere with the
housekeeping aminoacylation activity, suggesting that mitochondrial aaRSs, alike
cytosolic aaRSs are involved in other processes than translation. This opens new
research lines.

3.1 Introduction

Mammalian mitochondria possess a small circular genome, coding for 13 polypeptide chains only. These are subunits of the respiratory chain complexes involved in
the oxidative phosphorylation of ADP into ATP. The synthesis of these 13 subunits
requires a complete mitochondrial translation machinery. Interestingly, the RNA
components of this machinery are encoded by the mitochondrial genome. This is
the case of 11 mRNAs (leading to the 13 proteins), 2 rRNA, and 22 tRNAs representing the minimal and sufficient set to read all codons (Anderson et al. 1981).
All protein components are coded by the nuclear genome, synthesized in the cytosol, and imported into the organelle. They include the full sets of ribosomal proteins, as well as tRNA maturation and modification enzymes, aminoacyl-tRNA
synthetases (aaRSs), and also translation initiation, elongation, and termination
factors. The dual genetic origin of the macromolecules constituting the translation machinery has raised numerous questions as to their properties, mechanism,
and specificities of their partnerships, regulation of expression and mechanisms of
co-evolution. Indeed, the mitochondrial genome evolves 15–20 times more rapidly
than the nuclear genome (Brown et al. 1979; Castellana et al. 2011), generating
highly variable sequences so that the RNAs coded by the mammalian mitochondrial genomes are peculiar. All have lost some information as compared to their
bacterial homologs. For instance, mRNAs miss $3'$ and $5'$ untranslated regions,
ribosomal RNAs are significantly shorter than bacterial counterparts and tRNAs
present a range of peculiarities, from the absence of a few nucleotide signature
motifs to the absence of full structural domains (Willkomm and Hartmann 2006).

Herein we first focus on human mitochondrial tRNAs (mt-tRNAs) and aaRSs
(mt-aaRSs) and on their partnerships, to illustrate the present knowledge on

peculiar structural, biophysical, and functional properties of two partner macro-molecules encoded by two different genomes. Peculiarities of human mt-tRNA genes as highlighted by recent powerful bioinformatics approaches will be sum-marized and compared to mitochondrial tRNA genes in metazoan. Analysis of tRNAs will illustrate large sequence and structural variability, low thermodynamic stability, and high structural flexibility of this family of RNAs. Current knowledge on human mt-aaRSs will be summarized from genes to structural and functional properties of the corresponding proteins. The partnerships of aaRSs with their sub-strates will be described not only in terms of aminoacylation properties but also in terms of thermodynamics of substrate binding, a parameter that allowed for clear distinction from bacterial aminoacylation systems. As will be highlighted, the enlarged plasticity of the mitochondrial enzymes as compared to that of aaRS from other origins might be the result of an evolutionary adaptation of the nuclear-encoded protein to the rapidly evolving mitochondria-encoded RNAs. The human mitochondrial aspartylation system, studied in our laboratory, will serve as case study all along the review. After this first part, illustrating the status of wild-type macromolecules and thus referring to "order" in mitochondrial translation, "disor-der" will be considered as well as a molecular perturbation as well as a mitochon-drial pathology.

In the last two decades, mt-tRNA genes were recurrently reported as hosting point mutations linked to a variety of neuromuscular and neurodegenerative disor-ders. More recently, nuclear genes coding for mt-aaRSs also became the center of attention, with mutations causative of further disorders. The present understanding of their impact on the molecular level of "disorder" induced within tRNAs and aaRSs molecules will be summarized. While many mutations do interfere with the housekeeping aminoacylation reaction, several others have no detectable effect on it. This is in favor of the existence of additional mt-tRNA and mt-aaRS functions or properties that are altered by these mutations. These supplementary functions and properties need still to be determined. New research lines in this direction will be suggested.

3.2 Mammalian Mitochondrial tRNAs

3.2.1 Structural Properties of Mammalian Mitochondrial tRNAs

The gene content of the human mitochondrial genome is similar to what found in other metazoan mitochondria. It encodes for 22 mt-tRNAs (Anderson et al. 1981), one for each of the 20 amino acid specificities, and two additional ones for leu-cine and serine, respectively. This minimal set of tRNAs is sufficient for reading of all codons despite the genetic code is different in mitochondria from nuclear genomes. Already at the stage of bovine and human mitochondrial DNA sequenc-ing (Anderson et al. 1981) peculiar structural properties of tRNAs were noticed, marking significant differences with so-called canonical tRNAs present in bacteria or in the cytosol of eukaryotic cells. Most striking was the absence of a possible full cloverleaf structure for some of the expressed gene products, noticeably the absence of a complete structural domain for tRNASer1 (the D-arm of the cloverleaf), and, despite the presence of a potential cloverleaf for the other tRNAs, the absence of typical signature motifs in many sequences, motifs so far known as being conserved in all tRNAs. Mammalian mt-tRNAs fall into two groups according to the location of the corresponding gene on either of the mt-DNA coding strand. "Heavy" tRNAs (eight cases) are G-rich, while "Light" tRNAs (14 cases) are A, U, and C rich, leading to a series of typical structural characteristics such as biased base-pair content (Helm et al. 2000). Further, the full set of mammalian (and metazoan) mt-tRNAs do have a short variable region (Fig. 3.1) while canonical tRNAs also include structures with large variable regions (Giegé et al. 2012). Bioinformatic alignments to additional mammalian mt-tRNA genes confirmed and extended these properties and allowed for a fine-tuned description of detailed structural characteristics of each of the 22 tRNA families (Helm et al. 2000). As an outcome, mammalian mt-tRNAs structural properties range from canonical to highly degenerated types. Figure 3.1a indicates the most common degenerations of the classical cloverleaf structure in human mitochondrial sequences, particularly the loss of D/T loop interactions. Mt-tRNAs are expressed in cells to very low levels as compared to their cytosolic counterparts (estimated as 1 mt-tRNA per 160 cytosolic tRNA), rendering access for biochemical investigation of these RNAs very limited (Enriquez and Attardi 1996). Detailed experimental secondary structures were however established on in vitro transcripts and revealed large structural flexibility, with alternative folds to the cloverleaf co-existing in equilibrium (Bonnefond et al. 2005b; Helm et al. 1998; Sohm et al. 2003). This is due to a severe bias in nucleotide content (A, U, and C-rich, and G-poor) especially for the 14 tRNAs coded by the light DNA strand, leading to very low numbers of stable G-C base-pairs, and at opposite, to high levels of weaker A-U and G-U pairs. To be noticed also that the thermodynamic stabilities calculated for the cloverleaf folds are twice as weak as compared to those of canonical tRNAs (Fender et al. 2012). Post-transcriptional modifications (only characterized in a limited number of sequences; Suzuki et al. 2011) were found to be the triggers for stabilization of the cloverleaf fold as demonstrated with the case study highlighting the role of m1A9 modification in human mt-tRNA^{Lys} (Helm and Attardi 2004; Helm et al. 1998; Motorin and Helm 2010). This modification hinders base pairing of residue 9 (and of neighboring nucleotides 8 and 10) in the 3'-end domain of the tRNA with residues 64 (and neighboring residues 65 and 63, respectively) in the T-stem. Post-transcriptional modifications in mammalian mt-tRNAs (only characterized in a limited number of sequences; Suzuki et al. 2011) remain however quantitatively far more limited as compared to other tRNAs (7 % of modified nucleotides as compared to 13 %) suggesting that they play crucial roles (Helm et al. 1999; Suzuki et al. 2011). Interestingly, a number of specific modifications, such as taurine-dependent modifications of anticodon nucleotides, were reported to be key actors in mitochondrial codon reading (Kirino et al. 2005).

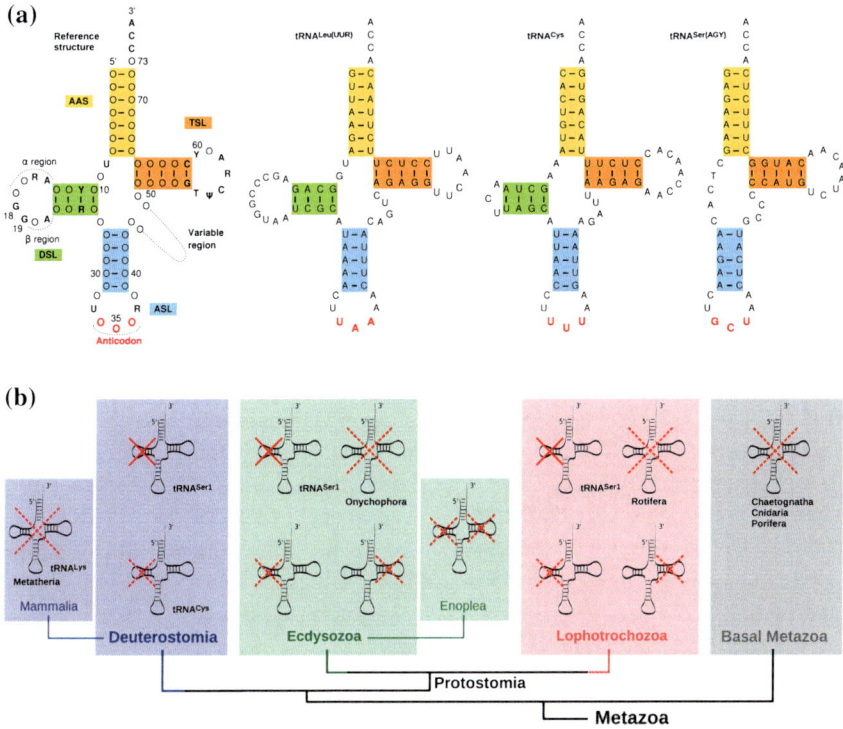

Fig. 3.1 Structural diversity and evolution of human mitochondrial tRNAs. **a**. Cloverleaf structures of three typical human mt-tRNA secondary structures as compared to the canonical cloverleaf reference structure (*left*). *AAS* amino acid acceptor stem, *DSL* D-stem and loop, *ASL* anticodon stem and loop, *TSL* T-stem and loop. The three examples presented, support the large range of structural profiles covered. tRNA$^{Leu(UUR)}$ is of classical type, with the full cloverleaf domains as well as the full set of conserved nucleotides involved in tertiary folding (see, in particular, the presence of residues G18, G19 in the D-loop, and of residues T54T55C56 in the T-loop, allowing for interaction between the corresponding domains). tRNA^{Lys} illustrates the family of mt-tRNAs still presenting the four domains of the cloverleaf, but with serious variations in the size of the D- and T-loops and their nucleotide content, missing the conserved elements, as well as the unusual nucleotides in the short connector between the acceptor and the D-stems. Finally, tRNA$^{Ser(AGY)}$ is the typical tRNA missing a full structural domain, namely the D-arm. **b**. Simplified view on the evolution of tRNA sets and tRNA secondary structures in Metazoa. Each panel highlights the most striking and representative structural deviation present within the considered evolutionay group. Accordingly, basal metazoan lost the full set of mt-tRNA genes and mammals, lost the gene for tRNA^{Lys} in methateria

The question as to the 3D fold of mammalian mt-tRNAs has retained the attention of many investigators. Indeed, whatever the secondary structural properties of these RNAs, it is expected that their 3D structures allow for the recognition by partner macromolecules, in particular the cognate aaRS for aminoacylation and the ribosome for codon reading and protein synthesis. An L-shaped structure based on an acceptor branch (T-arm and acceptor arm) and an anticodon branch (D-arm and anticodon arm) is thus expected as for tRNAs of all kingdoms of life (Giegé et al. 2012). This is considered despite the general absence of signature nucleotides (the so-called conserved and semi-conserved nucleotides) known to support such a fold in canonical tRNAs. Birefringence measurements, NMR, and chemical structure probing in solution were the leading approaches confirming the existence of L-shape-like folds for mt-tRNAs, bringing the two functional extremities, namely the acceptor 3′-end and the anticodon, at the expected distance from each other (reviewed in Giegé et al. 2012). The 3D structure of canonical tRNAs is based on a set of tertiary interactions between nucleotides at distance in the secondary structure, in particular on a network of interactions defining and stabilizing the "central core" of the L-shaped structure. Fine-tuned chemical structural probing on two case studies, human mt-tRNA^{Asp} and bovine mt-tRNA^{Phe} confirmed the existence of these networks (Messmer et al. 2009; Wakita et al. 1994). It remains to be verified if such networks can take place in all mt-tRNAs, an hypothesis that could not be supported by nucleotide conservation, but that will need a more detailed consideration of nucleotide partnership rules as tackled by Leontis and Westhof (Leontis et al. 2002). Initial data are available suggesting that at least all mammalian mt-tRNA^{Asp} would benefit from the set of tertiary network interactions despite nonconservation of involved nucleotides (Messmer et al. 2009). Extension of the analysis to the full set of mammalian mt-tRNAs is in progress. In regard of the elbow of the L, no hydrogen bonds could be detected between the D and T-loops, so that there may be no stabilization of the global L at this level. Accordingly, the angle formed by the two branches of the L may vary according to the tRNA, as also demonstrated by birefringence methods (Friederich and Hagerman 1997), and likely may vary according to its functional state, allowing for various structural adaptations to partner macromolecules (reviewed in Giegé et al. 2012). The role of post-transcriptional modifications in the global flexibility of the tRNA remains to be determined.

In summary, mammalian mt-tRNAs present a large diversity of structural types, ranging from canonical stable type for a few tRNA families only, to highly "bizarre" versions, characterized by an unusually high flexibility and plasticity of the full macromolecule, along an L-shaped-like global fold.

3.2.2 Comparison with Mitochondrial tRNA Genes from Other Metazoan

Due to their peculiarities, metazoan mt-tRNA genes are difficult to detect with the available search programs such as tRNAscan-SE (Lowe and Eddy 1997) and ARWEN (Laslett and Canbäck 2008), so that an approach using INFERNAL and covariance models was developed and implemented (Nawroki et al. 2009). A first systematic overview on nearly 2,000 metazoan RefSeq genomes (Pruitt et al. 2007) and their tRNA gene content allowed for interesting insights on the evolution of mt-tRNA genes (Jühling et al. 2012a). An overview on tRNA genes deprived of information coding for a full structural domain of the expressed RNA (D- or T-arm), and on loss of full mt-tRNA genes is illustrated in Fig. 3.1b. While mt-tRNA genes are encoded by mt-genomes over all metazoan clades, some organisms miss mt-tRNA genes and form exceptions. The loss of mt-tRNA genes was shown to be related with the loss of the corresponding aaRS genes in Cnidarians (e.g., jellyfish) (Haen et al. 2010). Members of Ceractinomorpha (sponges) (Wang and Lavrov 2008), Chaetognatha (arrow worms) (Miyamoto et al. 2010), and Rotifera (freshwater zooplankton) (Suga et al. 2008) also lost mt-tRNA genes. To be mentioned is the specific loss of tRNA^{Lys} genes in Metatheria (e.g., marsupials) where only a pseudo-gene remains (Dörner et al. 2001). It is assumed that in these organisms, by nuclear tRNAs (Alfonzo and Söll 2009; Duchêne et al. 2009).

Highly truncated tRNA gene sequences are known for Onychophora (velvet worms) (Braband et al. 2010), where unusual editing mechanisms are assumed to repair the tRNAs (Segovia et al. 2011) while other genes are completely missing. Other mt-tRNA genes lost only small sequence stretches so that the canonical cloverleaf of the tRNA can form without the need of repairing mechanisms. However, more than 90 % of mt-tRNAs share a four-arm cloverleaf structure, and the 10 % remaining mostly lost either the T-or the D-stem and developed replacement loops. The corresponding genes are found throughout all metazoan clades but are locally conserved. Hotspots of organisms where the full set of mt-tRNA genes have lost a complete domain are Ecdysozoa (arthropoda and nematoda) and Lophotrochozoa (molluscs and worms). In contrast, in Lepidosauria (turtles, snakes, and lizards) only the genes for mt-tRNA^{Cys} seem to spontaneously have lost the D-stem coding region along several parallel events (Macey et al. 1997; Seutin et al. 1994). Another well-known example of truncated mt-tRNA genes concerns Nematoda, where all mt-tRNAs lost their T-stem, except those of both tRNA^{Ser}, which instead lost their D-stem (Wolstenholme et al. 1994). While these truncated secondary structures in *C. elegans* are conserved within all other members of its family (round worms), a recent study on their sister group Enoplea detected even genes for "armless" tRNAs, namely sequences without both D- and T-stems (Jühling et al. 2012b). These sequences are conserved and are thereby the shortest functional tRNA sequences known so far (Jühling et al. 2012b). Compared to other metazoan mt-tRNA genes, mammalian genes lead to nearly canonical cloverleaf structures, and encode the full gene content with the exception of tRNA^{Lys}, which is missing in marsupials. However, hotspots of gene losses and of genes of truncated structures are distributed in other parts of the metazoan tree. Only the tRNA^{Ser1} gene, coding for a tRNA deprived of a D-domain, shows equal distribution all along the metazoans (Arcari and Brownlee 1980; de Bruijn et al. 1980).

3.3 Mitochondrial aaRSs

3.3.1 Genes and Proteins

Present-day mitochondrial genomes do not code for aaRSs, so that the full set of enzymes is necessarily nuclear encoded and that the expressed proteins become imported into mitochondria. A set of specific genes for mammalian mt-aaRSs has been recorded (Bonnefond et al. 2005a). This set distinguishes from the set of cytosolic synthetases for most enzymes, with the sole exceptions of LysRS and GlyRS, coded by a same gene but with distinguishing features allowing for dual location of the protein. The gene for LysRS undergoes alternative splicing so that the two final mature proteins distinguish by a few N-terminal amino acids only (Tolkunova et al. 2000). GlyRSs are generated from two translation initiation sites on the same gene, so that the two mature proteins are the same, but differ by the presence or the absence of a mitochondrial targeting sequence (MTS) (Mudge et al. 1998; Shiba et al. 1994). A further exception concerns mt-GlnRS for which no specific gene was found so far. As an alternative to the existence of a specific mt-GlnRS, the possible import of the cytosolic enzyme was proposed (Rinehart et al. 2005) as well as the existence of an indirect pathway based on misaminoacyla-tion of mt-tRNA^{Gln} with mt-GluRS, followed by transamidation of the charged glutamic acid into glutamine. Such a pathway exists both in yeast (Frechin et al. 2009a) and human mitochondria (Nagao et al. 2009). In yeast, the dual localiza-tion of GluRS is controlled by binding to Arc1p, a tRNA nuclear export cofactor that behaves as a cytosolic anchoring platform. When the metabolism of the yeast cell switches from fermentation to respiration, the expression of Arc1p is down-regulated and this increases the import of GluRS to satisfy a higher demand of mt glutaminyl-tRNA^{Gln} for mitochondrial protein synthesis (Frechin et al. 2009b).

Almost all mammalian mt-aaRSs have about the same size (once the MTS is removed upon import into mitochondria) and the same structural 2D organization as their homologs from the three kingdoms of life (Bonnefond et al. 2005a). The only exception is PheRS which is classically a tetramer ($\alpha2\beta2$) but only a mono-mer in mitochondria (Klipcan et al. 2008). Also, all mt-aaRSs contain the expected signature motifs for amino acid specificity and signature motifs of either class I or class II of aaRSs (Bonnefond et al. 2005a; Brindefalk et al. 2007; Sissler et al. 2005; Woese et al. 2000). The evolutionary origin of the different genes how-ever remains intriguing. Sequence alignments have indeed not identified sequence stretches or signature motifs that could originate from alpha-proteobacterial ances-tors, favoring various horizontal gene transfers events along evolution (Brindefalk et al. 2007).

During import into the mitochondria, the MTS is cleaved enzymatically. The detailed process is still under exploration for aaRSs but the trend for other pro-teins follows a two-step process. Once the polypeptide chain has crossed the mito-chondrial membranes and entered the mitochondrion, a first cleavage occurs. This cleavage might be definite. Alternatively, the partially matured protein then local-izes either in the matrix or anchors into (or at the surface of) the inner mitochon-drial membrane before another part of the MTS is cut off. Interestingly, ribosomes and elongation factors have been found close to the inner membrane suggesting that probably the entire machinery required for protein biosynthesis is located there, and accordingly aaRSs too. A particularity of MTSs is their nonconserved length, sequence and amino acid compositions (e.g., Chacinska et al. 2009). So far, no reliable consensus sequence was found that could be used to accurately predict the positions of maturation. This was highlighted by a wrongly predicted cleavage site of human mt-LeuRS, which led to a poor expression of a tentative mature protein in *E. coli* while a variant, shortened by further 39 N-terminal amino acids overproduced well in *E. coli* and was purified as an active enzyme while the one deprived of only 21 amino acids was insoluble (Bullard et al. 2000; Yao et al. 2003). Another well-documented example is human mt-AspRS. The predicted mature protein had a very low solubility when overexpressed in *E. coli* cells. Dynamic light scattering analyses revealed that aggregation proceeded during purification. A comparative analysis of a set of variants differing by their N-terminal sequences revealed that expression of the protein was actually enhanced when the N-terminus was extended by seven natural amino acids of the predicted mature N-terminus (Gaudry et al. 2012). The redesigned protein was highly soluble, monodisperse and functionally active in tRNA aminoacylation. It yielded crystals that were suitable for structure determination (Gaudry et al. 2012; Neuenfeldt et al. 2013). These results suggest that additional criteria should be taken into account for the prediction of the correct MTS cleavage sites and that the definition of the precise N-terminus of mature mt-aaRS should be determined experimentally.

3.3.2 Crystallographic Structures

As already highlighted, despite various evolutionary origins of the genes coding for mammalian mt-aaRSs, several enzymes have a same modular organization than their bacterial homologs. This is illustrated in Fig. 3.2 with three of the four crystal structures that have been determined so far for mt-aaRSs for exclusive mitochondrial location (additional crystallographic structures are available for human GlyRS (Cader et al. 2007; Xie et al. 2006) and for LysRS (Guo et al. 2008), aaRSs of dual cytosolic and mitochondrial location). Bovine mt-SerRS, human mt-TyrRS, and human mt-AspRS show an overall architecture close to that of their prokaryotic homologs (Bonnefond et al. 2007; Chimnaronk et al. 2005; Neuenfeldt et al. 2013). Mt-PheRS is again the exception. Instead of forming complex heterodimeric assemblies as bacterial, archaeal, and cytosolic enzymes, it forms a two-domain monomer, which only maintains the catalytic domain characteristic of class II. This human mitochondrial version is the smallest known aaRS (Yadavalli et al. 2009).

Along with the reduction of the tRNA pool (22 in human mitochondria) and the simplification of identity rules, several of these enzymes have adapted the way they recognize their tRNA substrate, especially when the latter display a non canonical 2D fold leading to higher flexibility. Positively charged patches at the surface of mt-SerRS were redistributed to bind tRNAs lacking an extended variable region, the hallmark and major identity element of prokaryotic, eukaryotic, and archaeal tRNAs[Ser] (Chimnaronk et al. 2005). The three other mt-aaRSs display a more electropositive tRNA-binding interface, which may favor interactions

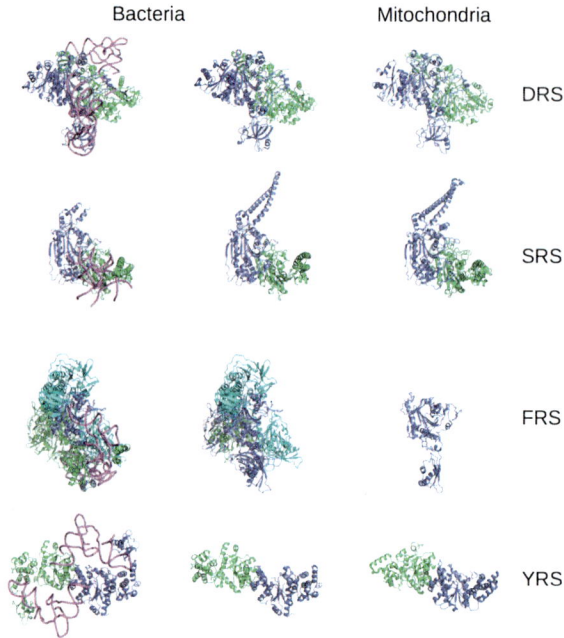

Fig. 3.2 Cyrstal structures of mammalian mitochondrial aaRSs (bovine or human) and of their bacterial homologs (*E. coli -Eco-* or *Thermus thermophiles -Tth-*). On the left, bacterial complexes (PDBids: 1C0A for *Eco*DRS/tRNA, 1SRS for *Tth*SRS/tRNA, 2IY5 for *Tth*FRS/tRNA and 1H3E for *Tth*YRS/tRNA) are shown with monomers A in blue, monomers B in green and tRNAs in pink, indicating the binding site of the cognate substrate. In the case of tetrameric bacterial FRS, monomers C and D are depicted in cyan. On the right, free forms of bacterial aaRSs (1EQR for *Eco*DRS, 1SRY for *Tth*SRS, 1B7Y for *Tth*FRS, 1H3F for *Tth*YRS without its C-terminal domain) and mitochondrial enzymes (4AH6 for *Hsa*DRS2, 1WLE for *Bta*SRS2, 3TUP for *Hsa*FRS2, and 2PID for *Hsa*YRS2) are represented in the same orientation and same color code. Except *Hsa*FRS which exhibits a totally different structural organization (monomer instead of heterotetramer), mitochondrial aaRSs have retained the overall architecture of their bacterial relatives. Names of organisms are abbreviated in a three-letter code (e.g., *Homo sapiens*: *Hsa*). Mitochondrial enzymes are referred to by the addition of the number 2

with the sugar-phosphate backbone of the substrate to compensate for a reduction of specific contacts with identity elements (Neuenfeldt et al. 2013). The ability of mt-TyrRS, mt-PheRS, and mt-AspRS to aminoacylate heterologuous tRNAs (Bonnefond et al. 2005a; Klipcan et al. 2012; Neuenfeldt et al. 2013) indicates a much higher substrate tolerance, which may be linked to an increased structural plasticity. For instance, mt-PheRS undergoes a large movement of its anticodon-binding domain upon tRNA binding, switching from a closed to an open conformation (Klipcan et al. 2012). The higher thermal sensitivity of human mt-AspRS as compared *E. coli* AspRS, its more open catalytic groove in the absence of tRNA and the amplitude of thermodynamic terms associated with tRNA binding are also in favor of a more dynamic structure (Neuenfeldt et al. 2013). Altogether, it appears that mt-aaRS properties have evolved to accompany the sequence and structure drift of mt-tRNAs. Enlarged intrinsic plasticity within a conserved architectural framework is one striking feature along this line. The underlying mechanisms enabling the crosstalk between nuclear and mitochondrial genomes remain to be explored.

3.4 tRNA/aminoacyl-tRNA Synthetase Partnerships in Mammalian Mitochondria

3.4.1 Aminoacylation of tRNAs, the Housekeeping Function of aaRSs

The partnership of tRNAs and aaRSs is a key event in translation. Each synthetase recognizes specifically its tRNA or family of isoaccepting tRNAs, and esterifies its 3′-CCA end with the specific amino acid. The charged tRNA enables delivery of the amino acid to the ribosome where translation takes place. The aminoacylation reaction involves a two-step process including first the activation of the specific amino acid into an adenylate in the presence of ATP, and second, the specific recognition of the cognate tRNA followed by transfer of the activated amino acid (Ibba et al. 2005). Deciphering the detailed mechanisms of these steps for bacterial, archeal, and eukaryotic cytosolic aminoacylation has retained the attention of a large number of research groups over several decades (Ibba et al. 2005). Analysis of mammalian mitochondrial aminoacylation systems is only at initial stages. Due to the dual origin of the two partner macromolecules and to the diverging structural properties of mt-tRNAs, the mechanisms of reciprocal recognition and of co-evolution of these macromolecules deserve much attention.

3.4.2 Mammalian Mitochondrial Synthetases have Low Catalytic Activities

Only a limited number of recombinant mammalian aaRSs have been obtained so far, allowing for biochemical and enzymatic characterization in vitro. As already reviewed elsewhere (Florentz et al. 2003; Suzuki et al. 2011) these enzymes present a 20- to 400- fold lower catalytic activity than their cytoplasmic and bacterial homologs. In the specific case of human mt-AspRS as compared to *E. coli* AspRS, the affinity for the substrate tRNA is higher by an order of magnitude as measured by isothermal titration calorimetry (ITC) while the affinity for an analog of the activated amino acid is of same level. However, the catalytic rate kcat for aminoacylation is 40-fold lower for the mt-aaRS (Neuenfeldt et al. 2013). The molecular reasons explaining the lower rate remain however elusive. Indeed, overimposition of the catalytic sites in the crystallographic structures of both enzymes, lead to an important overlap (less than 2 Å rmsd) not allowing to pinpoint intrinsic differences inline with a different catalytic activity (Fender et al. 2006; Neuenfeldt et al. 2013).

3.4.3 Identity Elements in Mitochondrial tRNAs are Limited

Specific recognition of tRNAs by aaRSs is driven by identity elements present in the tRNA (Giegé 2008; Giegé et al. 1998). These elements have been searched by mutagenic approaches on in vitro transcripts for a few mammalian mt-tRNAs (Florentz et al. 2003; Suzuki et al. 2011). Interestingly, while these sets are generally conserved along different organisms and even along kingdoms for a given amino acid specificity, they were found distinct in mt-tRNAs. A striking example concerns identity elements for aspartylation, one or the rare systems so far investigated. Major identity elements (elements for which strongest effects are observed upon mutation) are conserved all along evolution, as residues G73 (the so-called discriminator residue near the 3'- acceptor end), residue G10 in the D-stem, and residues G34, U35, and C36 forming the anticodon triplet. Transfer of this set of residues into host tRNAs of different specificities, converts these tRNAs into aspartic acid accepting species (Giegé et al. 1996). A mutagenic analysis performed on human mt-tRNA[Asp] revealed that only residues U35 and G36 are important elements for specific recognition and aspartylation by mt-AspRS while the other elements can be replaced by any other nucleotide without influencing the efficiency of tRNA recognition and aspartylation (Fender et al. 2006). Figure 3.3 illustrates this point. The striking non-importance of residue 73, otherwise highly conserved as G in tRNA[Asp] over all kingdoms of life, is a signature of mt-tRNA degeneration, and at the same time, of evolutionary adaptation of the synthetase. Deep insight into the structural environment of residue 73 in the catalytic site of human mt-AspRS reveals an enlarged space as compared to *E. coli* AspRS and other AspRS, allowing the fit of any of the four nucleotides, rather than the exclusive fit of a G residue at this position. A mutagenic analysis of the enzyme has confirmed this view (Fender et al. 2006). Another example of the peculiarity of identity elements in a mitochondrial system concerns human mt-tRNA[Tyr] (Bonnefond et al. 2005b). Base-pair G1-C72, forms an important identity element in archaeal and eukaryal tRNA[Tyr] (Bonnefond et al. 2005b). The mitochondrial tyrosine identity disobeys this rule, since mt-TyrRS is able to aminoacylate as well a tRNA with the G1-C72 pair as the opposite pair C1-G72. Other examples have been reviewed previously (Florentz et al. 2003) and will not be further discussed herein. They indicate that sequence analysis of mammalian mt-tRNAs lead to the conclusion that only a limited number of identity elements known for non mitochondrial aminoacylation systems are present (Florentz et al. 2003).

Despite the limited number of aminoacylation identity elements in mt-tRNAs, translation in mitochondria needs to be accurate so that the 13 synthesized proteins

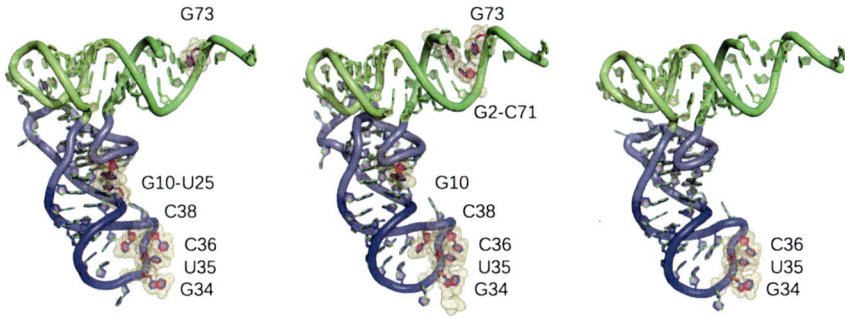

Fig. 3.3 Evolution of tRNA^{Asp} structure and identity elements. From the left to the right: tRNA^{Asp} from the yeast *Saccharomyces cerevisiae* (conformation in the complex with AspRS, PDBid: 1ASY), from *E. coli* (conformation in the complex with AspRS, PDBid: 1C0A), from human mitochondria (homology model—(Messmer et al. 2009)). The three molecules are shown in the same orientation with the acceptor stem in green, the D stem loop in light green, the anti-codon stem loop in dark blue, the variable loop and the T stem loop in light blue, respectively. Nucleotides defining the aspartate identity (Fender et al. 2012; Giegé et al. 1996) in each system are depicted in pink with a dotted surface. The human mitochondrial tRNA^{Asp} is characterized by a shortening of D and T loops, leading to an absence of bases interactions at the corner of the L scaffold, and by a reduced set of identity elements

(all subunits of respiratory chain complexes that are partners of more than 80 nuclear encoded subunits) are prepared without mistakes. It is hypothesized that the small competition created by 22 tRNAs only toward about the same number of aaRSs in the mitochondrial environment, as compared to more complex translation machineries in bacteria or eukaryotic cytosol with several hundreds of tRNAs for 20 aaRSs, can deal with a restricted number of identity elements. Further, selection by the elongation factor EF-Tu of properly charged tRNAs only may represent an additional process toward accurate protein synthesis (Nagao et al. 2007), as is the case in bacteria (LaRiviere et al. 2001).

3.4.4 Unprecedented Plasticity of Mitochondrial aaRSs and tRNAs

The discovery of mt-tRNAs that have lost critical structural information raised the question about how the nuclear-encoded mt-aaRSs have adapted to be able to deal with their partners? As already discussed, first insights were provided by resolution of crystal structures of mt-aaRSs (Bonnefond et al. 2007; Chimnaronk et al. 2005; Klipcan et al. 2008; Klipcan et al. 2012; Neuenfeldt et al. 2013). In the specific case of human mt-AspRS, a typical homodimeric bacterial-type AspRS, the 3D architecture is very close to that of the *E. coli* enzyme of same specificity, with the exception that it has a wider catalytic groove, a more electropositive surface potential, and an alternate interaction network at the subunits interface, a set of properties in line with support to facilitated tRNA partnership. An additional biophysical property, namely thermostability, illustrates further the originality of the protein. Comparative differential scanning fluorimetry analyses indicated that the mitochondrial protein is far less stable with regard to temperature than its bacterial homolog. It has a 12 °C lower melting point (Neuenfeldt et al. 2013). These properties are summarized in Table 3.1.

The partnership of a mt-aaRS (human mt-AspRS) with its substrates was investigated by ITC, an approach allowing for direct measurement of affinity (Kd) and of the thermodynamic parameters ΔH (variation in enthalpy), ΔS (variation in entropy), and ΔG (variation in free energy) (Neuenfeldt et al. 2013). Comparative analyses between human mt-AspRS and *E. coli* AspRS revealed a one order of magnitude higher affinity of the mitochondrial enzyme for cognate and noncognate tRNAs (cross binding studies of mt-AspRS with *E. coli* tRNA, and of *E. coli* AspRS with mt-tRNA^{Asp}), but with highly different entropy and enthalpy contributions (Table 3.1). Binding parameters of the cognate mitochondrial partners requires far larger enthalpic and entropic contributions than binding of the cognate bacterial partners, underlining reciprocal reorganization along complex formation. Such an adaptation is still possible when the mt aaRS meets the bacterial tRNA but is not possible in the opposite situation, namely when the bacterial enzyme and the mt-tRNA face each other. Thermodynamics thus contribute to explain the well-known unilateral aminoacylation of bacterial synthetases for bacterial tRNAs (Kumazawa et al. 1991). Interestingly, ITC measurements of small substrate binding, revealed that both enzymes bind a synthetic analog of the aspartyl-adenylate by a cooperative allosteric mechanism between the two subunits of the dimeric enzymes, but again with different thermodynamic contributions (Neuenfeldt et al. 2013) (Table 3.1).

Altogether, presently available structural, biophysical, and thermodynamic data support the view of so far unsuspected greater flexibility of mt-aaRS with respect to its bacterial homolog albeit a common architecture. This gain in plasticity may represent an evolutionary process that allows the nuclear-encoded proteins to adapt to the structurally degenerated RNAs from organelles. Evolutionary induced changes in intrinsic properties of proteins, may thus represent an alternative to other strategies, such as those reported for the mitochondrial ribosome, where the strong restriction in RNA sizes is compensated by extension of the number and size of the nuclear encoded proteins (Willkomm and Hartmann 2006). If mt-aaRS do have partner proteins that might also contribute to improved recognition of degenerated tRNAs remains an open question.

3.5 Human Mitochondrial tRNA and Synthetases in Pathologies

3.5.1 *Mitochondrial tRNAs and Human Pathologies*

In the last two decades, a large number of human neuromuscular and neurodegenerative disorders have been reported as correlated to point mutations in the mt-DNA encoded genes, with a large prevalence of mutations in tRNA genes

Table 3.1 Major structural and functional differences between dimeric mt and bacterial aspartyl-tRNA synthetases in favor of a greater plasticity of the organelle enzyme

Aspartyl tRNA synthetase			
Dimer surface	20 additional basic residues		(18 Lys + 2 Arg)
	Electrostatic potential more positive		
Dimer interface	70 versus 60 H bonds and 28 versus 20 salt bridges		
	About 25 % less specific interactions per Å^2		
Thermal stability (T_m)	Alone		37 °C versus 50 °C
	Bound to cognate tRNA		40 °C versus 50 °C
	Bound to AspAMS		45 °C versus 55 °C
Thermodynamic parameters			
Cognate tRNA^{Asp}	Kd	0.26 versus 3.1	μM
	ΔH	−20.3 versus −13.0	kcal/mol
	ΔT	+ 11.2 versus + 5.5	kcal/mol
	ΔG	−9.1 versus −7.5	kcal/mol
Noncognate tRNA^{Asp}	Kd	0.24 versus 22	μM
	ΔH	−14.0 versus −30.8	kcal/mol
	ΔT	+ 5.0 versus +24.4	kcal/mol
	ΔG	−9.0 versus −6.4	kcal/mol
AspAMS (monomer 1)	Kd	129 versus 29	nM
	ΔH	−13.2 versus −5.5	kcal/mol
	ΔT	+ 3.9 versus −4.8	kcal/mol
	ΔG	−9.4 versus −10.2	kcal/mol
AspAMS (monomer 2)	Kd	17 versus 3	nM
	ΔH	−21.8 versus −8.2	kcal/mol
	ΔT	+ 10.6 versus −3.5	kcal/mol
	ΔG	−10.6 versus −11.6	kcal/mol
tRNA^{Asp}			
Nucleotide content	90 versus 66 % A, U, and C		
	16 versus 7 out of 21 A–U and G–U base pairs		
	D and T loops are not classical		
Structural stability	ΔG	−22 versus −42	kcal/mol

Both polypeptide chains share 43 % identity between their amino acid sequences
Data compiled from (Fender et al. 2012) and (Neuenfeldt et al. 2013)

(reviewed for example in Florentz and Sissler 2003; Suzuki et al. 2011; Yarham et al. 2010; Ylikallio and Suomalainen 2012). Among the mutations leading to "mitochondrial disorders", 232 are distributed all over the 22 tRNAs (Fig. 3.4; data from MITOMAP, a human mitochondrial genome database http://www.m itomap.org/MITOMAP). Most striking cases concern tRNA^{Lys}, tRNA$^{Leu(UUR)}$, and tRNA^{Ile}, which form "hot spots" for mutations. Mutations in these tRNAs are most frequently correlated with Myoclonus Epilepsy with Ragged Red Fibers (MERRF) (Shoffner et al. 1990) and Mitochondrial Encephalomyopathy with Lactic Acidosis and Stroke-like episodes (MELAS) (Goto et al. 1990), respectively. However, this does not indicate a peculiar mutational susceptibility of the three mitochondrial genes, but is more likely due to systematic and intensive

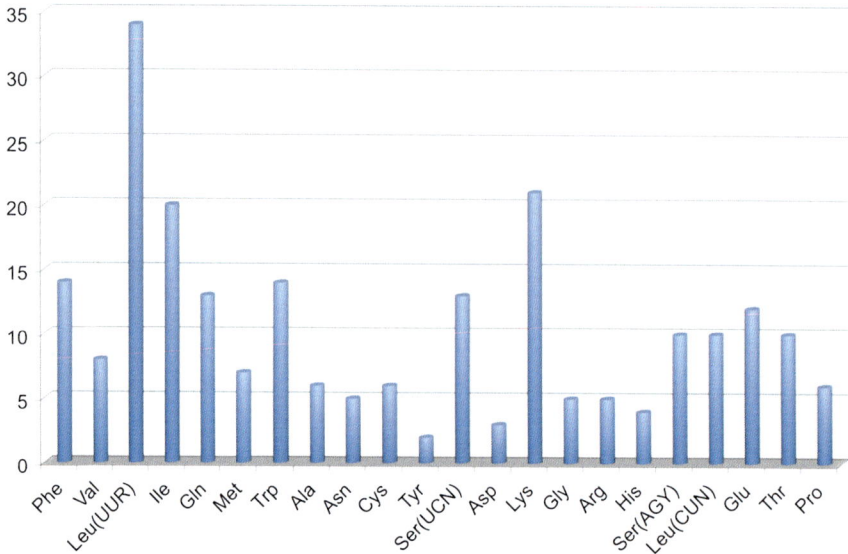

Fig. 3.4 Pathology-related mutations in the 22 mt- tRNA genes. tRNA genes are indicated by the three-letter code of the corresponding amino acid and are sorted according to their location on mt-DNA. Each column corresponds to the number of different mutations reported so far. To be noticed, three additional mutations are reported in the precursor of tRNA$^{Ser(UCN)}$ and one additional mutation at the junction between tRNA^{Gln} and tRNA^{Met}. Data are from MITOMAP, a database for mitochondrial genome mutations (http://www.mitomap.org/MITOMAP)

investigations of theses firstly reported examples of mt-tRNA genes correlated with human mitochondrial disorders.

The exponential rate of discovery of mutations in tRNA genes (3 mutations reported in 1990 and 232 mutations in 2012), together with the key role of tRNA in mitochondrial protein synthesis (linked to its global biology including gene expression, tRNA maturation, specific amino acid transfer, and regulation of these different functions), called for clarification of the molecular mechanisms for their pathogenicity. However, the relationships between genotypes and phenotypes appear very complicated since a given mutation can lead to a large variety of disorders of different severities (ranging from, e.g., limb weakness and exercise intolerance, to diabetes, leukoencephalopathy, encephalomyopathy, or fatal infantile cardiomyopathy, etc.). At the opposite, a given disorder can be linked to a variety of single point mutations in different tRNA genes. Also, each cell may contain hundreds to thousands copies of the mitochondrial genome, in a mixture of wild-type and mutated versions (heteroplasmic status). The variable distribution of affected tissues and the variable heteroplasmy levels lead to remarkable erratic and heterogenous clinical manifestations. Therefore, establishment of a mitochondrial disorder diagnosis can be difficult. It requires an evaluation of the family pedigree, in conjunction with a thorough assessment of clinical, imaging, and muscle biopsy analyses (McFarland et al. 2004). Also, a theoretical comparison of polymorphic (neutral mutations with no pathogenic manifestations) versus pathogenic mutations remains unsuccessful to identify simple basic features (at the levels of primary and secondary tRNA structures) that would make possible the prediction of pathogenicity of new mutations (Florentz and Sissler 2001; Yarham et al. 2011).

Numerous studies have attempted to unravel the molecular impacts of the mutations on the various properties of the affected tRNAs and lead so far to a mosaïcity of impacts. It is now clear that mutations can affect any step of the tRNA life cycle, either along tRNA biogenesis (maturation of 3'- or 5'- ends within the initial primary transcript, synthesis of the non coded CCA end, post-transcriptional modifications, folding and structure, stability), or tRNA function (aminoacylation, interaction with translation factors). Several reviews summarize the present view (Florentz and Sissler 2003; Rötig 2011; Ylikallio and Suomalainen 2012). In most cases, the effects of mutations are mild and affect either a single step of the tRNA life cycle or a combination of several of them. However, an initial impact is frequently observed on structural properties of affected tRNAs, followed by subsequent cascade effects on downstream functions (Florentz et al. 2003; Levinger et al. 2004; Wittenhagen and Kelley 2003). Therefore, any insight on the precise rules governing secondary and tertiary folding of the full set of human mt-tRNAs remains of high importance in order to comprehend the high sensitivity of these tRNAs to mutations perturbing their structure. Along these lines, pioneered experiments (Helm et al. 1998; Messmer et al. 2009), combined with the implementation of dedicated database (Pütz et al. 2007), and the development of bioinformatics tools (Bernt et al. 2012; Jühling et al. 2012a) are opening the path toward a solid knowledge on mt-tRNA 3D structures.

Finally, the housekeeping function of tRNAs, namely their capacity to become esterified by an amino acid, is not systematically affected in mutated variants, so that alternative functions of mt-tRNAs (Hou and Yang 2013; Mei et al. 2010) or alternative partnerships have to be considered (Jacobs and Holt 2000; Giegé et al. 2012).

3.5.2 Mitochondrial aaRSs and Pathologies

Lately, case-by-case reports linking mutations in nuclear genes coding for mitochondrial translation machinery proteins to pathologies (such as mutations in genes for elongation factor, tRNA modification enzymes, and ribosomal proteins), opened the way to "mitochondrial translation disorders" (Jacobs 2003). A new breakthrough took place in 2007 with the discovery in patients with cerebral white matter abnormalities of unknown origin of a first set of mutations present in *DARS2*, the nuclear gene coding for mt-AspRS (Scheper et al. 2007). These abnormalities were part of childhood-onset disorder called Leukoencephalopathy with Brain stem and Spinal cord involvement and Lactate elevation (LBSL; van der Knaap et al. 2003). Since this first discovery, mutations in eight additional mt-aaRS-encoding genes have been reported. They hit mt-ArgRS (Edvardson et al. 2007), mt-TyrRS (Riley et al. 2010), mt-SerRS (Belostotsky et al. 2011), mt-HisRS (Pierce et al. 2011), mt-AlaRS

(Götz et al. 2011), mt-MetRS (Bayat et al. 2012), mt-GluRS (Steenweg et al. 2012) and mt-PheRS (Elo et al. 2012) (Table 3.2). These recent correlations with human pathologies and the exponential description of reported cases, suggest as evidence that all mt-aaRS genes are likely affected by pathology-related mutations (that remain yet unveil), leading to a new family of disorders named according to the incriminated proteins namely "mt-aaRS disorders".

A detailed description and analysis of the full set of mutations in human mt-aaRS genes and their molecular and phenotypic implications has been reviewed (Konovalova and Tyynismaa 2013; Schwenzer et al. 2013). Here, the general outcomes are summarized. Table 3.2 recalls the main features characterizing the 65 nowadays-reported mutations in mt-aaRS genes. These include the type of pathogenic manifestation, familial pedigree, and affected tissues, as well as the number and types of mutations in each gene, their heterozygous versus homozygous, as well as the molecular impact on the synthetase and the final molecular impact on respiratory chain complexes. As a major outcome, it appears that whatever the mutation, no common combination of molecular steps correlates the mutations with the phenotypic expressions. Interestingly, the molecular impact of the mutations is not necessarily at the level of the housekeeping function of the synthetase, namely aminoacylation. Pathology-related mutations may have either a direct effect on the mitochondrial translation machinery by impacting one or several steps of mt-aaRS biogenesis and/or functioning. They may alternatively have an indirect effect by impacting ensuing steps and/or subsequent products activities [translation of the 13 mt-DNA-encoded subunits of respiratory chain complexes, respiratory chain complexes activities, and ATP synthesis]. Also, despite a dominant effect on brain and neuronal system is observed, sporadic manifestations are as well occurring in skeletal muscle, kidney, lung and/or heart. Along these lines, the selective vulnerability of tracts within the nervous system in case of mutations leading to splicing defects, for instance, is explained by tissue-specific differences in the concentration of the splicing factors (reduced in neural cell) (Edvardson et al. 2007; van Berge et al. 2012). However, the tissue-specificity of disorders remains an intriguing question. It is worth to establish the steady-state levels of various components of the mitochondrial translation machinery in different tissues, and correlate these levels with mitochondrial activity. This approach has been initiated by the evaluation of mRNA levels of the full set of human mt-aaRSs in 20 different human tissues (Fig. 3.5). A striking landscape of mRNA levels is observed highlighting tissue-specific differences by several orders of magnitude. There is no correlation between the various levels of mRNA and the amino acid content of the 13 mt-encoded proteins: leucine content is highest (14.4 %) followed by isoleucine, serine, and threonine (7 %), while arginine, aspartate, cysteine, glutamine, glutamate, and lysine contents is below 3 % (Schwenzer et al. 2013). We suggest that the low levels of aaRS mRNAs in brain, muscle and heart, lead to limiting mt translation activity in these tissues. Even a subtle change in mitochondrial translation efficiency may be detrimental in these tissues of high-energy demand. These data also suggest that mt-aaRS expressed to high levels may be involved in other functions than exclusively translation as is the case for cytosolic synthetases (Park et al. 2008).

Table 3.2 Human mt-aaRSs involved in mitochondrial disorders

	mt-AlaRS	mt-ArgRS	mt-AspRS	mt-GluRS	mt-HisRS	mt-MetRS	mt-pheRS	mt-SerRS	mt-TyrRS
Pathogenic manifestation[a]	CMP	PCH	LBSL	LBSL	PS	ARSAL	Encephalopathy	HUPRA	MLASA
Consanguinity of the parents	no	yes/no	yes/no	yes/no	no	no	yes/no	yes/no	yes/no
Affected tissues	Heart, brain, skeletal muscle	Brain	Brain, spinal cord	Brain	Ovarian sensori-neural system	Brain	Brain ± liver	Kidney, lung, muscle	Blood, skelatal muscle
Number of mutations	2	10	27	15	3	?	4	1	2
Non sense (frameshift/stop)	0	1	8	3	0	0	0	0	0
Missense	2	7	14	11	2	0	4	1	2
Deletion/insertion	0	2	5	1	1	Large insertion	0	0	0
Other	0	0	0	0	0	Gene duplication	0	0	0
Genetics compound	Heterozygous	Heterozygous/homozygous	Heterozygous/homozygous	Heterozygous/homozygous	Heterozygous	Heterozygous/homozygous	Heterozygous/homozygous	Homozygous	Homozygous
Negatively affected molecular event									
AaRS encoding mRNA expression/processing	nd	yes/no	yes/no	nd	no	no	nd	nd	nd
AaRS expression/stability	nd	yes/no	yes/no	nd	yes/no	yes	nd	nd	yes/no
AaRS import/oligomerziation/structure	nd	nd	yes/no	nd	yes/no	nd	yes/no	nd	no

(continued)

Table 3.2 (continued)

	mt-AlaRS	mt-ArgRS	mt-AspRS	mt-GluRS	mt-HisRS	mt-MetRS	mt-pheRS	mt-SerRS	mt-TyrRS
Aminoacylation activity (predicated)	yes	yes/no	yes/no	nd	yes	no	yes	yes	yes/no
Impact on the respiratory chain complex									
Global impact on translation	no	nd	no	yes	nd	yes	yes	nd	yes
Impact on RC activity[b]	yes	yes	yes	yes	nd	yes	yes	yes	yes

[a] Pathogenic manifestations are presented under the following acronyms. *CMP* Infantile Mitochondrial Cardimyopathy, *PCH* Ponto Cerebellar Hypoplasia, *LBSL* Leukoencephalopathy with Brain stem and Spinal cord involvement and Lactate elevation, *PS* Perrault Syndrome, *ARSAL* Autosomal Recessive Spastic Ataxia with Leukoencephalopathy, *HUPRA* HyperUricemia, Pulmonary hypertensions and Renal failure in infancy and Alkalosis, *MLASA* Myopathy, Lactic Acidosis and Sideroblastic Anemia. Consanguineous state of the parents and affected tissues are recalled. The number and type of mutations as well as their genetic compound are given. Molecular effects and impacts on the respiratory chain complexes are displayed

[b] All possible molecular effects on either aaRS biogenesis and/or function, or on translation and/or activity of the respiratory chain complexes, have not necessarily been investigated for all reported cases. For a more detailed view, please refer to (Schwenzer et al. 2013). For negatively affected molecular events, "yes" means that all tested mutations show this defect; "no" means that all tested mutations show no defect; "yes/no" means that some show a defect some do not. The table displays a mean picture: (i) a defect in mRNA expression and in aminoacylation does not necessarily refer to a same mutation; (ii) impacts on translation and aminoacylation activity can originate from separate mutations; and (iii) RC defects do not refer to tissue specificity: translation defect may correspond to, e.g., fibroblasts while aminoacylation activity was measured in muscle cells. *nd* stands for not determined

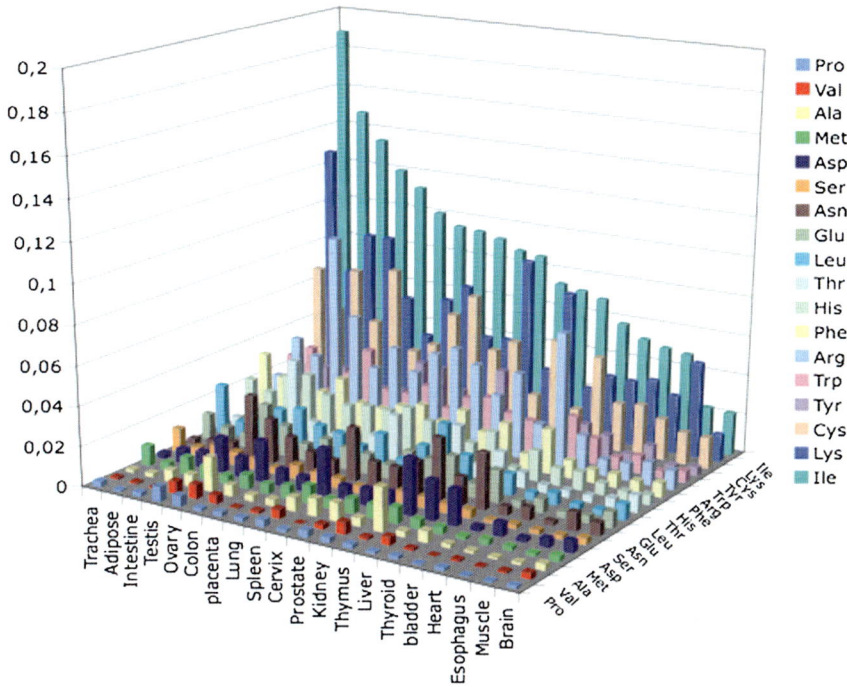

Fig. 3.5 Dosage of mt-aaRS-encoding mRNAs in different human tissues. The amount of mRNAs coding for 18 mt-aaRSs is determined by quantitative PCR on cDNAs prepared from total mRNA extracted from 20 human tissues. The mRNA for LysRS codes both for the cytosolic and mt enzyme. Results obtained for GlyRS, highly expressed, are missing from the graph (cytosolic and mt GlyRSs are both encoded by a single nuclear gene). No gene for mt-GlnRS has been reported so far. Values are normalized against the standard expression of GAPDH. They are mean values out of at least three independent experiments, from which the standard deviation is close to 50 %

To conclude, links between the activity of a given mt-aaRS along mitochondrial translation on one hand and ATP production on the other hand, involved a number of issues that need to be further explored. Those issues should take also into account the possibility that aminoacylation may turn out to be not the sole function of mt-aaRSs in a living cell and that these enzymes may also participate in other processes and/or be implicated in various fine-tuning mechanisms as is the case for cytosolic aaRSs. Indeed, various bacterial and eukaryal aaRSs were found to have many additional functions (e.g., Guo and Schimmel 2013). It becomes thus necessary to determine all potential interacting components of mt-aaRSs and to study their dynamic location within the organelle. In other works, the functional network of mt-aaRSs and its regulation needs to be tackled. New routes toward understanding of the molecular impacts of point mutations in nuclear mt-aaRS genes outside the frame of mitochondrial translation should become opened along these lines.

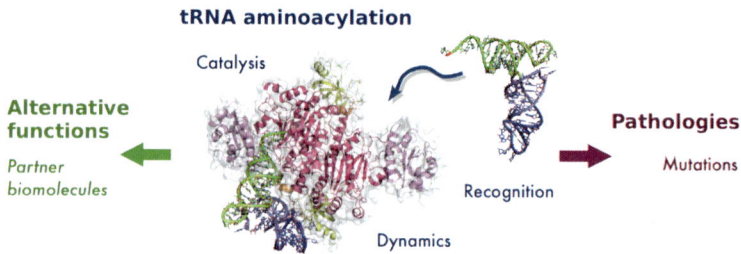

tRNA aminoacylation

Catalysis

Alternative functions

Partner biomolecules

Pathologies

Mutations

Recognition

Dynamics

Fig. 3.6 Mammalian mitochondrial tRNAs and aminoacyl-tRNA synthetases are key macromolecules in mitochondrial translation. They are also hot spots for a growing number of human pathologies, some of which being related to defects in the housekeeping functions, some not. Numerous questions both on fundamental knowledge ("order") and toward understanding of the molecular events underlying pathologies ("disorder") remain open. Both stimulate the development of new research lines outside the strict frame of mitochondrial translation

3.6 Conclusion and Perspectives

Mammalian mt-tRNAs and mt-aaRSs are known as key actors of mitochondrial translation, leading to the synthesis of 13 essential mitochondrial inner membrane proteins and subunits of respiratory chain complexes. The fact that these two families of partner macromolecules are coded by either of the two cellular genomes, the nuclear genome coding for the mt-aaRSs, and the mitochondrial genome coding for the mt-tRNAs, highlights different, but however connected, evolutionary pathways between RNAs and partner proteins allowing for tRNA aminoacylation. Despite important structural degeneration of mt-tRNAs linked to the high mutation rate of the mitochondrial genome, the conservation of "canonical" structural properties of mt-aaRSs include subtle but strong molecular adaptation so that the partnership between both macromolecules is maintained for accurate mitochondrial translation (Fig. 3.6). Detailed analysis of mammalian mt-tRNAs revealed new structural rules for RNAs, and bioinformatics compilations confirmed and enlarged the structural diversity of mt-tRNAs to the shortest versions ever discovered, calling for additional structural adaptation of functional RNAs. Investigation of mammalian mt-tRNAs and aaRSs are however still in an initial stage. Only a few systems have been characterized along a limited number of aspects. Open questions include for example, the common structural properties of mt-tRNAs and of synthetases all along metazoan mitochondria. Which are the identity elements in mt-tRNAs for specific aminoacylation by cognate mt-aaRS? Are these conserved or idiosynchratic? How far can a given set be degenerated and still allow for specificity? Post-transcriptional modification patterns of tRNAs and possible post-translational modifications of synthetases remain to become determined on full scale. The importance of these modifications is of crucial interest not only in structural stabilization and in the housekeeping codon reading function of mt-tRNAs, but also in alternative functions of mt-aaRSs. Post-translational modifications are indeed triggers to alternative functions of cytosolic aaRSs (e.g., Kim et al.

2012; Ofir-Birin et al. 2013). The related question of alternate function of mt-aaRSs is of key importance, especially in regard of understanding the molecular mechanisms of related disorders. Such functions, unrelated to aminoacylation, correspond to an emerging field of discoveries for cytosolic aaRSs. Examples include TyrRS involved in receptor-mediated signaling pathways associated with angiogenesis (Wakasugi et al. 2002a), TrpRS activated as an angiostatic factor (Wakasugi et al. 2002b), Glu-ProRS involved in the inflammatory response (Jia et al. 2008), LysRS plays a role in HIV-I packaging (Kleiman and Cen 2004) of GlyRS as anti-tumorigenic agent (Park et al. 2012).

The existence of an always growing panel of human disorders correlated to point mutations in either mt-tRNA genes or mt-aaRS genes, leads to investigations as to the molecular impacts of the mutations (Fig. 3.6). While a number of approaches pin-pointed a mosaicity of impacts on human mt-tRNA structure and function, and more recently on biophysical and functional properties of a small set of mt-aaRS, all linked to the housekeeping activity of both macromolecules in mitochondrial translation, a new field of investigation is emerging. Several of the pathology-related mutations are not located in the catalytic site of the protein, do not affect protein synthesis, and are thus indicative of new properties of mt-tRNAs and mt-aaRSs outside translation. Some hints on alternative functions have already been reported. Any route along this new topic deserves interest and should now be considered, as is currently the case for cytosolic tRNAs and synthetases (Guo and Schimmel 2013; Guo et al. 2010a, 2010b). Importantly, many other actors of the mammalian mitochondrial translation machinery also remain to be explored more systematically, not only for fundamental and evolutionary understanding but also because of their growing implication in human pathologies (e.g., Rötig 2011; Taylor and Turnbull 2005; Watanabe 2010).

Acknowledgments We thank Richard Giegé for critical reading of the manuscript and Gert Scheper and Koen de Groot for help in the qPCR experiments. Numerous contributions on mt-tRNAs and aaRS could not be mentioned because of space limitation and we apologize for this. Financial support came from Centre National de la Recherche Scientifique (CNRS), Université de Strasbourg, ANR MITOMOT (ANR-09-BLAN-0091-01/03), French National Program 'Investissements d'Avenir' (Labex MitoCross) administered by the "Agence National de la Recherche", and referenced ANR-10-IDEX-002-02; French-German PROCOPE program (DAAD D/0628236, EGIDE PHC 14770PJ), and German Academic Exchange Service (DAAD D/10/43622) for a doctoral fellowship.

References

Alfonzo JD, Söll D (2009) Mitochondrial tRNA import—the challenge to understand has just begun. Biol Chem 390:717–722

Anderson S, Bankier AT, Barrel BG, de Bruijn MHL, Coulson AR, Drouin J, Eperon JC, Nierlich DP, Roe BA, Sanger F, Schreier PH, Smith AJH, Staden R, Young IG (1981) Sequence and organization of the human mitochondrial genome. Nature 290:457–465

Arcari P, Brownlee GG (1980) The nucleotide sequence of a small (3S) seryl-tRNA (anticodon GCU) from beef heart mitochondria. Nucleic Acids Res 8:5207–5212

Bayat V, Thiffault I, Jaiswal M, Tétreault M, Donti T, Sasarman F, Bernard G, Demers-Lamarche J, Dicaire MJ, Mathieu J, Vanasse M, Bouchard JP, Rioux MF, Lourenco CM, Li Z, Haueter

C, Shoubridge EA, Graham BH, Brais B, Bellen HJ (2012) Mutations in the mitochondrial methionyl-tRNA synthetase cause a neurodegenerative phenotype in flies and a recessive ataxia (ARSAL) in humans. PLoS Biol 10:e1001288

Belostotsky R, Ben-Shalom E, Rinat C, Becker-Cohen R, Feinstein S, Zeligson S, Segel R, Elpeleg O, Nassar S, Frishberg Y (2011) Mutations in the mitochondrial seryl-tRNA synthetase cause hyperuricemia, pulmonary hypertension, renal failure in infancy and alkalosis, HUPRA syndrome. Am J Hum Genet 88:193–200

Bernt M, Donath A, Jühling F, Externbrink F, Florentz C, Fritzsch G, Pütz J, Middendorf M, Stadler PF (2012) MITOS: improved de novo metazoan mitochondrial genome annotation. Mol Phylogenet Evol [Epub ahead of print]

Bonnefond L, Fender A, Rudinger-Thirion J, Giegé R, Florentz C, Sissler M (2005a) Toward the full set of human mitochondrial aminoacyl-tRNA synthetases: characterization of AspRS and TyrRS. Biochemistry 44:4805–4816

Bonnefond L, Frugier M, Giegé R, Rudinger-Thirion J (2005b) Human mitochondrial TyrRS disobeys the tyrosine idenity rules. RNA 11:558–562

Bonnefond L, Frugier M, Touzé E, Lorber B, Florentz C, Giegé R, Sauter C, Rudinger-Thirion J (2007) Crystal structure of human mitochondrial tyrosyl-tRNA synthetase reveals common and idiosyncratic features. Structure 15:1505–1516

Braband A, Cameron SL, Podsiadlowski L, Daniels SR, Mayer G (2010) The mitochondrial genome of the onychophoran Opisthopatus cinctipes (Peripatopsidae) reflects the ancestral mitochondrial gene arrangement of Panarthropoda and Ecdysozoa. Mol Phylogenet Evol 57:285–292

Brindefalk B, Viklund J, Larsson D, Thollesson M, Andersson SG (2007) Origin and evolution of the mitochondrial aminoacyl-tRNA synthetases. Mol Biol Evol 24:743–756

Brown WM, George M, Wilson AC (1979) Rapid evolution of animal mitochondrial DNA. Proc Natl Acad Sci USA 76:1967–1971

Bullard J, Cai Y-C, Spremulli L (2000) Expression and characterization of the human mitochondrial leucyl-tRNA synthetase. Biochem Biophys Acta 1490:245–258

Cader MZ, Ren J, James PA, Bird LE, Talbot K, Stammers DK (2007) Crystal structure of human wildtype and S581L-mutant glycyl-tRNA synthetase, an enzyme underlying distal spinal muscular atrophy. FEBS Lett 581:2959–2964

Castellana S, Vicario S, Saccone C (2011) Evolutionary patterns of the mitochondrial genome in Metazoa: exploring the role of mutation and selection in mitochondrial protein coding genes. Genome Biol Evol 3:1067–1079

Chacinska A, Koehler CM, Milenkovic D, Lithgow T, Pfanner N (2009) Importing mitochondrial proteins: machineries and mechanisms. Cell 1387:628–644

Chimnaronk S, Gravers Jeppesen M, Suzuki T, Nyborg J, Watanabe K (2005) Dual-mode recognition of noncanonical tRNAs(Ser) by seryl-tRNA synthetase in mammalian mitochondria. EMBO J 24:3369–3379

de Bruijn MH, Schreier PH, Eperon IC, Barrell BG, Chen EY, Armstrong PW, Wong JF, Roe BA (1980) A mammalian mitochondrial serine transfer RNA lacking the "dihydrouridine" loop and stem. Nucleic Acids Res 8:5213–5222

Dörner M, Altmann M, Pääbo S, Mörl M (2001) Evidence for import of a lysyl-tRNA into marsupial mitochondria. Mol Biol Cell 12:2688–2698

Duchêne AM, Pujol C, Maréchal-Drouard L (2009) Import of tRNAs and aminoacyl-tRNA synthetases into mitochondria. Curr Genet 55:1–18

Edvardson S, Shaag A, Kolesnikova O, Gomori JM, Tarassov I, Einbinder T, Saada A, Elpeleg O (2007) Deleterious mutation in the mitochondrial arginyl-transfer RNA synthetase gene is associated with pontocerebellar hypoplasia. Am J Hum Genet 81:857–862

Elo JM, Yadavalli SS, Euro L, Isohanni P, Götz A, Carroll CJ, Valanne L, Alkuraya FS, Uusimaa J, Paetau A, Caruso EM, Pihko H, Ibba M, Tyynismaa H, Suomalainen A (2012) Mitochondrial phenylalanyl-tRNA synthetase mutations underlie fatal infantile Alpers encephalopathy. Hum Mol Genet 21:4521–4529

Enriquez JA, Attardi G (1996) Analysis of aminoacylation of human mitochondrial tRNAs. Methods Enzymol 264:183–196

Fender A, Gaudry A, Jühling F, Sissler M, Florentz C (2012) Adaptation of aminoacylation rules to mammalian mitochondria. Biochimie 94:1090–1097

Fender A, Sauter C, Messmer M, Pütz J, Giegé R, Florentz C, Sissler M (2006) Loss of a primordial identity element for a mammalian mitochondrial aminoacylation system. J Biol Chem 281:15980–15986

Florentz C, Sissler M (2001) Disease-related *versus* polymorphic mutations in human mitochondrial tRNAs: where is the difference? EMBO Rep 2(6):481–486

Florentz C, Sissler M (2003) Mitochondrial tRNA aminoacylation and human diseases. In: Lapointe J, Brakier-Gingras L (eds) Translation mechanisms. Landes Bioscience, Georgetown, pp 129–143

Florentz C, Sohm B, Tryoen-Tóth P, Pütz J, Sissler M (2003) Human mitochondrial tRNAs in health and disease. Cell Mol Life Sci 60:1356–1375

Frechin M, Duchêne A-M, Becker HD (2009a) Translating organellar glutamine codons: A case by case scenario? RNA Biol 6:31–34

Frechin M, Senger B, Brayé M, Kern D, Martin RP, Becker HD (2009b) Yeast mitochondrial Gln-tRNA(Gln) is generated by a GatFAB-mediated transamidation pathway involving Arc1p-controlled subcellular sorting of cytosolic GluRS. Genes Dev 23:1119–1130

Friederich MW, Hagerman PJ (1997) The angle between the anticodon and aminoacyl acceptor stems of yeast tRNA(Phe) is strongly modulated by magnesium ions. Biochemistry 36:6090–6099

Gaudry A, Lorber B, Messmer M, Neuenfeldt A, Sauter C, Florentz C, Sissler M (2012) Redesigned N-terminus enhances expression, solubility, and crystallisability of mitochondrial enzyme. Protein Eng Des Sel 25:473–481

Giegé R (2008) Toward a more complete view of tRNA biology. Nat Struct Mol Biol 15:1007–1014

Giegé R, Florentz C, Kern D, Gangloff J, Eriani G, Moras D (1996) Aspartate identity of transfer RNAs. Biochimie 78:605–623

Giegé R, Jühling F, Pütz J, Stadler P, Sauter C, Florentz C (2012) Structure of transfer RNAs: similarity and variability. Wiley Interdiscip Rev RNA 3:37–61

Giegé R, Sissler M, Florentz C (1998) Universal rules and idiosyncratic features in tRNA identity. Nucleic Acids Res 26:5017–5035

Goto Y, Nonaka I, Horai S (1990) A mutation in the tRNA$^{Leu(UUR)}$ gene associated with the MELAS subgroup of mitochondrial encephalomyopathies. Nature 348:651–653

Götz A, Tyynismaa H, Euro L, Ellonen P, Hyötyläinen T, Ojala T, Hämäläinen RH, Tommiska J, Raivio T, Oresic M, Karikoski R, Tammela O, Simola KO, Paetau A, Tyni T, Suomalainen A (2011) Exome sequencing identifies mitochondrial alanyl-tRNA synthetase mutations in infantile mitochondrial cardiomyopathy. Am J Hum Genet 88:635–642

Guo M, Ignatov M, Musier-Forsyth K, Schimmel P, Yang XL (2008) Crystal structure of tetrameric form of human lysyl-tRNA synthetase: Implications for multisynthetase complex formation. Proc Natl Acad Sci USA 105:2331–2336

Guo M, Schimmel P (2013) Essential nontranslational functions of tRNA synthetases. Nat Chem Biol 9:145–153

Guo M, Schimmel P, Yang X-L (2010a) Functional expansion of human tRNA synthetases achieved by structural inventions. FEBS Lett 584:434–442

Guo M, Yang XL, Schimmel P (2010b) New functions of aminoacyl-tRNA synthetases beyond translation. Nat Rev Mol Cell Biol 11:668–674

Haen KM, Pett W, Lavrov DV (2010) Parallel loss of nuclear-encoded mitochondrial aminoacyl-tRNA synthetases and mtDNA-encoded tRNAs in Cnidaria. Mol Biol Evol 27:2216–2219

Helm M, Attardi G (2004) Nuclear control of cloverleaf structure of human mitochondrial tRNA(Lys). J Mol Biol 337:545–560

Helm M, Brulé H, Degoul F, Cepanec C, Leroux J-P, Giegé R, Florentz C (1998) The presence of modified nucleotides is required for cloverleaf folding of a human mitochondrial tRNA. Nucleic Acids Res 26:1636–1643

Helm M, Brulé H, Friede D, Giegé R, Pütz J, Florentz C (2000) Search for characteristic structural features of mammalian mitochondrial tRNAs. RNA 6:1356–1379

Helm M, Florentz C, Chomyn A, Attardi G (1999) Search for differences in post-transcriptional
modification patterns of mitochondrial DNA-encoded wild-type and mutant human tRNA^{Lys}
and tRNA$^{Leu(UUR)}$. Nucleic Acids Res 27:756–763

Hou YM, Yang X (2013) Regulation of cell death by transfer RNA. Antioxid Redox Signal [Epub
ahead of print]

Ibba M, Francklyn C, Cusack S (2005) The aminoacyl-tRNA synthetases. Landes Biosciences,
Georgetown

Jacobs HT, Holt IJ (2000) The np 3243 MELAS mutation: damned if you aminoacylate, damned
if you don't. Hum Mol Genet 1:463–465

Jacobs HT (2003) Disorders of mitochondrial protein synthesis. Hum Mol Genet 12:R293–301

Jia J, Arif A, Ray PS, Fox PL (2008) WHEP domains direct noncanonical function of glutamyl-
Prolyl-tRNA synthetase in translational control of gene expression. Mol Cell 29:679–690

Jühling F, Pütz J, Bernt M, Donath A, Middendorf M, Florentz C, Stadler PF (2012a) Improved
systematic tRNA gene annotation allows new insights into the evolution of mitochondrial tRNA structures and into the mechanisms of mitochondrial genome rearrangements.
Nucleic Acids Res 40:2833–2845

Jühling F, Pütz J, Florentz C, Stadler PF (2012b) Armless mitochondrial tRNAs in Enoplea
(Nematoda). RNA Biol 9:1161–1166

Kim DG, Choi JW, Lee JY, Kim H, Oh YS, Lee JW, Tak YK, Song JM, Razin E, Yun SH, Kim S
(2012) Interaction of two translational components, lysyl-tRNA synthetase and p40/37LRP,
in plasma membrane promotes laminin-dependent cell migration. FASEB J 26:4142–4159

Kirino Y, Goto Y, Campos Y, Arenas J, Suzuki T (2005) Specific correlation between the wobble
modification deficiency in mutant tRNAs and the clinical features of a human mitochondrial
disease. Proc Natl Acad Sci USA 102:7127–7132

Kleiman L, Cen S (2004) The tRNALys packaging complex in HIV-1. Int J Biochem Cell Biol
36:1776–1786

Klipcan L, Levin I, Kessler N, Moor N, Finarov I, Safro M (2008) The tRNA-induced
conformational activation of human mitochondrial phenylalanyl-tRNA synthetase.
Structure 16:1095–1104

Klipcan L, Moor N, Finarov I, Kessler N, Sukhanova M, Safro MG (2012) Crystal structure of
human mitochondrial PheRS complexed with tRNA(Phe) in the active "open" state. J Mol
Biol 415:527–537

Konovalova S, Tyynismaa H (2013) Mitochondrial aminoacyl-tRNA synthetases in human disease. Mol Genet Metab [Epub ahead of print]

Kumazawa Y, Himeno H, Miura K, Watanabe K (1991) Unilateral aminoacylation specificity
between bovine mitochondria and eubacteria. J Biochem 109:421–427

LaRiviere FJ, Wolfson AD, Uhlenbeck OC (2001) Uniform binding of aminoacyl-tRNAs to elongation factor Tu by thermodynamic compensation. Science 294:165–168

Laslett D, Canbäck B (2008) ARWEN: a program to detect tRNA genes in metazoan mitochondrial nucleotide sequences. Bioinformatics 24:172–175

Leontis NB, Stombaugh J, Westhof E (2002) The non-Watson-Crick base pairs and their associated isostericity matrices. Nucleic Acids Res 30:3497–3531

Levinger L, Mörl M, Florentz C (2004) Mitochondrial tRNA 3′ end metabolism and human disease. Nucleic Acids Res 32:5430–5441

Lowe TM, Eddy SR (1997) tRNAscan-SE: a program for improved detection of transfer RNA
genes in genomic sequence. Nucleic Acids Res 25:955–964

Macey JR, Larson A, Ananjeva NB, Papenfuss TJ (1997) Replication slippage may cause parallel
evolution in the secondary structures of mitochondrial transfer RNAs. Mol Biol Evol 14:30–39

McFarland R, Elson JL, Taylor RW, Howell N, Turnbull DM (2004) Assigning pathogenicity to
mitochondrial tRNA mutations: when 'definitely maybe' is not good enough. Trends Genet
20:591–596

Mei Y, Yong J, Liu H, Shi Y, Meinkoth J, Dreyfuss G, Yang X (2010) tRNA binds to cytochrome
c and inhibits caspase activation. Mol Cell 37:688–698

Messmer M, Pütz J, Suzuki T, Suzuki T, Sauter C, Sissler M, Florentz C (2009) Tertiary net-work in mammalian mitochondrial tRNAAsp revealed by solution probing and phylogeny. Nucleic Acids Res 37:6881–6895

Miyamoto H, Machida RJ, Nishida S (2010) Complete mitochondrial genome sequences of the three pelagic chaetognaths Sagitta nagae, Sagitta decipiens and Sagitta enflata. Comp Biochem Physiol Part D Genomics Proteomics 5:65–72

Motorin Y, Helm M (2010) tRNA stabilization by modified nucleotides. Biochemistry 49:4934–4944

Mudge SJ, Williams JH, Eyre HJ, Sutherland GR, Cowan PJ, Power DA (1998) Complex organi-sation of the 5′-end of the human glycine tRNA synthetase gene. Gene 209:45–50

Nagao A, Suzuki T, Katoh T, Sakaguchi Y, Suzuki T (2009) Biogenesis of glutaminyl-mt tRNAGln in human mitochondria. Proc Natl Acad Sci USA 106:16209–16214

Nagao A, Suzuki T, Suzuki T (2007) Aminoacyl-tRNA surveillance by EF-Tu in mammalian mitochondria. Nucleic Acids Symp Ser (Oxf) 51:41–42

Nawroki EP, Kolbe DL, Eddy SR (2009) Infernal 1.0: Inference of RNA Alignments. Bioinformatics 25:1335–1337

Neuenfeldt A, Lorber B, Ennifar E, Gaudry A, Sauter C, Sissler M, Florentz C (2013) Thermodynamic properties distinguish human mitochondrial aspartyl-tRNA synthetase from bacterial homolog with same 3D architecture. Nucleic Acids Res 41:2698–2708

Ofir-Birin Y, Fang P, Bennett SP, Zhang HM, Wang J, Rachmin I, Shapiro R, Song J, Dagan A, Pozo J, Kim S, Marshall AG, Schimmel P, Yang XL, Nechushtan H, Razin E, Guo M (2013) Structural switch of lysyl-tRNA synthetase between translation and transcription. Mol Cell 49:30–42

Park MC, Kang T, Jin D, Han JM, Kim SB, Park YJ, Cho K, Park YW, Guo M, He W, Yang XL, Schimmel P, Kim S (2012) Secreted human glycyl-tRNA synthetase implicated in defense against ERK-activated tumorigenesis. Proc Natl Acad Sci USA 109:E640–E647

Park SG, Schimmel P, Kim S (2008) Aminoacyl tRNA synthetases and their connections to dis-ease. Proc Natl Acad Sci USA 105:11043–11049

Pierce SB, Chisholm KM, Lynch ED, Lee MK, Walsh T, Opitz JM, Li W, Klevit RE, King MC (2011) Mutations in mitochondrial histidyl tRNA synthetase HARS2 cause ovarian dys-genesis and sensorineural hearing loss of Perrault syndrome. Proc Natl Acad Sci USA 108:6543–6548

Pruitt KD, Tatusova T, Maglott DR (2007) NCBI reference sequences (RefSeq): a curated non-redundant sequence database of genomes, transcripts and proteins. Nucleic Acids Res 5(Database issue):D61–D65

Pütz J, Dupuis B, Sissler M, Florentz C (2007) Mamit-tRNA, a database of mammalian mito-chondrial tRNA primary and secondary structures. RNA 13:1184–1190

Riley LG, Cooper S, Hickey P, Rudinger-Thirion J, McKenzie M, Compton A, Lim SC, Thorburn D, Ryan MT, Giegé R, Bahlo M, Christodoulou J (2010) Mutation of the mitochondrial tyrosyl-tRNA synthetase gene, YARS2, causes myopathy, lactic acidosis, and sideroblastic anemia–MLASA syndrome. Am J Hum Genet 87:52–59

Rinehart J, Krett B, Rubio M-AT, Alfonzo JD, Söll D (2005) Saccharomyces cerevisiae imports the cytosolic pathway for Gln-tRNA synthesis into the mitochondion. Genes Dev 19:583–592

Rötig A (2011) Human diseases with impaired mitochondrial protein synthesis. Biochim Biophys Acta 1807:1198–1205

Scheper GC, van der Klok T, van Andel RJ, van Berkel CG, Sissler M, Smet J, Muravina TI, Serkov SV, Uziel G, Bugiani M, Schiffmann R, Krageloh-Mann I, Smeitink JA, Florentz C, Coster RV, Pronk JC, van der Knaap MS (2007) Mitochondrial aspartyl-tRNA synthetase deficiency causes leukoencephalopathy with brain stem and spinal cord involvement and lactate elevation. Nat Genet 39:534–539

Schwenzer H, Zoll J, Florentz C, Sissler M (2013) Pathogenic implications of human mitochon-drial aminoacyl-tRNA synthetases. In: KIM (ed) Topics in current chemistry-aminoacyl-tRNA synthetases: Applications in chemistry, Biology and Medicine. Springer (in press)

Segovia R, Pett W, Trewick S, Lavrov DV (2011) Extensive and evolutionarily persistent mitochondrial tRNA editing in Velvet Worms (phylum Onychophora). Mol Biol Evol 28:2873–2881

Seutin G, Lang BF, Mindell DP, Morais R (1994) Evolution of the WANCY region in amniote mitochondrial DNA. Mol Biol Evol 11:329–340

Shiba K, Schimmel P, Motegi H, Noda T (1994) Human glycyl-tRNA synthetase. Wide divergence of primary structure from bacterial counterpart and species-specific aminoacylation. J Biol Chem 269:30049–30055

Shoffner J, Lott M, Lezza AMS, Seibel P, Ballinger SW, Wallace DC (1990) Myoclonic epilepsy and ragged red fiber disease (MERRF) is associated with mitochondrial DNA tRNA[Lys] mutation. Cell 61:931–937

Sissler M, Pütz J, Fasiolo F, Florentz C (2005) Mitochondrial aminoacyl-tRNA synthetases. In: Ibba M, Francklyn C, Cusack S (eds), Aminoacyl-tRNA synthetases, chapter 24, pp 271–284. Landes Biosciences, Georgetown

Sohm B, Frugier M, Brulé H, Olszak K, Przykorska A, Florentz C (2003) Towards understanding human mitochondrial leucine aminoacylation identity. J Mol Biol 328:995–1010

Steenweg ME, Ghezzi D, Haack T, Abbink TE, Martinelli D, van Berkel CG, Bley A, Diogo L, Grillo E, Te Water Naudé J, Strom TM, Bertini E, Prokisch H, van der Knaap MS, Zeviani M (2012) Leukoencephalopathy with thalamus and brainstem involvement and high lactate 'LTBL' caused by EARS2 mutations. Brain 135:1387–1394

Suga K, Mark Welch DB, Tanaka Y, Sakakura Y, Hagiwara A (2008) Two circular chromosomes of unequal copy number make up the mitochondrial genome of the rotifer Brachionus plicatilis. Mol Biol Evol 25:1129–1137

Suzuki T, Nagao A, Suzuki T (2011) Human mitochondrial tRNAs: biogenesis, function, structural aspects, and diseases. Annu Rev Genet 45:299–329

Taylor RW, Turnbull DM (2005) Mitochondrial DNA mutations in human disease. Nat Rev Genet 6:389–402

Tolkunova E, Park H, Xia J, King MP, Davidson E (2000) The human lysyl-tRNA synthetase gene encodes both the cytoplasmic and mitochondrial enzymes by means of an unusual splicing of the primary transcript. J Biol Chem 275:35063–35069

van Berge L, Dooves S, van Berkel CG, Polder E, van der Knaap MS, Scheper GC (2012) Leukoencephalopathy with brain stem and spinal cord involvement and lactate elevation is associated with cell-type-dependent splicing of mtAspRS mRNA. Biochem J 441:955–962

van der Knaap MS, van der Voorn P, Barkhof F, Van Coster R, Krägeloh-Mann I, Feigenbaum A, Blaser S, Vles JS, Rieckmann P, Pouwels PJ (2003) A new leukoencephalopathy with brainstem and spinal cord involvement and high lactate. Ann Neurol 53:252–258

Wakasugi K, Slike BM, Hood J, Ewalt KL, Cheresh DA, Schimmel P (2002a) Induction of angiogenesis by a fragment of human tyrosyl-tRNA synthetase. J Biol Chem 277:20124–20126

Wakasugi K, Slike BM, Hood J, Otani A, Ewalt KL, Friedlander M, Cheresh DA, Schimmel P (2002b) A human aminoacyl-tRNA synthetase as a regulator of angiogenesis. Proc Natl Acad Sci USA 99:173–177

Wakita K, Watanabe Y-I, Yokogawa T, Kumazawa Y, Nakamura S, Ueda T, Watanabe K, Nishikawa K (1994) Higher-order structure of bovine mitochondrial tRNA[Phe] lacking the 'conserved' GG and TYCG sequences as inferred by enzymatic and chemical probing. Nucleic Acids Res 22:347–353

Wang X, Lavrov DV (2008) Seventeen new complete mtDNA sequences reveal extensive mitochondrial genome evolution within the Demospongiae. PLoS ONE 3:e2723

Watanabe K (2010) Unique features of animal mitochondrial translation systems. The non-universal genetic code, unusual features of the translational apparatus and their relevance to human mitochondrial diseases. Proc Jpn Acad Ser B Phys Biol Sci 86:11–36

Willkomm DK, Hartmann RK (2006) Intricacies and surprises of nuclear-mitochondrial co-evolution. Biochem J 399:e7–e9

Wittenhagen LM, Kelley SO (2003) Impact of disease-related mitochondrial mutations on tRNA structure and function. Trends Biochem Sci 28:605–611

Woese CR, Olsen GJ, Ibba M, Söll D (2000) Aminoacyl-tRNA synthetases, the genetic code, and the evolutionary process. Microbiol and Mol Biol Reviews 64:202–236

Wolstenholme DR, Okimoto R, Mcfarlane JL (1994) Nucleotide correlations that suggest tertiary interactions in the TV-replacement loop-containing mitochondrial tRNAs of the nematodes, *Caenorhabditis elegans* and *Ascaris suum*. Nucleic Acids Res 22:4300–4306

Xie W, Schimmel P, Yang XL (2006) Crystallization and preliminary X-ray analysis of a native human tRNA synthetase whose allelic variants are associated with Charcot-Marie-Tooth disease. Acta Crystallograph Sect F Struct Biol Cryst Commun 62:1243–1246

Yadavalli SS, Klipcan L, Zozulya A, Banerjee R, Svergun D, Safro M, Ibba M (2009) Large-scale movement of functional domains facilitates aminoacylation by human mitochondrial phenylalanyl-tRNA synthetase. FEBS Lett 583:3204–3208

Yao YN, Wang L, Wu XF, Wang ED (2003) The processing of human mitochondrial leucyl-tRNA synthetase in the insect cells. FEBS Lett 534:139–142

Yarham JW, Al-Dosary M, Blakely EL, Alston CL, Taylor RW, Elson JL, McFarland R (2011) A comparative analysis approach to determining the pathogenicity of mitochondrial tRNA mutations. Hum Mutat 32:1319–1325

Yarham JW, Elson JL, Blakely EL, McFarland R, Taylor RW (2010) Mitochondrial tRNA mutations and disease. Wiley Interdiscip Rev RNA 1:304–324

Ylikallio E, Suomalainen A (2012) Mechanisms of mitochondrial diseases. Ann Med 44:41–59

Chapter 4
Mitochondrial Targeting of RNA and Mitochondrial Translation

Ivan Tarassov, Ivan Chicherin, Yann Tonin, Alexandre Smirnov,
Petr Kamenski and Nina Entelis

Abstract Mitochondrial translation depends on the macromolecular components imported from the cytosol, which include translation factors, ribosomal proteins, aminoacyl-tRNA synthetases, and a variable number of small noncoding RNAs. The lasts are essentially tRNAs, but other small RNAs, like mammalian 5S rRNA, are also concerned by the RNA mitochondrial targeting pathway. If their importance in mitochondrial translation was demonstrated in each case where it was addressed, the precise function of these molecules differs from one system to another: in many cases they complement lacking mtDNA encoded counterparts, in others can fulfill conditional functions, finally they can complement the lack of needed mitochondrial enzymatic activities. In any case, it appears that the innated capacity of mitochondria to import small RNA molecules is supplied by specific

I. Tarassov (✉) · I. Chicherin · Y. Tonin · A. Smirnov · P. Kamenski · N. Entelis
Molecular Genetics, Genomics, Microbiology (UMR 7156), Universty of Strasbourg
and CNRS, 21 René Descartes, 67084 Strsbourg, France
e-mail: i.tarassov@unistra.fr

I. Chicherin
e-mail: i.v.chicherin@gmail.com

Y. Tonin
e-mail: yann.tonin@wanadoo.fr

A. Smirnov
e-mail: Skard1@yandex.ru

P. Kamenski
e-mail: peter@protein.bio.msu.ru

N. Entelis
e-mail: n.entelis@unistra.fr

I. Chicherin · A. Smirnov · P. Kamenski
Molecular Biology Department, Biology Faculty of M. V., Lomonossov Moscow State
University, 119991 Moscow, Russia

A.-M. Duchêne (ed.), *Translation in Mitochondria and Other Organelles*,
DOI: 10.1007/978-3-642-39426-3_4, © Springer-Verlag Berlin Heidelberg 2013

additional protein, often performing their "second job" to deliver the needed RNA in the organelle. This mechanism, still not understood in details, remains the unique natural pathway of nucleic acids delivery in mitochondria, and is therefore of a significant interest as a tool permitting to target this organelle with potentially therapeutic molecules and thus addressing a very important bulk of human pathologies linked with dysfunctions of mitochondrial translation machinery.

4.1 Introduction

This chapter focuses the role of mitochondrially targeted cytosolic RNA molecules in the organellar translation. Mitochondria import the majority of macromolecules from the cytosol, including almost a thousand of proteins and a various number of small noncoding RNAs. Mitochondrial targeting of RNAs is nowadays to be considered as a quasi-ubiquitous process, found in almost all studied biological systems (for review, see Entelis et al. 2001c; Rubio and Hopper 2011; Salinas et al. 2008). In all studied cases, naturally imported RNAs are small and noncoding: they are essentially tRNAs, but larger RNAs (like 5S rRNA, MRP, or RNase P RNA components) or smaller ones (like microRNAs) were described as mitochondrially targeted. A major part of these molecules clearly refer to protein synthesis machinery (tRNAs, 5S rRNA, RNase P). So far, different cases of RNA import of even similar types of molecules may have different functional implications— sometimes they are clearly essential, sometimes only conditionally important, in some cases their implication in mitochondrial translation is still not evident. The first part of the chapter tries to review these different cases with the emphasis on still unclear functional issues or contradictory data.

Mitochondrial DNA mutations are the recognized source of neuromuscular pathologies (Ruiz-Pesini et al. 2007). This large group of "mitochondrial diseases" includes mutations in virtually all mtDNA encoded genes and often have a strong impact on mitochondrial translation (Smits et al. 2010). One of their main features is the phenomenon of heteroplasmy, meaning the simultaneous presence of mutant and wild-type genomes in the same cell, the level of heteroplasmy determining the pathogenic effect of the mutation. All these pathologies have no efficient classical therapy and many attempts are currently done to develop alternative gene therapy strategies. Two main approaches were proposed—the "allotopic" one was to express in the nucleus the lacking or therapeutically active molecule and to address into mitochondria; and the "antigenomic" one, which aims to decrease the heteroplasmy below the threshold pathogenic level by addressing into the organelle specific recombinant molecules (Smith et al. 2004). If both strategies may use imported protein, nucleic acids delivery into mitochondria appears as more attractive. Therefore, in the second part of the chapter, we shall focus here RNA targeting into mitochondria, since many recent studies were aimed to exploit this pathway as a tool to correct dysfunctions of the mitochondrial translation machinery.

4.2 Imported tRNAs and Mitochondrial Translation

4.2.1 Plants

The situation in higher plants appears as the most evident—up to now, in all cases mitochondrially imported RNA species were limited to tRNAs and, also in all cases, imported tRNA species strictly corresponded to those whose genes were absent in mtDNA (Dietrich et al. 1992; Kumar et al. 1996; Salinas et al. 2008; Schneider and Marechal-Drouard 2000). The mechanism of this pathway is actively studied. Aminoacyl-tRNA synthetases, another component of the translation machinery, were formerly implicated in tRNA mitochondrial targeting (Dietrich et al. 1996; Small et al. 1992), while at the mitochondria level, Voltage Dependent Anion Channel (VDAC) and/or TOM complex were involved in tRNA translocation across mitochondrial membranes (Salinas et al. 2006, 2008).

There still is no evident evolutionary explanation why a given tRNA gene was conserved in mtDNA and the other—not, as well why there is a so significant variation in the imported tRNAs among different plant species, ranging from few to more than a half of the mitochondrial set. One could assume therefore that the emergency of tRNA import was polyphyletic and occurred many times during evolution, maybe in a sporadic way (Kumar et al. 1996). The similarity of mitochondrial and nuclear genetic code in plants may, at some extent, explain this situation: indeed nuclear-encoded and mitochondrially addressed tRNAs not only can function both in cytosolic and organellar translation, but also cannot become potentially toxic nonsense/missense suppressor tRNAs in the cytosol.

The question arises if there exist any regulatory mechanism which would permit to optimize mitochondrial translation by differential tRNA import efficiency. A priori there is no limitation for that, taking into account that mitochondrial subset of tRNA is a minor one comparing to the cytosolic pool. Indeed, in the case of a lower plant cells, in *C. reinhardtii,* mitochondrial abundance of nuclear DNA encoded tRNA was found to be in correlation with mtDNA genes codon usage (Vinogradova et al. 2009), which may reflect the existence of a kind of retrograde control of RNA import or, alternatively, a controlled system of RNA degradation in the organelle.

4.2.2 Trypanosomes

The group of *Trypanosomatidae* represents an extreme case, where all or almost all tRNAs are nuclearly encoded (Hauser and Schneider 1995; Rubio et al. 2000; Schneider 1994; Schneider and Marechal-Drouard 2000; Simpson et al. 1989). The need of these molecules for mitochondrial translation appears as evident, since no mtDNA-coded tRNA genes were ever identified while mitochondrial translation of the kinetoplast DNA (trypanosomal equivalent of mtDNA) is rather standard. The situation on these protozoans was recently exhaustively described in (Schneider 2011), therefore, we focus here only few translation-related issues.

In such trypanosomatids as *Trypanosoma brucei* or *Leishmania tarentolae*, cytosolic and mitochondrial pools of tRNAs are not identical, the exception concerning the initiator tRNA^{Met} and the tRNA^{Sec}, both found only in the cytosolic compartment of the cell (Geslain et al. 2006; Tan et al. 2002b). This fact refers directly to the absence of functional need of these tRNA species in the mitochondrial compartment. Mitochondrial initiation of translation does not require special eukaryotic initiator tRNA since a portion of mitochondrially imported cytosolic elongator tRNA^{Met} is formylated by specific tRNA^{Met}-formyltransferase (Charriere et al. 2005; Tan et al. 2002a). This formylated elongator tRNA is then recognized by the mitochondrial bacterial type initiation factor. This is just an example of the higher extent of mitochondrial translation flexibility, which is observed also in other species, as described below. Also, as in many other cases, it is difficult to explain how such mechanism was retained in evolution. It clearly demonstrates that many features of mitochondrial translation may have sporadic origins, and therefore being species/genera-specific. On the other hand, it also illustrate strictly classical functional explanation of RNA import in terms of translation processes, without involving "alternative functions" of imported RNA molecules, as for tRNA^{Sec}, whose presence in trypanosomatid mitochondria would be obsolete, in the absence of selenoproteins.

4.2.3 Yeast

In ascomycetes yeast *Saccharomyces cerevisiae* the situation with imported tRNAs is all but simple, taking into account that mtDNA appeared to encode all the set of tRNAs, in theory able to fulfill translational activity (Foury et al. 1998). So far two cases of tRNA import were independently described. The first concerns one out of two cytosolic tRNA^{Lys}, tRNA$^{Lys}_1$ (tRNA$^{Lys}_{CUU}$, or tRK1). This cytosolic tRNA was partially found in the mitochondrial compartment of the cell in 1979 (Martin et al. 1979), while the second tRNA^{Lys} (RNA$^{Lys}_{UUU}$, tRK2) resided only in the cytosol. The mechanism of tRK1 delivery into yeast mitochondria was studied in details and involved several targeting factors (Enolase-2 as an RNA chaperone and the cytosolic precursor of mitochondrial lysyl-tRNA synthetase pre-Msk1p) and the TOM/TIM translocators (Entelis et al. 2006; Kamenski et al. 2007b; Kolesnikova et al. 2010; Tarassov et al. 1995), issues recently reviewed elsewhere (Rubio and Hopper 2011; Salinas et al. 2008; Schneider 2011). Up to more recent studies its function in the mitochondria was not clear. Indeed, mtDNA codes for another tRNA^{Lys} (tRK3), which, due to its modification of the wobble position (cmnm^{5}S^2 U) was predicted to decode both AAA and AAG lysine codons in the organelle. Several alternative functions (like participating in splicing or mtDNA replication priming) were proposed for mitochondrial tRK1 (reviewed in Entelis et al. 2001c), so far these hypotheses never found experimental evidence. Finally, a peculiar mechanism was uncovered. It was shown that mitochondrially encoded tRK3 anticodon wobble base was undermodified (underthiolated) at higher temperatures, and, as a consequence loses the ability to efficiently decode the very rare mitochondrial AAG codons, while the imported CUU anticodon bearing tRK1 complements this deficiency (Kamenski et al. 2007a; Tarassov et al. 2007) (Fig. 4.1). This mechanism was the first case of mitochondrial translational

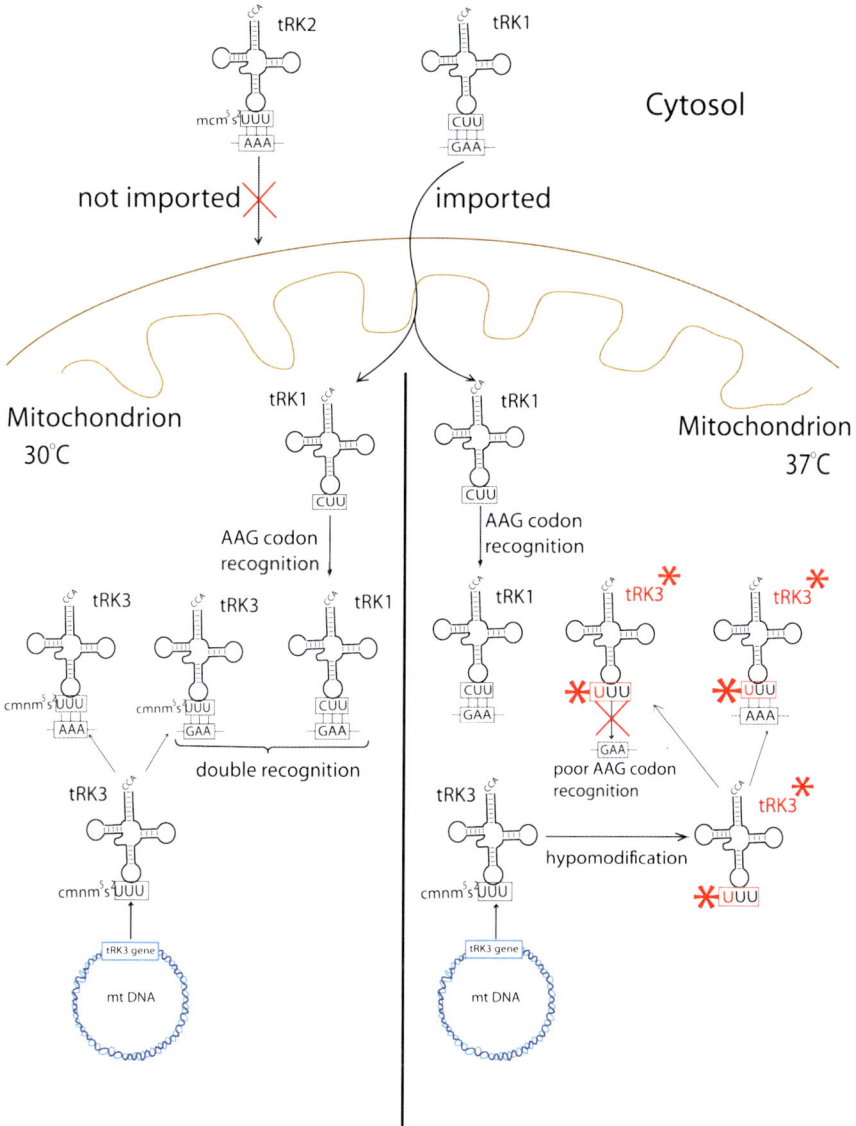

Fig. 4.1 tRNA import into yeast mitochondria is a mechanism of translational adaptation to stress conditions. At the *left* side of the scheme, normal conditions of growth (30 °C), when both lysine codons can be read by mtDNA encoded tRK3. At the *right* side—at higher temperature tRK3 wobble position of the anticodon in undermodified (underthiolated) which prevents efficient decoding of the AAG lysine codon, while imported tRK1 becomes essential for mitochondrial translation

adaptation to stress conditions. So far many questions remains to be solved. tRK1 is imported in mitochondria in a constitutional manner, and not only at higher temperature, while its function in mito-translation becomes detectable only at elevated one. So it is not clear if it participates in translation in a permanent way and if it is more or equally efficient to decode AAG codons in normal conditions. In this context it is worth to stress that tRK1 is aminoacylated by the cytosolic lysyl-tRNA synthetase (KARS), imported into mitochondria in aminoacylated form and cannot then be reacylated in the organelle by the mature mitochondrial synthetase (MSK1), the last one able to charge only tRK3 (Tarassov et al. 1995). This obviously limits the efficiency of tRK1 use in mitochondrial translation. It is still not evident if there exist any regulatory mechanisms governing this targeting. Indeed, it was demonstrated that differential deregulation of the proteasomal degradation of proteins (ubiquitin/26S proteasome system, UPS) may have various effects on tRK1 import (either increasing of decreasing it) (Brandina et al. 2007), which may reflect existence of such a regulation. So, once more, the function of the imported tRNA species was strictly classical while no evidence of any other "alternative" function was ever found.

The second described case of tRNA import in yeasts concerns $tRNA^{Gln}$ species. Formation of Glutaminyl-tRNAs can, in principle, proceed either by a specific glutaminyl-tRNA synthetase (QRS), as in eukaryotic cells and many bacteria, or by misaminoacylation of the corresponding $tRNA^{Gln}$ by the nondiscriminating glutamyl-tRNA synthetase (ERS) with subsequent transamidation of the aminoacyl residue in Glu-$tRNA^{Gln}$ to generate Gln-$tRNA^{Gln}$, as it is found in the majority of bacteria and all the archea (Feng et al. 2005; Ibba and Soll 2004). The last pathway is also used in mitochondria, so far in yeast amidotransferase activity has not been characterized for long time and, to explain how the glutamine decoding system could function, a new mechanism involving tRNA import was proposed (Rinehart et al. 2005) (Fig. 4.2a). Following this scheme, mtDNA encoded $tRNA^{Gln}_{UUG}$, due to a specific wobble base modification (mcm^5S^2U or $cmnm^5S^2U$, cited differently in different reports), can decode only CAA glutamine codons (hypothesis based on earlier studies, reviewed in Yokoyama and Nishimura 1995). The two isoacceptor $tRNAs^{Gln}$ (anticodons CUG and UUG) would be imported from the cytosol, along with the cytosolic QRS, which proceeds the aminoacylation in the "eukaryotic" fashion permitting decoding of the other two glutamine codons (CAG and CAA). This hypothesis responds to the question how glutamination may occur when no amidotransferase is present. It is also in agreement with the idea that mitochondrial enzyme, mERS, cannot generate Glu-$tRNA^{Gln}$, the required intermediate for the transamidation pathway (Rinehart et al. 2005). However, more recently, a new heterotrimeric complex with amidotransferase activity (GatA/GatF/GatB) was characterized, which, once assembled in mitochondria, can modify the mtDNA encoded $tRNA^{Glu}$ misacylated by nondiscriminating cytosolyc ERS, partially imported into the organelle (Fig. 4.2b) (Frechin et al. 2009). Such a mechanism eliminates the functional need of $tRNA^{Gln}$ import, and indeed, the latter study failed to detect any considerable amount of cytosolic $tRNAs^{Gln}$ in yeast mitochondria. Furthermore, the incapacity of the modified U in the wobble position of the mitochondrial

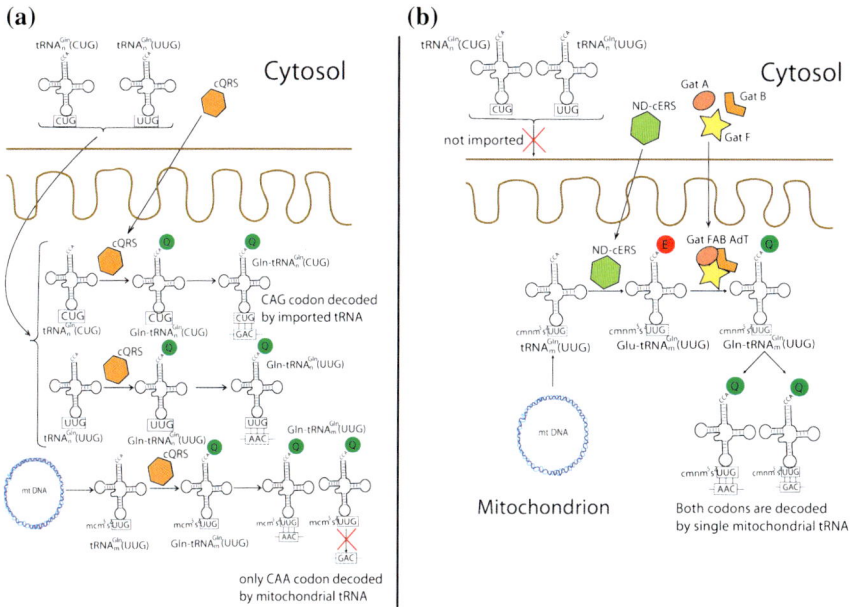

Fig. 4.2 tRNA^{Gln} in yeast mitochondria. **a** The model proposing simultaneous import of both two cytosolic tRNA^{Gln} and the cytosolic QRS. **b** The model proposing the generation of mitochondrial tRNA^{Gln} by the nondiscriminating cytosolic ERS and tripartite amidotransferase (Gat FAB AdT), both imported into mitochondria

tRNA^{Gln}$_{UUG}$, to pair with both A and G in the codon's first position is not supported by many reports (for example, Kurata et al. 2008; Umeda et al. 2005). Still remains the possibility that both mechanisms may coexist, but the former one being minor in terms of efficiency.

4.2.4 Mammalians

For a number of years it was assumed that mammalian mitochondria do not import any tRNA (Enriquez and Attardi 1996a, b; Enriquez et al. 1996), since no significant amounts of cytosolic tRNAs were detected in these organelles by classical hybridization methods. So far, a more recent study suggested that such pathway may, however, exist. It was proposed that a system which resembles to that proposed for yeast (see above) to import tRNAs^{Gln} into mitochondria would also exist in human cells (Rubio and Hopper 2011; Rubio et al. 2008) basing on the same reasoning as for the yeast system, the authors suggested that the modified wobble U34 of the mtDNA encoded tRNAs^{Gln} would affect decoding CAG codon, according to commonly agreed codon-anticodon recognition rules (Agris et al. 2007; Sprinzl and Vassilenko 2005). So, to fulfil complete set of decoding, mitochondria must import corresponding nuclear-encoded tRNA^{Gln} isoacceptors (CUG and UUG anticodons) from the cytosol (see the scheme corresponding to the yeast system, Fig. 4.2a). Indeed, these two cytosolic tRNAs were detected in purified rat and human mitochondria by a more sensitive RT-PCR approach (Rubio et al. 2008). These data are in agreement with the fact that mammalian mitochondria were already described as able to import tRNAs in artificial way (Entelis et al. 2001b; Kolesnikova et al. 2002), and also import other RNA molecules in vivo (Magalhaes et al. 1998; Smirnov et al. 2008a; Yoshionari et al. 1994), so their innate capacity to internalize RNA is out of doubt. So far, the fact that these tRNAs are detectable only by RT-PCR indicates on tiny amounts of them in mitochondria. Furthermore, the transamidation pathway, allowing to complement the absence of mitochondrial QRS, was described (Nagao et al. 2009). This last study permitted to demonstrate that human mitochondrial ERS can misaminoacylate the mtDNA encoded tRNA^{Gln} by a glutamate, which is modified by a heterotrimeric transamidase hGatCAB imported into mitochondria. This mechanism makes the functional issue of the cytosolic tRNAs^{Gln} import not evident. The possibility remains open that both pathways are coexisting but the tRNA-based having a minor functional input.

In the context of this chapter, one can also cite the results of a systematic analysis of human mitochondrial transcriptome, where not only cytosolic tRNA^{Gln} was also detected, but also tRNA^{Lys} and tRNA^{Leu} ones (Mercer et al. 2011). However, there is no hint for the moment if this was due to contamination problems or corresponds to the in vivo situation, since no plausible function for cytosolic tRNA^{Lys} and tRNA^{Leu} in human mitochondrial compartment was ever assigned.

4.3 Import of 5S rRNA in Mammalians and Mitochondrial Translation

Found for the first time in beef liver mitochondria, nuclear DNA encoded 5S rRNA appeared to be nearly the most abundant ribonucleic acid species inside the organelles (Yoshionari et al. 1994; Magalhaes et al. 1998; Entelis et al. 2001a). Its presence in mitochondrial matrix was somewhat unexpected, since 5S rRNA genes have disappeared from mitochondrial genomes of animal cells, and this component was believed to be replaced by new proteins, so typical for mitochondrial ribosomes (Smits et al. 2007). Furthermore, no 5S rRNA was detected in large-scale purified mammalian mitoribosomes (Sharma et al. 2003). On the other hand, a recent series of works, focused on function of this small but important molecule, showed clearly that 5S rRNA, being an extremely conservative ribosomal component, administers the clockwork-like mechanism of all main functional sites of ribosomal machine (Smith et al. 2001; Kiparisov et al. 2005; Kouvela et al. 2007). This fact suggests that the total loss of 5S rRNA cannot possibly occur, if only the mitoribosome has not suffered some substantial structural reorganization of its protein component before (what is probable only in very dynamic systems, exemplified by some protists). How did eukaryotic cells solve the problem of 5S rRNA loss from mitochondrial genomes? One of possible ways, as mentioned above, consists in direct substitution of this RNA with new ribosomal proteins forming the central protuberance (yeast, trypanosomatids) (Sharma et al. 2009; Smirnov et al. 2008a). Still, the obvious existence of cytosolic 5S rRNA import pathway in mammalian cells, as well as recent data discussed below, suggest that in the latter case, the role of the lost component may be performed by its cytosolic ortholog.

5S rRNA targeting into mammalian mitochondria mechanisms were recently described. This pathway includes RNA protein interaction involving at least two cytosolic precursor of mitochondrial proteins, preMRP-L18, and rhodanese and involves, as in the case of yeast tRNA mitochondrial import, structural rearrangements of the imported RNA molecule (Smirnov et al. 2008b, 2010, 2011) (Fig. 4.3). In the context of this chapter, it is worth to note that preMRP-L18 protein belongs to the L5-family of proteins which are direct interactants with 5S rRNAs in all known ribosomes. As a matter of fact, no MRP-L18 homologue was ever found in yeast mitochondria where no 5S rRNA is imported. This fact is to be considered as an indirect evidence on the involvement of the imported

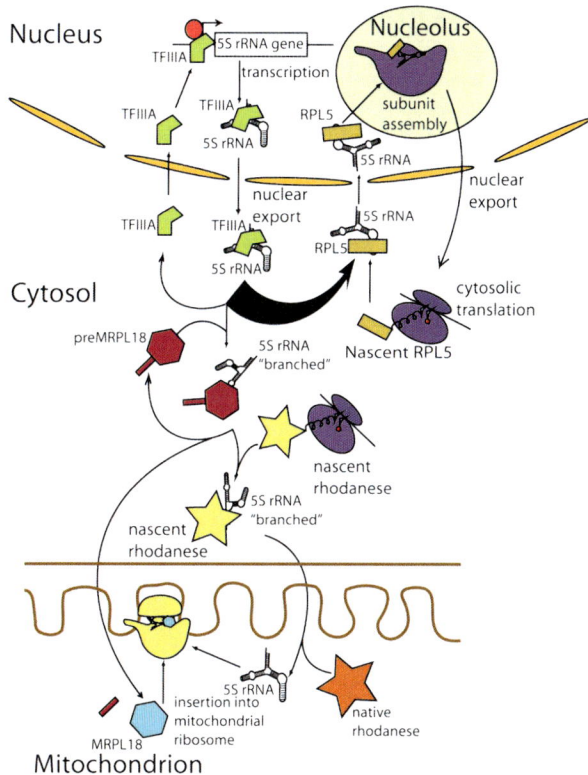

Fig. 4.3 5S rRNA import and cellular translation in mammalians. The intracellular distribution of 5S is ruled by a succession of RNA–protein interaction and chaperone events, which withdraw a minor portion of the cytosolic pool of the RNA from the RPL5 dependent nuclear re-export and further cytosolic translation to address it toward mitochondrial matrix where it is hypothesized to become a part of mitochondrial translation machinery

RNA molecule in mitochondrial ribosome, and is not in direct contradiction with the previously described "5S rRNA-less mitoribosome" lacking also many ribosomal proteins described thereafter (Dennerlein et al. 2010; Kolanczyk et al. 2010; Richter et al. 2010; Rorbach et al. 2008; Soleimanpour-Lichaei et al. 2007; Wanschers et al. 2012). Further analysis revealed that, indeed, a sub-population of mitochondrially localized 5S rRNA co-purified with mitochondrial ribosomes, being essentially associated with monosomal fraction, but not with the fraction of the large ribosomal subunit. This result may mean that the association of 5S rRNA with the mitochondrial ribosomes is leaky and the RNA molecule is easily lost upon subunits dissociation, which would explain why it was not previously detected. This idea was also in a good agreement with the fact that inhibition of 5S rRNA import by siRNA-downregulation of targeting factors (either rhodanese or MRP-L18) strongly affected mitochondrial translation (Smirnov et al. 2011). It is clear that additional functional and structural studies are needed to validate the "ribosomal" function of the imported rRNA, but all currently available data are in favor of it. Once more, as in the case of tRNA import in yeast mitochondria, the primary function of the imported RNA seems to be exploited in the organelle, while no evidence for any other alternative function was found.

4.4 Import of RNase P RNA in Mammalians and Mitochondrial Translation

Another case of RNA import that might be indirectly related to mitochondrial translation concerns the RNA component of 5'-end tRNA processing enzyme, RNase P in mammalians. RNases P were long time considered as obligatory containing a ribozyme (Joyce 1989). Since many species (but not all) do not possess corresponding genes in their mitochondrial genomes, it was suggested that they import this 200–300 bases long RNA from the cytosol. In the case of mammalian cells, the first experimental evidence of the presence of RNase P RNA in the mitochondria was reported more than 20 years ago (Doersen et al. 1985), which corroborated with a further study (Puranam and Attardi 2001). So far, the amount of this RNA associated with mitochondria was extremely low, which raised a debate about the existence of the RNase P RNA import pathway, along with that of MRP RNA, a related RNA species [mitochondrial RNA processing (MRP) enzyme RNP complex, proposed to participate in the initiation of mtDNA replication] (Chang and Clayton 1989; Topper et al. 1992). Furthermore, more recently RNA-free (only proteinaceous) mitochondrial RNase P was characterized in several species, including mammalians (Pavlova et al. 2012; Gutmann et al. 2012; Gobert et al. 2010; Holzmann et al. 2008; Rossmanith and Potuschak 2001). This discovery made unclear the role of the tiny amounts of RNase P RNA (H1 RNA) in mammalian mitochondria. So far, another study suggested a new possible function of this RNA. Indeed, RNA import into mammalian mitochondria, including three RNAs hypothesized to be targeted into the organelle, namely 5S rRNA, MRP RNA, and RNase P (RNA H1), was reported to depend on a protein associated with the inner mitochondrial membrane or intermembrane space, PNPT1 (PNPase, polynucleotide phosphorylase). Inhibition of PNPT1 resulted in a decrease of RNA import. tRNA processing was also analyzed and it was demonstrated that in conditions of PNPT1 downregulation single tRNA genes produced correctly maturated tRNA, while processing of tandem ones was affected. It was then proposed that RNA-containing RNase P may be required only for maturation of the laters (Wang et al. 2010). This defect is expected to affect not only the very tRNA maturation (and therefore its functionality in translation), but also mRNA one, since tRNA secondary structures are thought to serve as signals of primary multicistronic transcript cleavage (Ojala et al. 1981). More recently, analysis of a case of pathological mutations in PNPT1 gene (causing a respiratory chain deficiency), demonstrated that when PNPase was reduced, translation in mitochondria was also affected, while overexpression of a wild-type gene fully restored the deficiency (Vedrenne et al. 2012). In the context of this chapter, it is noteworthy to mention that in PNPT1 mutants not only RNA H1 import appears as affected (von Ameln et al. 2012), but also 5S rRNA one (Vedrenne et al. 2012), which may be the main reason for translational deficiency. At this moment, the exact meaning of the H1 RNA presence in mammalian mitochondria is still subject to debate, although coexistence of both RNA-containing and RNA-free RNases P remains a plausible possibility.

4.5 Import of Micro RNAs and Mitochondrial Translation

Several recent studies, essentially systematic ones, have found microRNAs (miR-NAs) in mitochondria of liver cells, myoblasts, HeLa cells etc. (Bandiera et al. 2011; Barrey et al. 2011; Bian et al. 2010; Kren et al. 2009). If the question of their origin is still open, at least for some of them nuclear coding and mitochondrial import was hypothesized. The mechanisms of this targeting were still not studied. This type of small noncoding RNA molecules participate usually in regulation of mRNA translation and stability (Fabian et al. 2010). There is still very little information on possible involvement of imported miRNAs in mitochondrial translational regulation. So far, it was already reported that miR-181c miRNA species would be nuclear-encoded, mitochondrially targeted, and regulating mitochondrial cytochrome oxidase (COX) subunit 1 mRNA translation (Das et al. 2012). One can expect that further examples of such regulation may appear in the future. In this context it is also crucial to determine with precision the real origin of these RNAs, since the mitochondrial targeting for molecules having homology with mtDNA is sometimes thorough to prove and requires specific model systems of in vitro and in vivo import, as it was done for tRNAs or 5S rRNAs for yeast, plant, or mammalian systems. Besides the mechanistic aspect of potential microRNA import into mitochondria, it will also be challenging to identify the exact functions of a given imported microRNA in mitochondrial translation. Indeed, since in other biological systems they were essentially implicated in cap-dependent translation regulation (Fabian et al. 2010), which is not existing in mitochondria, one can reasonably expect the existence of other, still not discovered mechanisms.

4.6 Import of RNA as a Tool to Correct Dysfunctions in Mitochondrial Translation

Mutations in the mtDNA are an important cause of human diseases, which affect essentially (but not exclusively) the nervous system and the muscles. Diseases associated with deleterious mutations of the mtDNA are often severe neuromuscular disorders, such as for example the syndromes MERRF, MELAS, CPEO, KSS, or LHON. Each of these diseases is to be considered as "rare" (<1:5,000–10,000), but their global socio-medical input is very important and often underestimated. Among the over 350 mtDNA pathogenic mutations characterized to-date in the relatively small human mitochondrial genome (16,569 base pairs), more than 140 are located in tRNA genes, over 70 in protein-coding genes, and a few in the rRNA genes, the rest 120 being deletions of variable size (Ruiz-Pesini et al. 2007). Most of these mutations affect mitochondrial translation, either in specific (one mRNA) or nonspecific (nonfunctional tRNAs or rRNAs) manner. To-date, these diseases do not have efficient curative therapies. The gene therapy approach therefore represents an open possibility, which may be followed in two directions. The first one, commonly referred to as "allotopic expression" (Smith et al. 2004), is to express the gene of interest in the nucleus and to target its product into mitochondria where it would replace its mutant counterpart. The second direction, referred to as "anti- genomic" (Taylor et al. 2001), is to inhibit the replication of mutant mtDNA molecules, thus favoring the replication of wild-type genomes. This approach is based on the heteroplasmy of most deleterious mtDNA mutations, i.e., within each cell mutated mtDNA molecules coexist with wild-type ones. As clinical symptoms are detected only above a relatively high proportion of mutant mtDNA, even partial inhibition of the propagation of mutant versions of mtDNA is expected to permit a significant rescue of mitochondrial functions. Here below we describe how both strategies were experimentally tested by exploiting the RNA mitochondrial pathway, in each case observing improvement of mitochondrial translation affected by mtDNA mutations.

4.6.1 Allotopic Strategy

The first description of successful allotopic strategy using RNA import refers to the yeast model, where a nonsense mutation in mtDNA encoded COX2 subunit was functionally complemented by an engineered mitochondrially targeted cytosolic tRNA (Kolesnikova et al. 2000). This study, validating the idea of exploiting the RNA import for correction of mtDNA mutations effects, demonstrated that eukaryotic type tRNAs can be functional in mitochondrial translation. A similar approach was further applied to two different pathogenic mutations in human mitochondrial tRNAs genes. The first one concerned a point mutation A8344 > G localized in the tRNA^{Lys} gene, often associated with the MERRF syndrome (Myoclonic Epilepsy with Red Ragged Fibers). In this study (Kolesnikova et al. 2004), we used the fact that human mitochondria are able to import the yeast tRNAs^{Lys} (tRK1, see above) or their mutant derivatives (Entelis et al. 2002, 2001b; Kolesnikova et al. 2002). Several versions of specifically designed importable tRNA versions were indeed shown to partially correct all mitochondrial functions affected by the MERRF mutation. In the context of this chapter, it is noteworthy that this mutation in the tRNA^{Lys} affects modification of the anticodon, which leads to a recognition of AAA or AAG lysine codon with different efficiencies (the last one is less efficiently read) and consecutive abortive or inefficient translation (Kirino et al. 2004; Yasukawa et al. 2001). Mitochondrially targeted cytosolic tRNA^{Lys} was shown to reverse this defect of translation and, as the functional consequence, to improve other mitochondrial functions, both in model immortalized cybrid cells or in patient derived primary fibroblasts (Kolesnikova et al. 2004).

The second, more recent successful attempt to apply allotopic strategy using RNA import concerns another point mutation in the mitochondrial tRNA^{Leu} gene, A3243 > G, causing the MELAS disease (Mitochondrial Encephalopathy, Lactic Acidosis, Stroke-like episodes). In this case no naturally imported tRNA^{Leu} was available neither in human nor in yeast cells, so a set of chimeric tRNAs with switched aminoacylation identity was created based on tRK importable versions. It was found that introduction of leucinylation identity elements [essentially localized in the anticodon (Giegé et al. 1998)] in the context of yeast cytosolic tRNAs^{Lys} did not affect their capacity to be mitochondrially targeted into yeast, but more important, also into human mitochondria. It was then demonstrated that the imported recombinant tRNA was able to be correctly aminoacylated by the Leucine (either in the cytosol or in the organelle) and to participate in mitochondrial translation. Transient or stable expression of the importable tRNA in model cybrid cells bearing the MELAS mutation lead to a significant rescue of mitochondrial translation, and consequently other mitochondrial functions (respiration, electron transfer complexes activity, etc.) (Karicheva et al. 2011).

Several alternative strategies, but which also can be termed as "allotopic", were developed by another team, studying the RNA mitochondrial import mechanisms in a trypanosomatid *Leishmania tropica*. An important part of the results of this team (Goswami et al. 2006) were subject of controversy which is still not resolved (Schekman 2010), therefore we shall not discuss here the mechanistic aspects of this pathway, but in the context of the current chapter, is would be worthy to mention recent development of this study leading to propose a therapeutic use of RNA import. All these studies exploit the mitochondrial inner membrane anchored RIC (RNA Import Complex) to deliver RNAs with therapeutic capacities into mitochondria (Adhya 2008). If RIC was described as facilitating tRNA delivery into *Leishmania* mitochondria, it was also proposed that it can be internalized by cultured human cells in a caveolin-1-dependent manner and to be then inserted in human mitochondrial membranes (Mahata et al. 2006). This insertion was described to promote import of cytosolic tRNAs into human mitochondria, which permitted to complement translation defects caused either point mtDNA mutation in either tRNA^{Lys} (MERRF) or 2 kb deletion (KSS, Kearns-Seyre syndrome). A related method was more recently used to address *any* type of mtDNA mutation, comprising single and multiple deletions, or even more generally many types of mitochondrial dysfunction, as altered ROS production. To this end the RIC reconstituted from recombinant proteins was complexed with synthetic transcripts covering significant or all coding portions of mammalian mtDNA, by introducing into these transcripts a previously identified "import signature" (a short hairpin structure present in all imported *Leishmania* tRNAs), and addressing it into affected cells or tissues (Jash and Adhya 2011, 2012; Mahato et al. 2011; Jash et al. 2011).

A recent development of allotopic strategy described exploited the newly identified RNA import mitochondrial factor, PNPT1 (PNPase, polynucleotide phosphorylase). This inner membrane localized protein was formerly shown to be the part of RNA mitochondrial delivery machinery and its specificity relayed to recognition of a conserved hairpin structure in the imported H1 RNA molecule (Wang et al. 2010). This "import signature", referred to as RP sequence, was fused to either tRNA^{Lys} or tRNA^{Leu} expressed in nucleus, which promoted their import into human mitochondria and corrected in a significant manner translation defects due to MERRF or MELAS mutations, correspondingly (Wang et al. 2012) (Fig. 4.4). As a matter of fact, the same RP import signature permitted to address into mitochondria much larger RNA molecules in the PNPase-dependent manner, like COX2 mRNA expressed in the nucluues (Wang et al. 2012).

Allotopic strategy can also be imagined by using the artificially induced protein-driven RNA import, which is not strictly corresponding to the natural RNA import pathway focused in the chapter, still being worth to be mentioned here. Indeed, covalent (Seibel et al. 1995; Vestweber and Schatz 1989) or noncovalent (Sieber et al. 2011b) complexing of nucleic acids with mitochondrially imported proteins or even peptides can promote their co-internalization into the organelles, and thus can be used for mitochondrial addressing functionally active molecules, as tRNAs, or even mRNAs, into mitochondria bearing pathogenic mutations.

4.6.2 Antigenomic Strategy

Reducing the heteroplasmy level of mtDNA mutations below the pathogenic threshold was attempted by many strategies, including addressing into mitochondria peptide nucleic acids (PNAs), specific restriction enzymes, using mitochondrial substrates and, on the level of the organism, specific diets (reviewed in Smith and Lightowlers 2010; Smith et al. 2004), all of these approaches are valuable but still have many technical limitations. The first validation of RNA import as a tool for

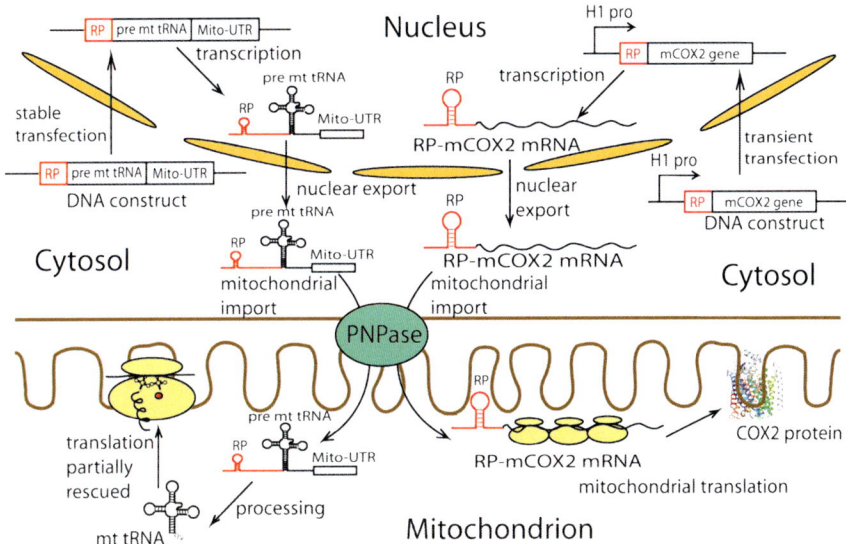

Fig. 4.4 Two allotopic gene therapy models exploiting *PNPase* -promoted RNA import. At the *left* side is schematically presented the tRNA import into human mitochondria, which permits complementation of tRNA mutations in mtDNA (MERRF, MELAS). RP, RNA import signature of RNase P H1 RNA fused to the pre-tRNA, which targets the RNA into mitochondrial. At the *right* side—a similar approach but when RP sequence was fused to an mRNA (COX2 respiratory subunit)

antigenomic strategy was recently described by our team (Comte et al. 2013), the model concerned a large 8 kb long deletion from base 8363 to 15438 and including several structural genes and 6 tRNA ones. This heteroplasmic deletion provokes the KSS disease and is characterized by a significant, though not very dramatic, changes in mitochondrial translation pattern, which may be explained by a 2–3 lower copy number of the genes localized within the deletion. The rationale of the approach was to target mitochondria with short oligoribonucleotides complementary to the new sequence generated by the mutation which could affect replication of only mutant mtDNA molecule. Two types of RNA molecules were used as vectors to address "therapeutic" oligoribonucleotides into mitochondria; the first one was based on 5S rRNA, where the so-called β-domain, dispensable for mitochondrial targeting (Smirnov et al. 2008b), was replaced by the oligonucleotides corresponding to either the H- or L-strands of mtDNA region surrounding the deletion (Fig. 4.5). The second RNA-vector was based on mini-RNAs resulting form SELEX experiments held on importable tRNA sequences, which represent a simple two-hairpin structures where the linker can be replaced by oligonucleotides of interest (Kolesnikova et al. 2010). It was demonstrated that both types of chimeric RNA molecules were addressed into mitochondria of cultured human cells and that their import was accompanied by a reproducible decrease of the ratio between mutant and wild-type mtDNA. This shift of heteroplasmy (transient in the case of transient transfection and stable when

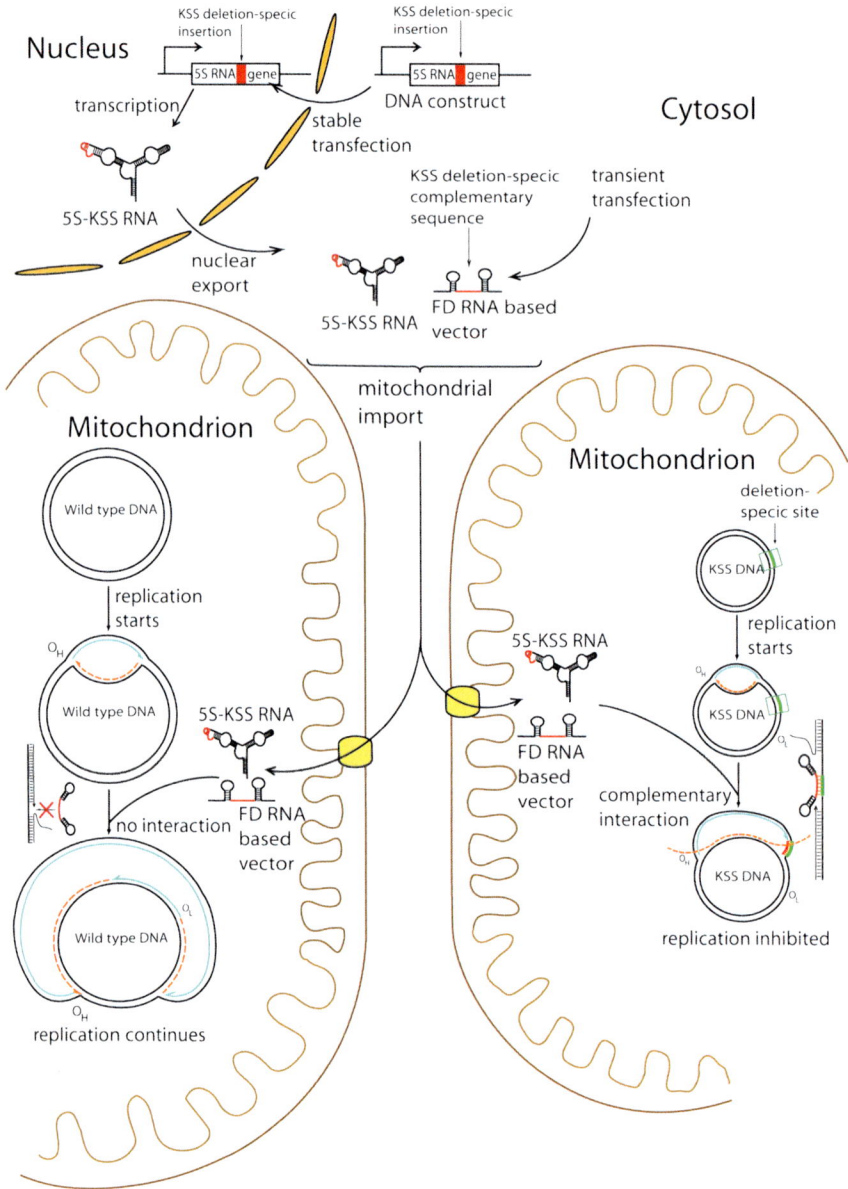

Fig. 4.5 Antigenomic gene therapy models exploiting RNA mitochondrial import. In both cases, a large deletion associated with the KSS disease was addressed. At the *upper left* side—the use of natural import of 5S rRNA, whose recombinant versions were stably expressed in cybrid cells. The antigenomic oligonucleotide replacing the 5S rRNA β-domain is in *red*. At the *upper right* side—FD short tRNA derivatives were used for transient transfection of cybrid cells. At the *bottom*, the schema illustrates the specificity of anti-replication effect of the recombinant mitochondrially targeted RNAs: no effect on wild-type mtDNA replication (*left*) and inhibition of KSS mutant mtDNA (*right*)

recombinant RNAs were stably expressed) resulted in a restoration of a normal mitochondrial translation patters, thus producing a curative effect (Comte et al. 2013).

4.7 Concluding Remarks

To summarize the main issues of this chapter, one can remark that RNA targeting into mitochondria is, in the majority of studied cases, intimately linked to mitochondrial translation, either in direct (tRNAs, rRNA) or nondirect (regulation of translation by miRNA) manner. As it was stressed many times (Sieber et al. 2011a; Schneider 2011; Rubio and Hopper 2011; Salinas et al. 2008), RNA import, even sharing a number of common features among species, still is considered as emerged many times upon evolution. One can suggest that innate capacity of mitochondria to import short RNAs was used to either complement or to optimize the organellar biosynthetic apparatus. It is interesting to note that if the proteins participating in this pathway vary enormously among species and often are "moonlighting" (performing their secondary jobs, i.e., enolase, aminoacyl-tRNA synthetases, rhodanese, ribosomal protein MRP-L18, PNPase, etc.), each time when the imported RNA function inside the organelle was investigated in details, it was rather a classical one in translation (tRNA, 5S rRNA) and not alternative. The functionality of these imported molecules and the flexibility of the overall process make the RNA import pathway an extremely promising tool for the future gene therapy of mitochondrial pathologies, and the first model systems described here above are well illustrating this idea. Indeed, RNA mitochondrial import might be altered in order to either send in the organelles fully functional translation-related molecules to replace the mutant ones or to promote a reduction of mutant mtDNA to correct translation defects.

Acknowledgments   The authors thank all the members of the UMR 7156 GMGM mitochondrial team (Strasbourg, France), H. Becker (Strasbourg, France), R. N. Lightowlers (Newcastle, U.K.), and J. Herrmann (Kaiserslautern, Germany) for helpful discussions. The studies of our team cited in the chapter were supported by the CNRS, University of Strasbourg, ANR (Agence Nationale de Recherche), AFM (Association Française cotre les Myopathies), FRM (Fondation de Recherche Médicale), and the National Program Investissement d'Avenir (LabEx MitoCross).

IC and PK are supported by Russian Foundation for Basic Research and Russian Ministry of Education and Science (Federal programme "Scientific stuff of innovative Russia").

References

Adhya S (2008) Leishmania mitochondrial tRNA importers. Int J Biochem Cell Biol 40(12):2681–2685

Agris PF, Vendeix FA, Graham WD (2007) tRNA's wobble decoding of the genome: 40 years of modification. J Mol Biol 366(1):1–13

Bandiera S, Ruberg S, Girard M, Cagnard N, Hanein S, Chretien D, Munnich A, Lyonnet S, Henrion-Caude A (2011) Nuclear outsourcing of RNA interference components to human mitochondria. PLoS ONE 6(6):e20746

Barrey E, Saint-Auret G, Bonnamy B, Damas D, Boyer O, Gidrol X (2011) Pre-microRNA and mature microRNA in human mitochondria. PLoS ONE 6(5):e20220

Bian Z, Li LM, Tang R, Hou DX, Chen X, Zhang CY, Zen K (2010) Identification of mouse liver mitochondria-associated miRNAs and their potential biological functions. Cell Res 20(9):1076–1078

Brandina I, Smirnov A, Kolesnikova O, Entelis N, Krasheninnikov IA, Martin RP, Tarassov I (2007) tRNA import into yeast mitochondria is regulated by the ubiquitin-proteasome system. FEBS Lett 581(22):4248–4254

Chang DD, Clayton DA (1989) Mouse RNAase MRP RNA is encoded by a nuclear gene and contains a decamer sequence complementary to a conserved region of mitochondrial RNA substrate. Cell 56(1):131–139

Charriere F, Tan TH, Schneider A (2005) Mitochondrial initiation factor 2 of Trypanosoma brucei binds imported formylated elongator-type tRNA(Met). J Biol Chem 280(16):15659–15665

Comte C, Tonin Y, Heckel-Mager AM, Boucheham A, Smirnov A, Aure K, Lombes A, Martin RP, Entelis N, Tarassov I (2013) Mitochondrial targeting of recombinant RNAs modulates the level of a heteroplasmic mutation in human mitochondrial DNA associated with Kearns Sayre Syndrome. Nucleic Acids Res 41(1):418–433

Das S, Ferlito M, Kent OA, Fox-Talbot K, Wang R, Liu D, Raghavachari N, Yang Y, Wheelan SJ, Murphy E, Steenbergen C (2012) Nuclear miRNA regulates the mitochondrial genome in the heart. Circ Res 110(12):1596–1603

Dennerlein S, Rozanska A, Wydro M, Chrzanowska-Lightowlers ZM, Lightowlers RN (2010) Human ERAL1 is a mitochondrial RNA chaperone involved in the assembly of the 28S small mitochondrial ribosomal subunit. Biochem J 430(3):551–558

Dietrich A, Weil JH, Marechal-Drouard L (1992) Nuclear-encoded transfer RNAs in plant mitochondria. Annu Rev Cell Biol 8:115–131

Dietrich A, Marechal-Drouard L, Carneiro V, Cosset A, Small I (1996) A single base change prevents import of cytosolic tRNA(Ala) into mitochondria in transgenic plants. Plant J 10(5):913–918

Doersen CJ, Guerrier-Takada C, Altman S, Attardi G (1985) Characterization of an RNase P activity from HeLa cell mitochondria. Comparison with the cytosol RNase P activity. J Biol Chem 260(10):5942–5949

Enriquez JA, Attardi G (1996a) Analysis of aminoacylation of human mitochondrial tRNAs. Methods Enzymol 264:183–196

Enriquez JA, Attardi G (1996b) Evidence for aminoacylation-induced conformational changes in human mitochondrial tRNAs. Proc Natl Acad Sci U S A 93(16):8300–8305

Enriquez JA, Perez-Martos A, Lopez-Perez MJ, Montoya J (1996) In organello RNA synthesis system from mammalian liver and brain. Methods Enzymol 264:50–57

Entelis NS, Kolesnikova OA, Dogan S, Martin RP, Tarassov IA (2001a) 5 S rRNA and tRNA import into human mitochondria. Comparison of in vitro requirements. J Biol Chem 276(49):45642–45653

Entelis NS, Kolesnikova OA, Dogan S, Martin RP, Tarassov IA (2001b) 5 S rRNA and tRNA import into human mitochondria. Comparison of in vitro requirements. J Biol Chem 276(49):45642–45653

Entelis NS, Kolesnikova OA, Martin RP, Tarassov IA (2001c) RNA delivery into mitochondria. Adv Drug Deliv Rev 49(1–2):199–215

Entelis N, Kolesnikova O, Kazakova H, Brandina I, Kamenski P, Martin RP, Tarassov I (2002) Import of nuclear encoded RNAs into yeast and human mitochondria: experimental approaches and possible biomedical applications. Genet Eng (N Y) 24:191–213

Entelis N, Brandina I, Kamenski P, Krasheninnikov IA, Martin RP, Tarassov I (2006) A glycolytic enzyme, enolase, is recruited as a cofactor of tRNA targeting toward mitochondria in Saccharomyces cerevisiae. Genes Dev 20:1609–1620

Fabian MR, Sonenberg N, Filipowicz W (2010) Regulation of mRNA translation and stability by microRNAs. Annu Rev Biochem 79:351–379

Feng L, Sheppard K, Tumbula-Hansen D, Soll D (2005) Gln-tRNAGln formation from Glu-tRNAGln requires cooperation of an asparaginase and a Glu-tRNAGln kinase. J Biol Chem 280(9):8150–8155

Foury F, Roganti T, Lecrenier N, Purnelle B (1998) The complete sequence of the mitochondrial genome of Saccharomyces cerevisiae. FEBS Lett 440(3):325–331

Frechin M, Senger B, Braye M, Kern D, Martin RP, Becker HD (2009) Yeast mitochondrial Gln-tRNA(Gln) is generated by a GatFAB-mediated transamidation pathway involving Arc1p-controlled subcellular sorting of cytosolic GluRS. Genes Dev 23(9):1119–1130

Geslain R, Aeby E, Guitart T, Jones TE, Castro de Moura M, Charriere F, Schneider A, Ribas de Pouplana L (2006) Trypanosoma seryl-tRNA synthetase is a metazoan-like enzyme with high affinity for tRNASec. J Biol Chem 281(50):38217–38225

Giegé R, Sissler M, Florentz C (1998) Universal rules and idiosyncratic features in tRNA identity. Nucleic Acids Res 26:5017–5035

Gobert A, Gutmann B, Taschner A, Gossringer M, Holzmann J, Hartmann RK, Rossmanith W, Giege P (2010) A single Arabidopsis organellar protein has RNase P activity. Nat Struct Mol Biol 17(6):740–744

Goswami S, Dhar G, Mukherjee S, Mahata B, Chatterjee S, Home P, Adhya S (2006) A bifunctional tRNA import receptor from Leishmania mitochondria. Proc Natl Acad Sci U S A 103(22):8354–8359

Gutmann B, Gobert A, Giege P (2012) PRORP proteins support RNase P activity in both organelles and the nucleus in Arabidopsis. Genes Dev 26(10):1022–1027

Hauser R, Schneider A (1995) tRNAs are imported into mitochondria of Trypanosoma brucei independently of their genomic context and genetic origin. EMBO J 14(17):4212–4220

Holzmann J, Frank P, Loffler E, Bennett KL, Gerner C, Rossmanith W (2008) RNase P without RNA: identification and functional reconstitution of the human mitochondrial tRNA processing enzyme. Cell 135(3):462–474

Ibba M, Soll D (2004) Aminoacyl-tRNAs: setting the limits of the genetic code. Genes Dev 18(7):731–738

Jash S, Adhya S (2011) Suppression of reactive oxygen species in cells with multiple mitochondrial DNA deletions by exogenous protein-coding RNAs. Mitochondrion 11(4):607–614

Jash S, Adhya S (2012) Induction of muscle regeneration by RNA-mediated mitochondrial restoration. FASEB J 26(10):4187–4197

Jash S, Chowdhury T, Adhya S (2011) Modulation of mitochondrial respiratory capacity by carrier-mediated transfer of RNA in vivo. Mitochondrion 12(2):262–270

Joyce GF (1989) RNA evolution and the origins of life. Nature 338(6212):217–224

Kamenski P, Kolesnikova O, Jubenot V, Entelis N, Krasheninnikov IA, Martin RP, Tarassov I (2007a) Evidence for an adaptation mechanism of mitochondrial translation via tRNA import from the cytosol. Mol Cell 26(5):625–637

Kamenski P, Vinogradova E, Krasheninnikov I, Tarassov I (2007b) Directed import of macromolecules into mitochondria. Mol Biol 41:187–202

Karicheva OZ, Kolesnikova OA, Schirtz T, Vysokikh MY, Mager-Heckel AM, Lombes A, Boucheham A, Krasheninnikov IA, Martin RP, Entelis N, Tarassov I (2011) Correction of the consequences of mitochondrial 3243A > G mutation in the MT-TL1 gene causing the MELAS syndrome by tRNA import into mitochondria. Nucleic Acids Res 39(18):8173–8186

Kiparisov S, Petrov A, Meskauskas A, Sergiev PV, Dontsova OA, Dinman JD (2005) Structural and functional analysis of 5S rRNA in Saccharomyces cerevisiae. Mol Genet Genomics 274(3):235–247

Kirino Y, Yasukawa T, Ohta S, Akira S, Ishihara K, Watanabe K, Suzuki T (2004) Codon-specific translational defect caused by a wobble modification deficiency in mutant tRNA from a human mitochondrial disease. Proc Natl Acad Sci U S A 101(42):15070–15075

Kolanczyk M, Pech M, Zemojtel T, Yamamoto H, Mikula I, Calvaruso MA, van den Brand M, Richter R, Fischer B, Ritz A, Kossler N, Thurisch B, Spoerle R, Smeitink J, Kornak U, Chan D, Vingron M, Martasek P, Lightowlers RN, Nijtmans L, Schuelke M, Nierhaus KH, Mundlos S (2010) NOA1 is an essential GTPase required for mitochondrial protein synthesis. Mol Biol Cell 22(1):1–11

Kolesnikova OA, Entelis NS, Mireau H, Fox TD, Martin RP, Tarassov IA (2000) Suppression of mutations in mitochondrial DNA by tRNAs imported from the cytoplasm. Science 289(5486):1931–1933

Kolesnikova O, Entelis N, Kazakova H, Brandina I, Martin RP, Tarassov I (2002) Targeting of tRNA into yeast and human mitochondria: the role of anticodon nucleotides. Mitochondrion 2(1–2):95–107

Kolesnikova OA, Entelis NS, Jacquin-Becker C, Goltzene F, Chrzanowska-Lightowlers ZM, Lightowlers RN, Martin RP, Tarassov I (2004) Nuclear DNA-encoded tRNAs targeted into mitochondria can rescue a mitochondrial DNA mutation associated with the MERRF syndrome in cultured human cells. Hum Mol Genet 13(20):2519–2534

Kolesnikova O, Kazakova H, Comte C, Steinberg S, Kamenski P, Martin RP, Tarassov I, Entelis N (2010) Selection of RNA aptamers imported into yeast and human mitochondria. RNA 16(5):926–941

Kouvela EC, Gerbanas GV, Xaplanteri MA, Petropoulos AD, Dinos GP, Kalpaxis DL (2007) Changes in the conformation of 5S rRNA cause alterations in principal functions of the ribosomal nanomachine. Nucleic Acids Res 35(15):5108–5119

Kren BT, Wong PY, Sarver A, Zhang X, Zeng Y, Steer CJ (2009) MicroRNAs identified in highly purified liver-derived mitochondria may play a role in apoptosis. RNA Biol 6(1):65–72

Kumar R, Marechal-Drouard L, Akama K, Small I (1996) Striking differences in mitochondrial tRNA import between different plant species. Mol Gen Genet 252(4):404–411

Kurata S, Weixlbaumer A, Ohtsuki T, Shimazaki T, Wada T, Kirino Y, Takai K, Watanabe K, Ramakrishnan V, Suzuki T (2008) Modified uridines with C5-methylene substituents at the first position of the tRNA anticodon stabilize U.G wobble pairing during decoding. J Biol Chem 283(27):18801–18811

Magalhaes PJ, Andreu AL, Schon EA (1998) Evidence for the presence of 5S rRNA in mammalian mitochondria. Mol Biol Cell 9(9):2375–2382

Mahata B, Mukherjee S, Mishra S, Bandyopadhyay A, Adhya S (2006) Functional delivery of a cytosolic tRNA into mutant mitochondria of human cells. Science 314(5798):471–474

Mahato B, Jash S, Adhya S (2011) RNA-mediated restoration of mitochondrial function in cells harboring a Kearns Sayre Syndrome mutation. Mitochondrion 11(4):564–574

Martin R, Schneller JM, Stahl A, Dirheimer G (1979) Import of nuclear deoxyribonucleic acid coded lysine-accepting transfer ribonucleic acid (anticodon C-U-U) into yeast mitochondria. Biochemistry 18:4600–4605

Mercer TR, Neph S, Dinger ME, Crawford J, Smith MA, Shearwood AM, Haugen E, Bracken CP, Rackham O, Stamatoyannopoulos JA, Filipovska A, Mattick JS (2011) The human mitochondrial transcriptome. Cell 146(4):645–658

Nagao A, Suzuki T, Katoh T, Sakaguchi Y (2009) Biogenesis of glutaminyl-mt tRNAGln in human mitochondria. Proc Natl Acad Sci U S A 106(38):16209–16214

Ojala D, Montoya J, Attardi G (1981) tRNA punctuation model of RNA processing in human mitochondria. Nature 290(5806):470–474

Pavlova LV, Gossringer M, Weber C, Buzet A, Rossmanith W, Hartmann RK (2012) tRNA processing by protein-only versus RNA-based RNase P: kinetic analysis reveals mechanistic differences. ChemBioChem 13(15):2270–2276

Puranam RS, Attardi G (2001) The RNase P associated with HeLa cell mitochondria contains an essential RNA component identical in sequence to that of the nuclear RNase P. Mol Cell Biol 21(2):548–561

Richter R, Rorbach J, Pajak A, Smith PM, Wessels HJ, Huynen MA, Smeitink JA, Lightowlers RN, Chrzanowska-Lightowlers ZM (2010) A functional peptidyl-tRNA hydrolase, ICT1, has been recruited into the human mitochondrial ribosome. EMBO J 29(6):1116–1125

Rinehart J, Krett B, Rubio MA, Alfonzo JD, Soll D (2005) Saccharomyces cerevisiae imports the cytosolic pathway for Gln-tRNA synthesis into the mitochondrion. Genes Dev 19(5):583–592

Rorbach J, Richter R, Wessels HJ, Wydro M, Pekalski M, Farhoud M, Kuhl I, Gaisne M, Bonnefoy N, Smeitink JA, Lightowlers RN, Chrzanowska-Lightowlers ZM (2008) The human mitochondrial ribosome recycling factor is essential for cell viability. Nucleic Acids Res 36(18):5787–5799

Rossmanith W, Potuschak T (2001) Difference between mitochondrial RNase P and nuclear RNase P. Mol Cell Biol 21(23):8236–8237

Rubio MA, Hopper AK (2011) Transfer RNA travels from the cytoplasm to organelles. Wiley Interdiscip Rev RNA 2(6):802–817

Rubio MA, Liu X, Yuzawa H, Alfonzo JD, Simpson L (2000) Selective importation of RNA into isolated mitochondria from Leishmania tarentolae. RNA 6(7):988–1003

Rubio MA, Rinehart JJ, Krett B, Duvezin-Caubet S, Reichert AS, Soll D, Alfonzo JD (2008) Mammalian mitochondria have the innate ability to import tRNAs by a mechanism distinct from protein import. Proc Natl Acad Sci U S A 105(27):9186–9191

Ruiz-Pesini E, Lott MT, Procaccio V, Poole JC, Brandon MC, Mishmar D, Yi C, Kreuziger J, Baldi P, Wallace DC (2007) An enhanced MITOMAP with a global mtDNA mutational phylogeny. Nucleic Acids Res 35(Database issue):D823–D828

Salinas T, Duchene AM, Delage L, Nilsson S, Glaser E, Zaepfel M, Marechal-Drouard L (2006) The voltage-dependent anion channel, a major component of the tRNA import machinery in plant mitochondria. Proc Natl Acad Sci U S A 103(48):18362–18367

Salinas T, Duchene AM, Marechal-Drouard L (2008) Recent advances in tRNA mitochondrial import. Trends Biochem Sci 33(7):320–329

Schekman R (2010) Editorial expression of concern: a bifunctional tRNA import receptor from Leishmania mitochondria. Proc Natl Acad Sci U S A 107:9476

Schneider A (1994) Import of RNA into mitochondria. Trends Cell Biol 4:282–286

Schneider A (2011) Mitochondrial tRNA import and its consequences for mitochondrial translation. Annu Rev Biochem 80:1033–1053

Schneider A, Marechal-Drouard L (2000) Mitochondrial tRNA import: are there distinct mechanisms? Trends Cell Biol 10(12):509–513

Seibel P, Trappe J, Villani G, Klopstock T, Papa S, Reichmann H (1995) Transfection of mitochondria: strategy towards a gene therapy of mitochondrial DNA diseases. Nucleic Acids Res 23(1):10–17

Sharma MR, Koc EC, Datta PP, Booth TM, Spremulli LL, Agrawal RK (2003) Structure of the mammalian mitochondrial ribosome reveals an expanded functional role for its component proteins. Cell 115(1):97–108

Sharma MR, Booth TM, Simpson L, Maslov DA, Agrawal RK (2009) Structure of a mitochondrial ribosome with minimal RNA. Proc Natl Acad Sci U S A 106(24):9637–9642

Sieber F, Duchene AM, Marechal-Drouard L (2011a) Mitochondrial RNA import: from diversity of natural mechanisms to potential applications. Int Rev Cell Mol Biol 287:145–190

Sieber F, Placido A, El Farouk-Ameqrane S, Duchene AM, Marechal-Drouard L (2011b) A protein shuttle system to target RNA into mitochondria. Nucleic Acids Res 39(14):e96

Simpson AM, Suyama Y, Dewes H, Campbell DA, Simpson L (1989) Kinetoplastid mitochondria contain functional tRNAs which are encoded in nuclear DNA and also contain small minicircle and maxicircle transcripts of unknown function. Nucleic Acids Res 17(14):5427–5445

Small I, Marechal-Drouard L, Masson J, Pelletier G, Cosset A, Weil JH, Dietrich A (1992) In vivo import of a normal or mutagenized heterologous transfer RNA into the mitochondria of transgenic plants: towards novel ways of influencing mitochondrial gene expression? EMBO J 11(4):1291–1296

Smirnov A, Entelis N, Krasheninnikov I, P. MR, Tarassov I (2008a) 5S rRNA: peculiarities of structure, interactions with macromolecules and possible functions. Progress in Biological Chemistry (Russian, translated in English)

Smirnov A, Tarassov I, Mager-Heckel AM, Letzelter M, Martin RP, Krasheninnikov IA, Entelis N (2008) Two distinct structural elements of 5S rRNA are needed for its import into human mitochondria. RNA 14(4):749–759

Smirnov A, Comte C, Mager-Heckel AM, Addis V, Krasheninnikov IA, Martin RP, Entelis N, Tarassov I (2010) Mitochondrial enzyme rhodanese is essential for 5 S ribosomal RNA import into human mitochondria. J Biol Chem 285(40):30792–30803

Smirnov A, Entelis N, Martin RP, Tarassov I (2011) Biological significance of 5S rRNA import into human mitochondria: role of ribosomal protein MRP-L18. Genes Dev 25(12):1289–1305

Smith PM, Lightowlers RN (2010) Altering the balance between healthy and mutated mitochondrial DNA. J Inherit Metab Dis 34(2):309–313

Smith MW, Meskauskas A, Wang P, Sergiev PV, Dinman JD (2001) Saturation mutagenesis of 5S rRNA in Saccharomyces cerevisiae. Mol Cell Biol 21(24):8264–8275

Smith PM, Ross GF, Taylor RW, Turnbull DM, Lightowlers RN (2004) Strategies for treating disorders of the mitochondrial genome. Biochim Biophys Acta 1659(2–3):232–239

Smits P, Smeitink JA, van den Heuvel LP, Huynen MA, Ettema TJ (2007) Reconstructing the evolution of the mitochondrial ribosomal proteome. Nucleic Acids Res 35(14):4686–4703

Smits P, Smeitink J, van den Heuvel L (2010) Mitochondrial translation and beyond: processes implicated in combined oxidative phosphorylation deficiencies. J Biomed Biotechnol 2010:737385

Soleimanpour-Lichaei HR, Kuhl I, Gaisne M, Passos JF, Wydro M, Rorbach J, Temperley R, Bonnefoy N, Tate W, Lightowlers R, Chrzanowska-Lightowlers Z (2007) mtRF1a is a human mitochondrial translation release factor decoding the major termination codons UAA and UAG. Mol Cell 27(5):745–757

Sprinzl M, Vassilenko KS (2005) Compilation of tRNA sequences and sequences of tRNA genes. Nucleic Acids Res 33(Database issue):D139–D140

Tan TH, Bochud-Allemann N, Horn EK, Schneider A (2002a) Eukaryotic-type elongator tRNAMet of Trypanosoma brucei becomes formylated after import into mitochondria. Proc Natl Acad Sci U S A 99(3):1152–1157

Tan TH, Pach R, Crausaz A, Ivens A, Schneider A (2002b) tRNAs in Trypanosoma brucei: genomic organization, expression, and mitochondrial import. Mol Cell Biol 22(11):3707–3717

Tarassov I, Entelis N, Martin R (1995) Mitochondrial import of a cytoplasmic lysine-tRNA in yeast is mediated by cooperation of cytoplasmic and mitochondrial lysyl-tRNA synthetases. EMBO J 14:3461–3471

Tarassov I, Kamenski P, Kolesnikova O, Karicheva O, Martin RP, Krasheninnikov IA, Entelis N (2007) Import of nuclear DNA-encoded RNAs into mitochondria and mitochondrial translation. Cell Cycle 6(20):2473–2477

Taylor RW, Wardell TM, Smith PM, Muratovska A, Murphy MP, Turnbull DM, Lightowlers RN (2001) An antigenomic strategy for treating heteroplasmic mtDNA disorders. Adv Drug Deliv Rev 49(1–2):121–125

Topper JN, Bennett JL, Clayton DA (1992) A role for RNAase MRP in mitochondrial RNA processing. Cell 70(1):16–20

Umeda N, Suzuki T, Yukawa M, Ohya Y, Shindo H, Watanabe K (2005) Mitochondria-specific RNA-modifying Enzymes responsible for the biosynthesis of the wobble base in mitochondrial tRNAs. Implications for the molecular pathogenesis of human mitochondrial diseases. J Biol Chem 280(2):1613–1624

Vedrenne V, Gowher A, De Lonlay P, Nitschke P, Serre V, Boddaert N, Altuzarra C, Mager-Heckel AM, Chretien F, Entelis N, Munnich A, Tarassov I, Rotig A (2012) Mutation in PNPT1, which encodes a polyribonucleotide nucleotidyltransferase, impairs RNA import into mitochondria and causes respiratory-chain deficiency. Am J Hum Genet 91(5):912–918

Vestweber D, Schatz G (1989) DNA-protein conjugates can enter mitochondria via the protein import pathway. Nature 338(6211):170–172

Vinogradova E, Salinas T, Cognat V, Remacle C, Marechal-Drouard L (2009) Steady-state levels of imported tRNAs in Chlamydomonas mitochondria are correlated with both cytosolic and mitochondrial codon usages. Nucleic Acids Res 37(5):1521–1528

von Ameln S, Wang G, Boulouiz R, Rutherford MA, Smith GM, Li Y, Pogoda HM, Nurnberg G, Stiller B, Volk AE, Borck G, Hong JS, Goodyear RJ, Abidi O, Nurnberg P, Hofmann K, Richardson GP, Hammerschmidt M, Moser T, Wollnik B, Koehler CM, Teitell MA, Barakat

A, Kubisch C (2012) A mutation in PNPT1, encoding mitochondrial-RNA-import protein PNPase, causes hereditary hearing loss. Am J Hum Genet 91(5):919–927

Wang G, Chen HW, Oktay Y, Zhang J, Allen EL, Smith GM, Fan KC, Hong JS, French SW, McCaffery JM, Lightowlers RN, Morse HC 3rd, Koehler CM, Teitell MA (2010) PNPASE regulates RNA import into mitochondria. Cell 142(3):456–467

Wang G, Shimada E, Zhang J, Hong JS, Smith GM, Teitell MA, Koehler CM (2012) Correcting human mitochondrial mutations with targeted RNA import. Proc Natl Acad Sci U S A 109(13):4840–4845

Wanschers BF, Szklarczyk R, Pajak A, van den Brand MA, Gloerich J, Rodenburg RJ, Lightowlers RN, Nijtmans LG, Huynen MA (2012) C7orf30 specifically associates with the large subunit of the mitochondrial ribosome and is involved in translation. Nucleic Acids Res 40(9):4040–4051

Yasukawa T, Suzuki T, Ishii N, Ohta S, Watanabe K (2001) Wobble modification defect in tRNA disturbs codon-anticodon interaction in a mitochondrial disease. EMBO J 20(17):4794–4802

Yokoyama S, Nishimura S (1995) Modified nucleosides and codon recognition. In: RajBhandary U, Söll D (eds) tRNA: structure, biosynthesis and function. American Society for Microbiology Press, Washington, DC, pp 207–233

Yoshionari S, Koike T, Yokogawa T, Nishikawa K, Ueda T, Miura K, Watanabe K (1994) Existence of nuclear-encoded 5S-rRNA in bovine mitochondria. FEBS Lett 338(2):137–142

Chapter 5
Mechanisms and Control of Protein Synthesis in Yeast Mitochondria

Steffi Gruschke and Martin Ott

Abstract In the yeast *Saccharomyces cerevisiae*, eight proteins are encoded by the mitochondrial genome. Seven of them are core catalytic subunits of complexes III and IV of the respiratory chain and the ATP synthase and thus essential for oxidative phosphorylation (OXPHOS), while one protein is soluble and a constituent of the small subunit of mitochondrial ribosomes. The expression of these proteins is mainly controlled posttranscriptionally by so-called translational activators. These nuclear-encoded factors act on the 5′-untranslated region (UTR) of their specific client mRNA and stimulate translation. In addition, translational activators play multiple roles in regulation and organization of mitochondrial protein synthesis. The mitochondrial OXPHOS complexes are assembled from subunits encoded by both the nuclear and the mitochondrial DNA. During the biogenesis of OXPHOS complexes, translational activators help to coordinate cytosolic and mitochondrial translation by adjusting mitochondrial protein synthesis to levels that can successfully be assembled. This chapter summarizes the current knowledge about how mitochondrial protein synthesis in the model organism *Saccharomyces cerevisiae* is coordinated with OXPHOS complex assembly.

5.1 Mechanisms of Protein Synthesis in Yeast Mitochondria

5.1.1 The Mitochondrial Genome in Yeast

Mitochondria are key organelles of eukaryotic cells that participate in important metabolic processes like the TCA cycle, fatty acid oxidation, and amino acid

S. Gruschke · M. Ott (✉)
Department of Biochemistry and Biophysics, Center for Biomembrane Research,
Stockholm University, 106 91 Stockholm, Sweden
e-mail: martin.ott@dbb.su.se

A.-M. Duchêne (ed.), *Translation in Mitochondria and Other Organelles*,
DOI: 10.1007/978-3-642-39426-3_5, © Springer-Verlag Berlin Heidelberg 2013

degradation; they also play important roles in a variety of biosynthetic pathways and contribute to regulation of cellular signaling and apoptosis. Mitochondria evolved 2 billion years ago when an archaeal cell established a symbiotic relationship with aerobic bacteria (Sagan 1967; Gray 1989). In the course of evolution, most of the former bacterial genes were transferred to the nuclear DNA. Concomitantly, the organellar genetic code developed away from the standard genetic code, so that the codon usages differ significantly between both systems. The most obvious alteration is that the universal stop codon TGA is translated into tryptophan (Barrell et al. 1979; Fox 1979). Today, mitochondrial genomes of fungi and higher eukaryotes typically code only for a small number of genes. These include genes for proteins of OXPHOS complexes as well as tRNAs and rRNAs of the mitochondrial translation machinery. In the yeast *Saccharomyces cerevisiae*, eight proteins are encoded in the mitochondrial DNA (mtDNA), 24 tRNAs as well as the rRNA of the small (15S) and large (21S) subunit of the mitochondrial ribosome (Borst and Grivell 1978). Of the eight mitochondrially encoded proteins, seven represents very hydrophobic core subunits of OXPHOS complexes (cytochrome *b* of the bc_1 complex, Cox1, Cox2, and Cox3 of cytochrome oxidase and Atp6, Atp8, and Atp9 of the ATP synthase) located in the inner membrane of mitochondria, while one is a soluble protein and a component of the small mitochondrial ribosomal subunit (Fig. 5.1). These proteins are synthesized on mitochondrial ribosomes which are, unlike bacterial ribosomes, permanently associated to the inner membrane (Fiori et al. 2003; Ott et al. 2006; Prestele et al. 2009), thereby allowing co-translational membrane insertion of the hydrophobic OXPHOS subunits (Hell et al. 2001; Jia et al. 2003; Szyrach et al. 2003; Ott and Herrmann 2010).

5.1.2 The Mitochondrial Ribosome in Yeast and the Process of Translation

The central components of the translation machinery in mitochondria are mitochondrial ribosomes. Due to the endosymbiotic origin of the organelles, it has long been assumed that mitochondrial ribosomes closely resemble the bacterial particles. However, although this is true for certain aspects like catalytic properties or sensitivity against antibiotics, millions of years of evolution have rendered the organellar ribosomes strikingly different from their ancestors. This is especially evident from their structure, because mitochondrial ribosomes contain typically much more protein and less rRNA. Furthermore, mitochondrial ribosomes greatly differ between species. Whereas the protein–rRNA ratio of bacterial ribosomes is 1:2, mitochondrial ribosomes of *S. cerevisiae* and *S. pombe* denote a 1:1 ratio and the relative gain in protein mass is even more pronounced in mitochondrial ribosomes of mammals (2:1). The functional significance of this huge increase in protein content is not entirely clear, but a likely explanation is that these ribosomes require additional stabilization due to the greatly reduced ribosomal RNA sequences and losses in structurally important RNA folds (Mears et al. 2006).

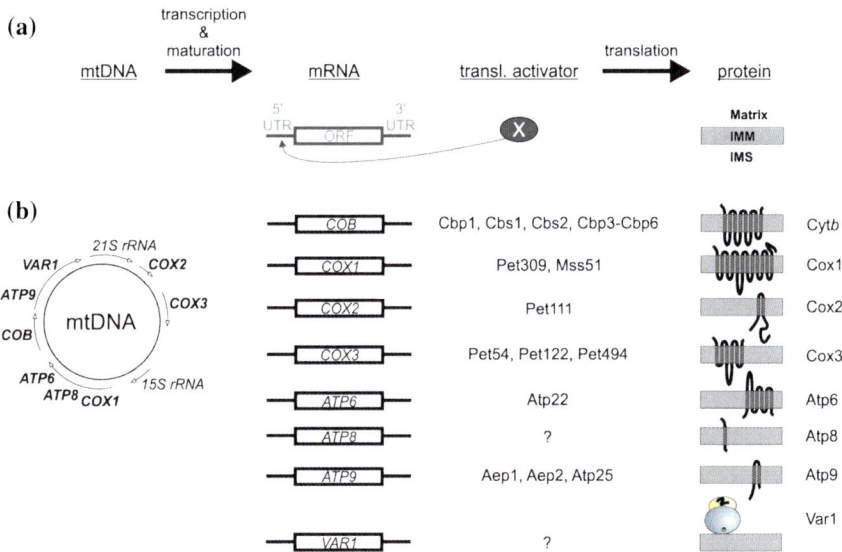

Fig. 5.1 Translational activators in yeast mitochondria. **a** General scheme of gene expression in yeast mitochondria. The mRNAs contain 5′- and 3′-untranslated regions (UTRs) flanking the respective open reading frame (ORF). Especially, the 5′-UTR is the target of specific translational activators (X), whose action is required to activate translation of their client mRNA. **b** Specific expression of genes encoded in the mitochondrial DNA with the help of translational activators. The mitochondrial genome (mtDNA) of *S. cerevisiae* encodes two ribosomal RNAs (*15S* and *21S* rRNA), eight proteins and 24 tRNAs (not shown). So far, translational activators for six of the eight mRNAs have been identified. The topologies of the mitochondrially encoded proteins in the inner membrane are depicted. Var1 is a component of the small ribosomal subunit. IMM, inner mitochondrial membrane. IMS, intermembrane space

The general process of protein synthesis in mitochondria involves many conserved translation factors and has mainly been analyzed by studying mammalian mitochondria, see also the Chap. 2. Whereas the elongation cycle of translation in mitochondria is assumed to closely resemble the bacterial process, termination of protein synthesis deviates from the ancestral system (Chrzanowska-Lightowlers et al. 2011; Kehrein et al. 2013). The initiation step of mitochondrial translation is only poorly understood. Unlike bacterial mRNAs or transcripts in the cytosol of eukaryotes, mitochondrial mRNAs do not contain Shine–Dalgarno sequences or 5′-cap structures that promote initiation of translation in these systems. Although in yeast nucleotides in the 15S rRNA of the small ribosomal subunit are complementary to sequences in mitochondrial mRNAs (Li et al. 1982), those regions do not fulfill a Shine–Dalgarno-like function as they are dispensable for translation (Costanzo and Fox 1988; Mittelmeier and Dieckmann 1995). Rather, sequences adjacent to the start AUG in the mRNA are implicated in this process. Accordingly, removal of the start AUG did not allow initiation of translation at alternative AUG codons within downstream sequences; changing the start AUG to AUA in *COX2* and *COX3* mRNAs only modestly reduced synthesis of Cox2 and Cox3 (Folley and Fox 1991; Mulero and Fox 1994; Bonnefoy and Fox 2000). Importantly, the 5'- and 3'-untranslated regions (UTRs) are, at least in yeast, the target of transcript-specific translational activators. These factors are mediating in a yet unknown manner the translation of their specific client mRNA.

5.1.3 The Concept of Translational Activators and Their Possible Functions

Translational activators (TAs) in yeast mitochondria have been studied for more than four decades (Fig. 5.1). The cytochrome oxidase subunits Cox2 and Cox3 were the first examples for which the concept of specific translational activation in yeast mitochondria was introduced. Early studies showed that deletion of the nuclear genes *PET111* and *PET494* specifically impairs the expression of the mitochondrial *COX2* and *COX3* gene, respectively (Cabral and Schatz 1978). In subsequent years, genetic screens revealed that mutants lacking one TA (and therefore one mitochondrially encoded protein) could regain respiratory growth by remodeling of sequences within the mitochondrial genome. Those rearrangements led to the generation of fusion genes or the exchange of regulatory regions with the result that affected transcripts acquired 5'-UTRs of other genes, making their expression independent from the authentic, missing TA but dependent on the factor controlling synthesis of the other gene (Muller et al. 1984). In yeast, expression of six of the eight mitochondrially encoded proteins depends on TAs; these factors are described below.

How exactly TAs exert their function in translation is not yet understood. Different molecular functions have been suggested, and because there is experimental evidence for all of them, TAs might not share one universal function but rather exert their roles at different steps of the translation process and in some cases have more than one function. Some TAs like Cbp1, Pet309, Pet111, Aep2, and Atp25 are required for stabilizing the transcripts they act on (Poutre and Fox 1987; Payne et al. 1991; Manthey and McEwen 1995; Ellis et al. 1999; Islas-Osuna et al. 2002; Zeng et al. 2008).

Another possibility is that TAs work by supporting initiation of translation by, e.g., assisting to load the mRNA correctly onto the ribosome as Shine–Dalgarno-like sequences are absent in mitochondrial messengers. Interactions of TAs with both the 5'-UTR and mitochondrial ribosomes would support such an idea and have been shown primarily by genetic study (McMullin et al. 1990; Haffter et al. 1991; Haffter and Fox 1992; Fox 1996). A direct binding of TAs to sequences within specific 5'-UTRs has been demonstrated in the cases of Pet111 and Pet122, where substitutions of amino acids in the TA could rescue adverse mutations in the 5'-UTR of *COX2* and *COX3*, respectively (Costanzo and Fox 1993; Mulero and Fox 1993a).

Translational activators participate in the organization of mitochondrial protein synthesis. Many TAs are peripheral or integral membrane proteins, and therefore localize the mRNAs to the matrix face of the inner membrane to facilitate interactions with the permanently membrane-associated translation machinery in mitochondria (McMullin and Fox 1993; Sanchirico et al. 1998). However, the organization of translation by TAs might be even more intricate than this. For example, Pet309 was found to be present in a complex of a molecular mass of about 900 kDa that also contained Cbp1, a protein required to stabilize the cytochrome *b* mRNA (Krause et al. 2004). Pet309 was independently shown to be in contact with the general mRNA metabolism factor Nam1 (Naithani et al. 2003). A third study revealed a Nam1 interaction with the yeast mitochondrial RNA polymerase (Rodeheffer et al. 2001). From all these findings, a model was suggested that links transcription, mRNA maturation, and protection as well as translation at the inner mitochondrial membrane (Krause et al. 2004). Furthermore, the specific TAs of Cox1, Cox2, and Cox3 interact with each other and thereby organize expression of the three cytochrome oxidase subunits in a way that allows efficient assembly of this respiratory chain complex (Naithani et al. 2003).

The last, and comparably well documented, function of TAs is the regulation of mitochondrial protein synthesis in response to the efficiency of respiratory chain assembly. Respiratory chain complexes and the ATP synthase are composed of subunits produced by two different genetic systems and the assembly of these complex machineries is a highly intricate event. To allow efficient assembly of the respiratory chain, the expression of mitochondrially and nuclear encoded subunits has to be coordinated. Translational activators participate in these regulatory circuits in the case of the bc_1 complex, cytochrome oxidase, and the ATP synthase in yeast. In general, when assembly of an OXPHOS complex is blocked due to missing nuclear-encoded structural subunits or the absence of specific assembly factors, sequestration of a TA in an assembly intermediate that contains a mitochondrially encoded subunit of the OXPHOS complex lowers the amounts of the TA available to stimulate translation (Fig. 5.2). By this, mitochondrial protein synthesis is adjusted to levels that can successfully be assembled into OXPHOS complexes. The coupling of synthesis and assembly is a conserved process as it had first been described for the biogenesis of chloroplast photosystems and there been termed "control of epistatic synthesis" (CES) (Wollman et al. 1999). The detailed mechanisms of how specific TAs act in such regulatory feedback loops in yeast mitochondria are outlined in the last section of this chapter.

5.1.4 Synthesis of Cytochrome b

Cytochrome *b* is the only subunit of the bc_1 complex which is encoded in the mitochondrial genome and translation of its mRNA (*COB* mRNA) is dependent on several factors (Rödel 1997). The *COB* gene is co-transcribed with an adjacent tRNA and the resulting bi-cistronic precursor has to be processed in a complex way (Christianson et al. 1983; Hollingsworth and Martin 1986; Chen and

(a) normal OXPHOS complex assembly

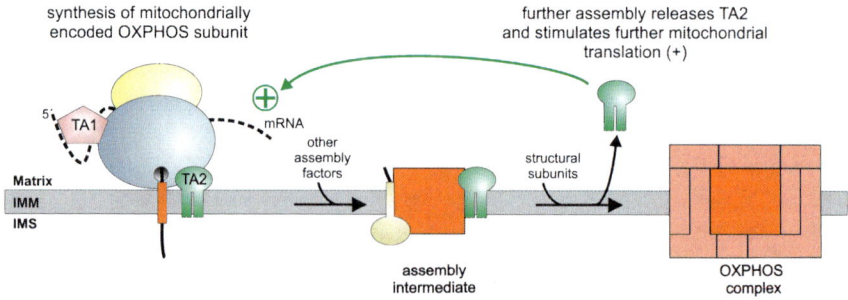

(b) impaired/inefficient OXPHOS complex assembly

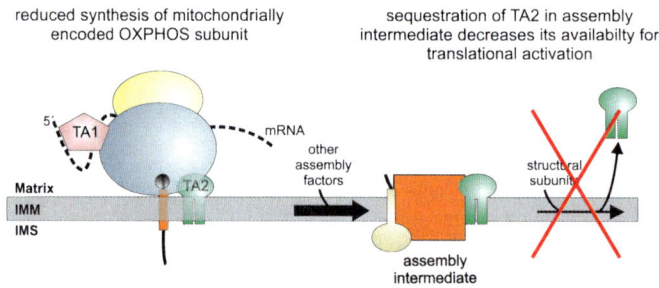

Fig. 5.2 The general principle of feedback loops regulating mitochondrial protein synthesis in response to the efficiency of OXPHOS complex assembly. Translation in mitochondria requires specific translational activators (TAs). Some TAs (TA2 in the scheme) have a dual role in activating translation of their client mRNA and in mediating assembly of the encoded protein into a respiratory chain complex. **a** Under normal circumstances when assembly is not disturbed, the TA2 is only transiently present in an assembly intermediate and released upon further assembly. It can then stimulate further translation of the regulated subunit (*dark orange*) at the mitochondrial ribosome (+). **b** If complex assembly is perturbed due to the absence of structural OXPHOS subunits or assembly factors encoded in the nucleus, TA2 cannot be released efficiently and is sequestered in the assembly intermediate. The TA is therefore not available for activating translation, which reduces of synthesis of the mitochondrially encoded OXPHOS subunit. By this, mitochondrial translation is adjusted to level that can be incorporated into OXPHOS complexes. IMM, inner mitochondrial membrane. IMS, intermembrane space

Martin 1988) The *COB* gene furthermore contains introns, some of which encode maturases that are required for the excision of introns within the same or other transcripts (Lazowska et al. 1980; Nobrega and Tzagoloff 1980; Dhawale et al. 1981; De La Salle et al. 1982). Both the unprocessed and the mature *COB* transcripts are shielded from exonucleolytic degradation by the factor Cbp1 (Weber and Dieckmann 1990; Mittelmeier and Dieckmann 1995; Islas-Osuna et al. 2002). This mRNA stabilization involves the interaction of Cbp1 with the 5'-UTR of cytochrome *b;* a single CCG triplet near the 5'-end is especially important (Chen and Dieckmann 1997). In addition to stabilization, Cbp1 is also directly necessary for translation of the *COB* mRNA (Islas-Osuna et al. 2002).

Four other TAs are specifically required for the synthesis of cytochrome *b*. Yeast cells lacking either the product of the *CBS1* or the *CBS2* gene cannot translate *COB* mRNA and accumulate the unprocessed pre-*COB* transcript (Rödel et al. 1985; Rödel 1986). Because Cbs1 and Cbs2 are needed for cytochrome *b* synthesis also in a strain where the *COB* gene does not contain any introns, they do not function in processing of pre-*COB* (Muroff and Tzagoloff 1990). The accumulation of the unprocessed transcript is presumably a secondary effect in Δ*cbs1* and Δ*cbs2* cells, because synthesis of the maturases encoded within introns requires translation of the pretranscript. Similar to the case of Cbp1, the *COB* 5′-UTR dictates the dependence on Cbs1 and Cbs2 (Rödel et al. 1985; Rödel and Fox 1987). The region within the 5′-UTR recognized by the two TAs lies in the sequence −232 to −4 relative to the start AUG at +1 (Mittelmeier and Dieckmann 1995). However, a direct interaction between either Cbs1 or Cbs2 and the 5′-untranslated region of *COB* mRNA was not yet shown, so it is not clear whether the factors act directly or indirectly through yet unknown components.

Additional factors involved in cytochrome *b* translation are the proteins Cbp3 and Cbp6, which form a functionally and structurally inseparable complex that binds to mitochondrial ribosomes in close proximity to the tunnel exit (Gruschke et al. 2011). In contrast to Δ*cbs1* and Δ*cbs2* cells, the pre-*COB* transcript is processed and matured similar to the wild type in the absence of Cbp3 or Cbp6 (Dieckmann and Tzagoloff 1985; Gruschke et al. 2011). Despite this, cytochrome *b* cannot accumulate in the mutants and the phenotype cannot be suppressed by a typical gene rearrangement within the mitochondrial genome (Tzagoloff et al. 1988; Gruschke et al. 2012; Kühl et al. 2012). This can be attributed to the fact that the Cbp3–Cbp6 complex exerts a second function in cytochrome *b* biogenesis; it is required for the stabilization and assembly of the newly synthesized protein. As soon as cytochrome *b* is fully synthesized, Cbp3–Cbp6 binds to the protein, the complex is released from the ribosome and after recruitment of the assembly factor Cbp4 cytochrome *b* is fed into the bc_1 complex assembly line (Gruschke et al. 2011; Gruschke et al. 2012). This dual role of Cbp3–Cbp6 in both translation of the *COB* mRNA and assembly of cytochrome *b* enables Cbp3–Cbp6 to act in a regulatory feedback circuit that adjusts the level of cytochrome *b* synthesis to the assembly efficiency of the bc_1 complex (Gruschke et al. 2012). This regulation is explained in more detail in the last section of the chapter.

5.1.5 Synthesis of Mitochondrially Encoded Cytochrome Oxidase Subunits

Three subunits of cytochrome oxidase (COX, complex IV) are encoded in the mitochondrial genome, namely Cox1, Cox2, and Cox3. The TA for *COX2*, Pet111, and one of the TAs for *COX3*, Pet494, served as the first examples for specific translational activation in mitochondria (Cabral and Schatz 1978). Due to the pioneering work of Tom Fox, Pet111 is the best-studied translational activator.

Like for other TAs, the phenotype of the *PET111* deletion mutant was found to be suppressible by the exchange of *COX2* regulatory regions to 5′-UTRs of other mitochondrial genes (Poutre and Fox 1987; Mulero and Fox 1993b). The 5′-UTR of the *COX2* mRNA is relatively short (54 bases), making mutagenetic analyses spanning the complete sequence easier than in other cases. By this means, it was shown that the sequence between −16 and −47 (relative to the translation start at +1) is sufficient to confer Pet111-mediated translational activation; within this region lies a predicted stem–loop structure between position −20 and −35 that is especially important (Dunstan et al. 1997). Like in the case of Cbp1 and the *COB* 5′-UTR, a direct interaction between this structural motif in the *COX2* 5′-UTR and Pet111 is very likely, because mutations in the protein can rescue base exchanges in the mRNA (Mulero and Fox 1993a). However, the functional role of Pet111 in translational activation of *COX2* on the molecular level has, as for the other TAs, not been elucidated yet.

Translational activation of *COX3* mRNA depends on three proteins, Pet54, Pet122, and Pet494 (Cabral and Schatz 1978; Muller et al. 1984; Costanzo and Fox 1986, 1988). All of these factors act on the 613 nucleotide long 5′-UTR of *COX3* mRNA (Costanzo and Fox 1988). In the case of Pet122, this interaction presumably is directed as a mutation within the protein can restore translation of an mRNA lacking a functionally important part of the 5′-UTR (Costanzo and Fox 1993). Furthermore, Pet54, Pet122, and Pet494 interact with each other at the inner mitochondrial membrane and thereby presumably help localizing synthesis of this cytochrome oxidase subunit to the membrane (McMullin and Fox 1993; Brown et al. 1994). In addition to a function in Cox3 synthesis, Pet54 has been proposed to play a role in maturation of the *COX1* mRNA (Valencik and McEwen 1991).

The third cytochrome oxidase subunit encoded in the mitochondrial genome is Cox1. Its synthesis depends on two proteins that are involved in posttranscriptional processes. Pet309 is required for translation of the *COX1* transcript, because yeast strains harboring an intronless *COX1* gene accumulate the mature mRNA, but fail to synthesize Cox1 (Manthey and McEwen 1995). Pet309 belongs to the class of PPR proteins that contain pentatricopeptide repeats, a motif involved in protein-RNA interactions (Lipinski et al. 2011). All of the seven PPRs of Pet309 are required for supporting translation of the *COX1* mRNA, suggesting direct interaction of the protein with the 5′-UTR of the messenger (Tavares-Carreon et al. 2008). Mss51 is the second protein involved in translation of the *COX1* transcript. Although initially thought to be required for splicing of the *COX1* precursor mRNA, experiments with strains harboring an intronless *COX1* gene showed that Mss51 rather functions as a translational activator (Faye and Simon 1983; Decoster et al. 1990). The *COX1* 5′-UTR again was shown to direct Mss51 dependence; however, the exchange of this regulatory region for that of another mitochondrial gene did not bypass the requirement for Mss51 (Perez-Martinez et al. 2003; Zambrano et al. 2007). The reason for this is a second posttranslational function of Mss51 in Cox1 biogenesis; it interacts with newly synthesized Cox1 and is part of cytochrome oxidase assembly intermediates (Perez-Martinez et al. 2003; Barrientos et al. 2004). Recent studies have shown that the synthesis of Cox1 is regulated in a highly complex manner in response to cytochrome oxidase assembly. In this process, Mss51 plays a key role by mediating both translation of the *COX1* mRNA and assembly of this respiratory chain complex (Fontanesi et al. 2008; Mick et al. 2011). It thus represents a second example of a TA mediating feedback modulation of mitochondrial protein synthesis in the context of OXPHOS complex assembly (see below).

5.1.6 Synthesis of Mitochondrially Encoded ATP Synthase Subunits

The mitochondrial genome of *S. cerevisiae* contributes three subunits to the formation of the ATP synthase, Atp6, Atp8 and Atp9, all of which are part of the membrane-integrated F_0 unit. The *ATP6* and *ATP8* genes are transcribed as one long precursor mRNA together with *COX1* (Simon and Faye 1984). After endonucleolytic cleavage of this pretranscript, maturation, and stabilization of the *ATP8/ATP6* bi-cistronic and/or the single mRNAs is accomplished by several nuclear encoded factors: Nca2, Nca3, and Nam1 (which is not only specific for Atp6 and Atp8) and Aep3 (Groudinsky et al. 1993; Camougrand et al. 1995; Pelissier et al. 1995; Ellis et al. 2004).

Translation of *ATP6* depends on the factor Atp22 (Zeng et al. 2007). Similar to other TAs, the absence of Atp22 can be overcome by a mitochondrial gene rearrangement leading to the generation of a *Cox1::ATP6* transcript. Translation of this mRNA is only dependent on Pet309 and Mss51, the TAs of *COX1*, but not on Atp22 (Zeng et al. 2007). In accordance, efficient synthesis of a mitochondrially encoded reporter gene was strictly dependent on the presence of Atp22 (Rak and Tzagoloff 2009). Although *ATP6* and *ATP8* are produced from a bi-cistronic transcript, *ATP22* deletion mutants specifically lack Atp6 but show normal translation rates of *ATP8* (Zeng et al. 2007). This suggests that a translational activator for *ATP8* still awaits identification.

The core component of the F_0 part of the ATP synthase is an oligomer of Atp9 subunits that forms the proton conducting channel. The *ATP9* gene is transcribed together with an adjacent tRNA and the *VAR1* gene and the polycistronic transcript is matured by endonucleolytic cleavage (Zassenhaus et al. 1984). The importance of the 5'-UTR of *ATP9* was recognized very early, as insertion of bases into this region impaired translation (Ooi et al. 1987). Three proteins influence translation of the *ATP9* mRNA: Aep1, which acts as a TA, Aep2 that is either required for the stabilization of the *ATP9* transcript or stimulating its translation and Atp25, which has a dual role in translation and assembly of ATP synthase (Payne et al. 1991, 1993; Ellis et al. 1999; Zeng et al. 2008). Atp25 is split into two halves and both portions function in mitochondria. The C-terminal half of the protein is conferring stability to the *ATP9* mRNA and expression of this part of the protein is sufficient to allow Atp9 synthesis in the *ATP25* deletion mutant

(Zeng et al. 2008). In the absence of the N-terminal half of Atp25; however, the translated Atp9 is not stably assembled into the Atp9-oligomer. This suggests that the N-terminal half of Atp25 is not dispensable for the biogenesis of Atp9 and might even mediate assembly of the Atp9 ring. Hence, Atp25 is a protein of dual function with a probability to modulate expression of *ATP9* in a feedback loop. However, this hypothesis has not yet been analyzed experimentally. Importantly, it was recently demonstrated that the synthesis of Atp6 and Atp8 is regulated in response to the assembly process of the ATP synthase complex (Rak and Tzagoloff 2009). This is described in detail below.

5.2 Nuclear Control of Protein Synthesis in Yeast Mitochondria

Translational activators are encoded in the nucleus. Soon after the discovery of Pet111 and Pet494 and the establishment of the concept of specific translational activation, regulation/control of the expression of these nuclear genes was investigated. Studies using a yeast strain with a chromosomal gene fusion consisting of the *COX3*-specific *PET494* and the *E. coli* β-galactosidase gene *lacZ* revealed that Pet494 is expressed at very low levels (Marykwas and Fox 1989). The TAs Pet122 and Pet111 are present in similarly low amounts (Fox 1996). The low abundance of translational activators implies that they are rate limiting for mitochondrial protein synthesis. This was confirmed for Pet494 by investigating diploid yeast strains homo- or heterozygous for the *PET494* locus or haploids carrying a high copy plasmid to overexpress the gene. Additionally, these strains harbored a mitochondrial genome that encodes the reporter construct *ARG8 ^{m}* (Steele et al. 1996). *ARG8* is a nuclear gene coding for a soluble enzyme, which is normally posttranslationally imported into mitochondria and involved in the biogenesis of arginine. The recoded version of the gene *ARG8 ^{m}* was integrated into the mtDNA of yeast deficient in the nuclear copy of *ARG8*. In the study of Steele et al., the open reading frame of *COX3* was substituted by the *ARG8 ^{m}* gene (*cox3::ARG8 ^{m}* mtDNA), making Arg8 synthesis dependent on *COX3*-specific translational activation. The available amount of Pet494 clearly correlated with the Arg8 expression rate, while *cox3::ARG8 ^{m}* expression only moderately correlated with Pet122 level (Steele et al. 1996).

The expression of many genes involved in respiration in yeast is modulated over a wide range of growth conditions and TAs seem to be no exception to this. *PET494* expression is subject to catabolite repression (Marykwas and Fox 1989). In the presence of glucose, the levels of the TA drop four to sixfold in comparison to cells grown on nonfermentable carbon sources. Furthermore, synthesis of Pet494 is regulated by oxygen, but in contrast to the transcriptional repression by glucose this is rather achieved on a translational level (Marykwas and Fox 1989). Interestingly, expression of *PET494* is heme independent. This is opposed to other respiratory genes that are responding to oxygen levels, where transcriptional upregulation under aerobic conditions is mediated by heme (Guarente and Mason 1983; Keng and Guarente 1987). The regulation of the expression of other TAs was not analyzed similarly detailed. However, expression of many genes necessary for respiration and especially subunits of the OXPHOS system is influenced by growth conditions (Guarente and Mason 1983; Lowry et al. 1983; Myers et al. 1987; Forsburg and Guarente 1989). Taken into account the feedback regulatory circles that were revealed in the last years, this regulation of nuclear gene expression can be considered as an example of how mitochondrial gene expression is influenced by carbon source, oxygen levels, or presence of heme .

5.3 Regulation of Mitochondrial Protein Synthesis in Response to Assembly of the Oxphos System

Both the nuclear as well as the mitochondrial protein synthesis machinery contribute subunits to the OXPHOS complexes. To ensure efficient assembly, these expression systems have to be coordinated temporally and spatially in a precise manner. In recent years, different groups have revealed how regulation of mitochondrial protein synthesis is accomplished and how the levels of mitochondrially encoded subunits are adjusted to allow an efficient OXPHOS assembly process (Fig. 5.2). The general principle is that TAs with dual functions are sequestered in OXPHOS assembly intermediates. When assembly proceeds normally, the TA is released to stimulate synthesis of its client protein. In contrast, when further assembly fails, the TA is trapped in the assembly intermediate and not available to activate new rounds of translation.

5.3.1 Regulation of Cytochrome b Synthesis

The yeast bc_1 complex is composed of nine nuclear-encoded subunits that are assembled around the core component cytochrome b, which is produced by the mitochondrial genetic system. Catalytically active are only the three proteins, cytochrome b, cytochrome c_1 (Cyt1), and the Fe/S protein Rip1, whereas the remaining seven subunits are accessory structural subunits. The step-wise assembly process involves four intermediates and has mainly been analyzed by the use of yeast strains lacking individual structural subunits of the bc_1 complex and their analysis by Blue Native polyacrylamide gel electrophoresis (BN PAGE) (Zara et al. 2007, 2009a, b; Gruschke et al. 2012; Smith et al. 2012). Assembly starts with synthesis and membrane insertion of cytochrome b, which is immediately bound by the Cbp3–Cbp6 complex (Fig. 5.3). Recruitment of the assembly factor Cbp4 results in assembly intermediate I that serves as a pool of unassembled cytochrome b even at steady state. Addition of the first two nuclear-encoded

(a) normal *bc₁* complex assembly or assembly impaired <u>after</u> formation of intermediate IV

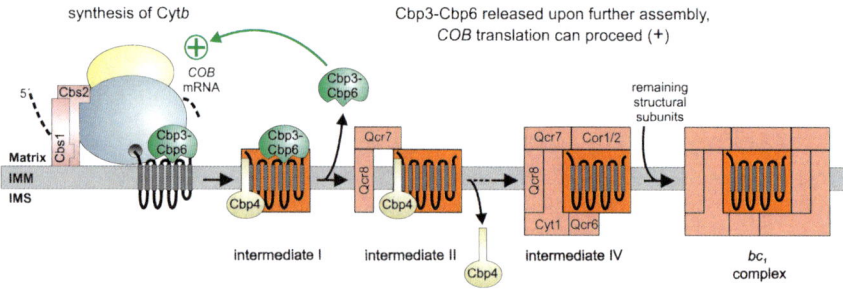

(b) *bc₁* complex assembly disturbed <u>before</u> formation of intermediate IV

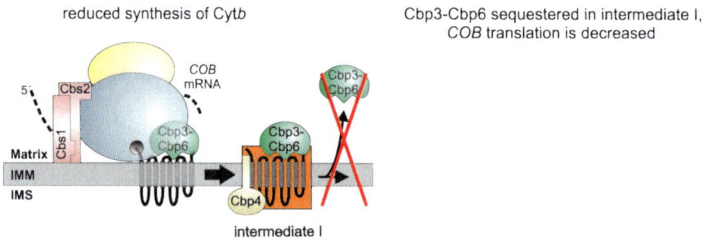

Fig. 5.3 Schematic representation of the regulatory feedback loop modulating cytochrome *b* synthesis in response to the assembly of the bc_1 complex. The Cbp3–Cbp6 complex exerts a dual role in the biogenesis of cytochrome *b*: In its ribosome-bound form it acts as a translational activator and together with Cbs1 and Cbs2 stimulates translation of the *COB* mRNA. It is also present as a nonribosome-bound form in association with cytochrome *b* and Cbp4, forming the first assembly intermediate of the bc_1 complex assembly line. A: Cytochrome *b* assembles through four intermediates into a functional bc_1 complex, three of which are depicted in the scheme. When assembly is undisturbed or can proceed at least until intermediate IV is formed, Cbp3–Cbp6 is released from intermediate I upon further assembly, can again activate *COB* mRNA translation at the ribosome (+) and cytochrome *b* synthesis is not affected. B: If complex assembly is disturbed before intermediate IV is formed, Cbp3–Cbp6 cannot be released efficiently and is sequestered in the accumulating intermediate I (*thick black arrow*). The complex is therefore not available for activating translation and consequently cytochrome *b* synthesis is reduced. IMM, inner mitochondrial membrane. IMS, intermembrane space

structural subunits Qcr7 and Qcr8 induces release of Cbp3–Cbp6, whereas Cbp4 stays attached. The cytochrome *b*-Cbp4-Qcr7-Qcr8 complex represents the second assembly intermediate and is further joined by the two core proteins Cor1 and Cor2. Addition of Cyt1 and the small acidic accessory subunit Qcr6 to intermediate III forms intermediate IV, which was previously described as the 500 kDa complex (Zara et al. 2009b). The incorporation of two accessory subunits (Qcr9 and Qcr10) and the last catalytic subunit (Rip1) completes formation of the bc_1 complex. When assembly is disturbed before intermediate IV can be generated, the synthesis of cytochrome *b* is reduced (Gruschke et al. 2012) (Fig. 5.3). This is caused by sequestration of Cbp3–Cbp6 in assembly intermediate I, which accumulates under these conditions and as a result the Cbp3–Cbp6 complex is not available at the mitochondrial ribosome to fulfill its function as a TA for the *COB* mRNA (Gruschke et al. 2012). This negative feedback can be overcome by over-expression of the Cbp3–Cbp6 complex, demonstrating its key role in this process. Formation of assembly intermediate IV seems to be a critical point in the pathway as disturbance of the last assembly step by either deletion of one of the structural subunits *QCR9*, *QCR10*, *RIP1* or required assembly factors (*MZM1*, *BCS1*) does not lead to reduced cytochrome *b* translation.

5.3.2 Regulation of Cox1 Synthesis

The cytochrome oxidase (COX, complex IV) is composed of 11 subunits in yeast, three of which are encoded in the mitochondrial genome. Two of these three subunits harbor redox-active heme and/or copper co-factors. Electrons flow from the Cu_A center of Cox2 to the heme *a* cofactor of Cox1, from where they are passed further to the active site of Cox1 composed of the Cu_B center and heme a_3. Heme a_3 binds molecular oxygen which serves as the final electron acceptor. The assembly of this OXPHOS complex is characterized very well and assisted by a considerable number of factors involved in co-factor acquisition, mediation of subunit interaction and feedback regulation (Fontanesi et al. 2006; Mick et al. 2011). Assembly of cytochrome oxidase is initiated from the central subunit of the complex, Cox1. Unassembled Cox1 with its redox-active cofactors is potentially harmful for cells as it may give rise to reactive oxygen species (Khalimonchuk et al. 2007). To ensure integrity of the cell, Cox1 synthesis has to be monitored precisely and adjusted to levels that can successfully be incorporated into COX. In recent years, it was found that Cox1 translation in yeast is subject to a complex feedback regulatory circle that achieves this fine tuning (Mick et al. 2011). The key role in this feedback loop is played by the dually functioning protein Mss51, which acts as a TA for *COX1* mRNA as well as a Cox1-assembly factor by bind-ing the newly synthesized protein (Perez-Martinez et al. 2003). Like Cbp3–Cbp6 in the case of the *bc₁* complex, Mss51 is sequestered in assembly intermediates that cannot be resolved when further assembly is blocked and thereby is precluded from activating new rounds of *COX1* translation (Fig. 5.4). Very recently, it was demonstrated that Mss51 contains two heme binding motifs in its N-terminus, thereby allowing it to act as a heme sensor and coordinate COX assembly with heme availability (Soto et al. 2012).

Besides Mss51, several other factors participate in COX assembly and the feed-back regulation mechanism (Fig. 5.4). The presence of Cox14 and Coa3/Cox25 is required to allow efficient interaction of Mss51 with newly synthesized Cox1, and thus they act as negative regulators of *COX1* synthesis by ensuring efficient seques-tration of Mss51 (Barrientos et al. 2004; Perez-Martinez et al. 2009; Mick et al. 2010; Fontanesi et al. 2011). *COX1* feedback regulation depends on the C-terminal region of Cox1 itself. Mutants lacking this part of the protein can synthesize and assemble Cox1 into a functional cytochrome oxidase, but do not exhibit assembly

(a) normal COX assembly and heme biosynthesis

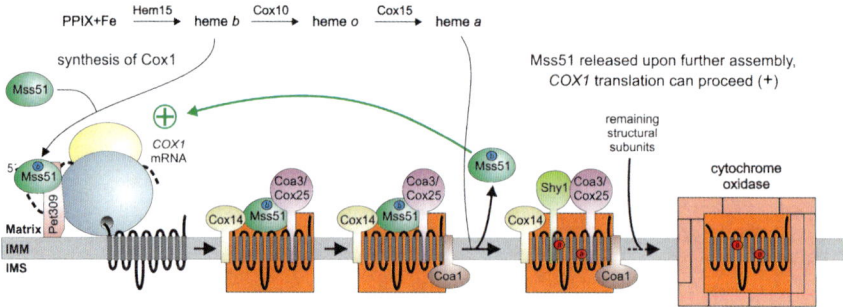

(b) disturbed COX assembly or heme biosynthesis

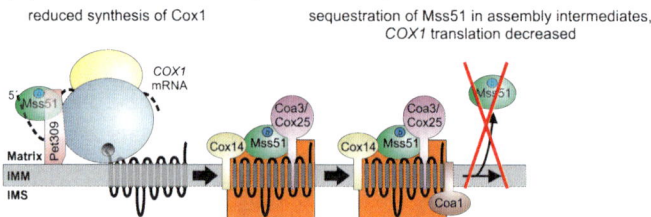

Fig. 5.4 Schematic representation of the regulatory feedback loop modulating Cox1 synthesis in response to the assembly of the COX complex. **a** Mss51 has two functions in the biogenesis of this OXPHOS complex; in concert with Pet309 it serves as a translational activator for *COX1* mRNA and, in addition, Mss51 is part of assembly intermediates, acting as a Cox1 chaperone. Mss51 is a heme-binding protein, which additionally allows to regulate *COX1* synthesis in response to the heme homeostasis of the cell. Hem15 (ferrochelatase) incorporates iron into the protophorphyrin IX ring (PPIX), thereby forming heme *b* from which subsequently heme *a*, one of the cofactors present in Cox1, is synthesized through the concerted action of Cox10 and Cox15. Together with hemylated Mss51, Cox14, Coa3/Cox25, and Coa1 are part of early Cox1 assembly intermediates. Presumably, at the step where Cox1 is hemylated and Shy1 enters the pathway, Mss51 is released and can again activate *COX1* mRNA translation (+). Incorporation of the remaining structural subunits releases the other assembly factors until formation of cytochrome oxidase is completed. **b** If heme biosynthesis or COX assembly is disturbed, Mss51 cannot function properly or is sequestered in assembly intermediates and therefore not available for activating translation; Cox1 synthesis consequently is reduced. IMM, inner mitochondrial membrane. IMS, intermembrane space

responsive reduction of Cox1 synthesis. It has been speculated that the molecular reason for this lies in the weakened interaction between Mss51 and Cox14 (Shingu-Vazquez et al. 2010). Coa1, Coa2, Shy1, and the mitochondrial Hsp70 chaperone Ssc1 are additional factors participating in COX assembly (Barrientos et al. 2002; Pierrel et al. 2007; Fontanesi et al. 2008; Pierrel et al. 2008; Fontanesi et al. 2010). The exact molecular composition of all COX assembly intermediates is, however, still under debate (McStay et al. 2012). A subcomplex consisting of Mss51, Cox14, Coa3/Cox25, and Coa1 bound to an oxidatively harmless, unhemylated form of Cox1 is stable in wild-type cells and presumably serves as a pool of assembly competent Cox1 (Khalimonchuk et al. 2010). In contrast, the Shy1-containing assembly intermediate comprises hemylated Cox1; however, Shy1 is most likely not required for hemylation *per se*, but rather stabilizes Cox1 in a conformation allowing the insertion of heme a_3. Although it is not entirely resolved yet, Mss51 presumably is released from Cox1 when Shy1 enters the assembly pathway. Mss51 can then again act as a TA and induce further Cox1 synthesis.

5.3.3 Regulation of Atp6/8 Synthesis

The ATP synthase is composed of three functionally and structurally distinct parts. The membrane-embedded F_0 part comprises Atp9 subunits, which form a ring-like structure, and the two proteins Atp6 and Atp8. These three proteins are encoded in the mitochondrial genome. The hydrophilic F_1 part is formed by a hexamer of alternating α and β subunits that mediate ATP synthesis and the central stalk, which is made up of subunits γ, δ, and ε. The stalk is in contact to the Atp9-ring as well as the $\alpha_3\beta_3$ hexamer. The third part of the enzyme is the peripheral stator stalk made up of four subunits, which is attached to both the $\alpha_3\beta_3$ oligomer and Atp6 in the membrane. By this, the $\alpha_3\beta_3$ hexamer, Atp6, and the stator form the stationary part of the enzyme. Driven by the electrochemical gradient across the membrane, protons flow back from the intermembrane space into the matrix at the interface between the Atp9 ring and Atp6, thereby rotating the Atp9 part and the central stalk stepwise and inducing conformational changes at the catalytic sites of the $\alpha_3\beta_3$ hexamer that drive ATP synthesis (Stock et al. 2000). The assembly process of the ATP synthase is not understood in every detail (Ackerman and Tzagoloff 2005; Rak et al. 2009). Early experiments indicated that assembly of the F_1 unit is independent from assembly of F_0 (Schatz 1968). The current idea is that ATP synthase assembly involves to distinct, but coordinately formed modules which are joined at the end (Rak et al. 2011). The main F_0 component, the Atp9 ring, is assembled from Atp9 monomers with the help of the N-terminal part of Atp25 and then interacts with the pre-assembled F_1 unit (Zeng et al. 2008; Rak et al. 2011). In parallel, a complex of Atp6, Atp8, and at least two stator stalk subunits is generated. Together with Atp6, the Atp9 ring forms the proton translocating channel of the enzyme complex. The joining of Atp6 with the Atp9 ring seems to occur at a rather late step of assembly and involves the assembly factor Atp10 and the inner membrane protein Oxa1 (Tzagoloff et al. 2004; Jia et al. 2007).

In 2009, Rak and Tzagoloff reported that translation of the *ATP8/ATP6* bi-cistronic mRNA is dependent on F_1 assembly (Fig. 5.5) (Rak and Tzagoloff 2009). Mutants lacking assembly factors required for the formation of the F_1 unit, Atp11 or Atp12, or the two main structural F_1 subunits α and β display reduced synthesis rates of Atp6 and Atp8. It was excluded that this was caused by an increased turnover of the newly translated proteins by analyzing expression of the *ARG8 ^{m}* reporter genes (*atp6::ARG8 ^{m}* or *atp8::ARG8 ^{m}*) that revealed impaired Arg8 synthesis. Overexpression of Atp22, the translational activator of *ATP6*, was able to

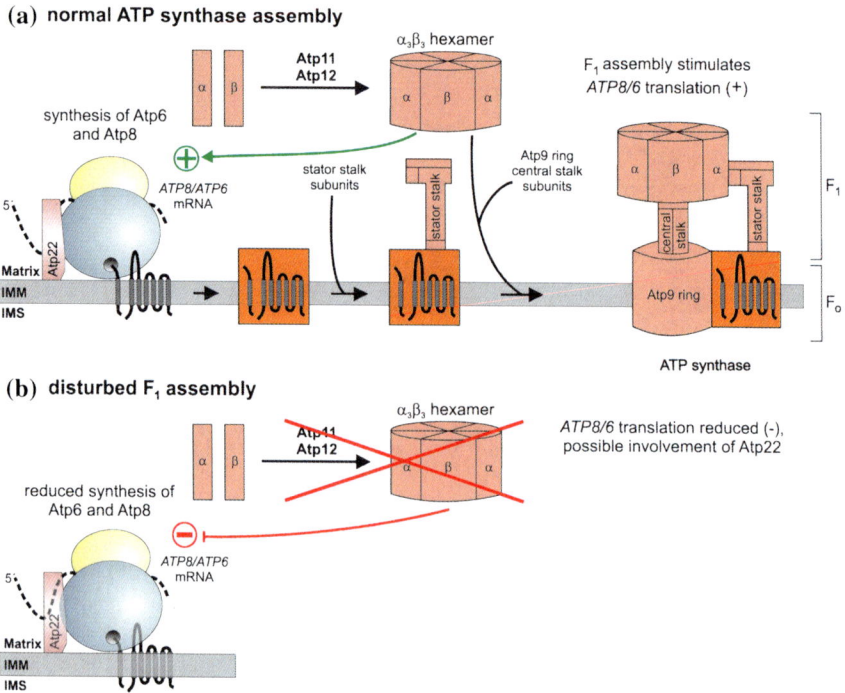

Fig. 5.5 Schematic representation of the regulatory feedback loop modulating Atp6/Atp8 synthesis in response to the assembly of the ATP synthase. A: The mitochondrially encoded ATP synthase subunits Atp6 and Atp8 are translated from a bi-cistronic mRNA with the help of the translational activator for *ATP6*, Atp22, and a yet unknown translational activator for *ATP8*. Atp6 and Atp8 are after their synthesis assembled with the stator stalk. The monomeric forms of the F_1 subunits α and β are prevented from aggregation and assembled into the $\alpha_3\beta_3$ hexamer by the assembly factors Atp11 and Atp12. After addition of the central stalk subunits, the F_1 part is joined to the Atp9 ring and the Atp6/8 module, forming the fully assembled ATP synthase. The successful assembly of the $\alpha_3\beta_3$ hexamer is required for efficient translation of the *ATP8/ATP6* mRNA (+). B: If F_1 formation is perturbed, synthesis of Atp6 and Atp8 is impaired (−). This regulation presumably involves Atp22, for details see text. IMM, inner mitochondrial membrane. IMS, intermembrane space

suppress the phenotype (Rak and Tzagoloff 2009). Currently, it is not clear how exactly this feedback regulation is mediated. It could be achieved either by the sequestration of Atp22 in some form of assembly intermediate (similar to the cases of Cbp3–Cbp6 and Mss51) or it could involve yet uncharacterized components. This system is physiological important, as it prevents the dissipation of the membrane potential in case the F_o part cannot efficiently be coupled to the F_1 complex. Although the F_1-dependent regulation of F_o biogenesis mechanistically differs from the feedback-regulated expression of *COB* or *COX1*, it provides another example of how mitochondrial translation is adjusted to the level of cytoplasmic protein synthesis.

5.4 Outlook

Biogenesis of OXPHOS complexes of dual genetic origin requires cross-talk of the two genetic systems involved. This regulation occurs at the level of mitochondrial protein synthesis, which is modulated by TAs that sense efficiency of assembly to down-regulate expression of their client protein when assembly fails. Despite the fact that specific translational activation of mitochondrial protein synthesis by nuclear genes is known since more than 40 years, we do not yet understand which exact molecular functions TAs exert during protein synthesis in the organelle. In addition to a presumably direct role in translation, mitochondrial TAs appear to be implicated in the organization of translation. The organization of cytochrome *b* biogenesis might serve as a good example to illustrate this: Cytochrome *b* is only efficiently synthesized when one of its TAs, the Cbp3–Cbp6 complex is present at the ribosomal tunnel exit (Gruschke et al. 2011). Because Cbp3–Cbp6 is also an essential assembly factor for cytochrome *b*, it is ensured that the newly synthesized protein experiences an optimally tailored environment for further assembly. Indeed, when cytochrome *b* is synthesized from an mRNA containing the 5′-UTR of another transcript, the proteins fails to accumulate robustly, while rates of synthesis of this ectopically expressed protein are indistinguishable from the authentic protein (Gruschke et al. 2012). Similar observations have also been reported previously for other ectopically expressed proteins (Sanchirico et al. 1998), suggesting that in mitochondria, each mRNA is translated by ribosomes that are specifically designed to optimally support biogenesis of the client protein (Gruschke and Ott 2010). It thus appears that mitochondrial protein synthesis is probably much more sophisticated organized than anticipated and that this system still harbors many exciting previously unidentified features.

Acknowledgments   We thank all members of our group for stimulating discussions. Our work is supported by the Swedish research council (VR), the Center for Biomembrane Research (CBR) at Stockholm University, the German research council (Research unit 967), Jaensson foundation, and the Carl Tryggers foundation.

References

Ackerman SH, Tzagoloff A (2005) Function, structure, and biogenesis of mitochondrial ATP synthase. Prog Nucleic Acid Res Mol Biol 80:95–133
Barrell BG, Bankier AT, Drouin J (1979) A different genetic code in human mitochondria. Nature 282:189–194
Barrientos A, Korr D, Tzagoloff A (2002) Shy1p is necessary for full expression of mitochondrial *COX1* in the yeast model of Leigh's syndrome. EMBO J 21:43–52
Barrientos A, Zambrano A, Tzagoloff A (2004) Mss51p and Cox14p jointly regulate mitochondrial Cox1p expression in *Saccharomyces cerevisiae*. EMBO J 23:3472–3482
Bonnefoy N, Fox TD (2000) In vivo analysis of mutated initiation codons in the mitochondrial *COX2* gene of *Saccharomyces cerevisiae* fused to the reporter gene *ARG8* ^{m} reveals lack of downstream reinitiation. Mol Gen Genet 262:1036–1046

Borst P, Grivell LA (1978) The mitochondrial genome of yeast. Cell 15:705–723

Brown NG, Costanzo MC, Fox TD (1994) Interactions among three proteins that specifically activate translation of the mitochondrial *COX3* mRNA in *Saccharomyces cerevisiae*. Mol Cell Biol 14:1045–1053

Cabral F, Schatz G (1978) Identification of cytochrome *c* oxidase subunits in nuclear yeast mutants lacking the functional enzyme. J Biol Chem 253:4396–4401

Camougrand N, Pelissier P, Velours G, Guerin M (1995) *NCA2*, a second nuclear gene required for the control of mitochondrial synthesis of subunits 6 and 8 of ATP synthase in *Saccharomyces cerevisiae*. J Mol Biol 247:588–596

Chen JY, Martin NC (1988) Biosynthesis of tRNA in yeast mitochondria. An endonuclease is responsible for the 3′-processing of tRNA precursors. J Biol Chem 263:13677–13682

Chen W, Dieckmann CL (1997) Genetic evidence for interaction between Cbp1 and specific nucleotides in the 5′ untranslated region of mitochondrial cytochrome *b* mRNA in *Saccharomyces cerevisiae*. Mol Cell Biol 17:6203–6211

Christianson T, Edwards JC, Mueller DM, Rabinowitz M (1983) Identification of a single transcriptional initiation site for the glutamic tRNA and *COB* genes in yeast mitochondria. Proc Natl Acad Sci USA 80:5564–5568

Chrzanowska-Lightowlers ZM, Pajak A, Lightowlers RN (2011) Termination of protein synthesis in mammalian mitochondria. J Biol Chem 286:34479–34485

Costanzo MC, Fox TD (1986) Product of *Saccharomyces cerevisiae* nuclear gene *PET494* activates translation of a specific mitochondrial mRNA. Mol Cell Biol 6:3694–3703

Costanzo MC, Fox TD (1988) Specific translational activation by nuclear gene products occurs in the 5′ untranslated leader of a yeast mitochondrial mRNA. Proc Natl Acad Sci USA 85:2677–2681

Costanzo MC, Fox TD (1993) Suppression of a defect in the 5′ untranslated leader of mitochondrial *COX3* mRNA by a mutation affecting an mRNA-specific translational activator protein. Mol Cell Biol 13:4806–4813

De La Salle H, Jacq C, Slonimski PP (1982) Critical sequences within mitochondrial introns: pleiotropic mRNA maturase and cis-dominant signals of the box intron controlling reductase and oxidase. Cell 28:721–732

Decoster E, Simon M, Hatat D, Faye G (1990) The *MSS51* gene product is required for the translation of the *COX1* mRNA in yeast mitochondria. Mol Gen Genet 224:111–118

Dhawale S, Hanson DK, Alexander NJ, Perlman PS, Mahler HR (1981) Regulatory Interactions between Mitochondrial Genes - Interactions between 2 Mosaic Genes. Proc Natl Acad Sci USA 78:1778–1782

Dieckmann CL, Tzagoloff A (1985) Assembly of the mitochondrial membrane system. *CBP6*, a yeast nuclear gene necessary for synthesis of cytochrome *b*. J Biol Chem 260:1513–1520

Dunstan HM, Green-Willms NS, Fox TD (1997) In vivo analysis of *Saccharomyces cerevisiae* *COX2* mRNA 5′-untranslated leader functions in mitochondrial translation initiation and translational activation. Genetics 147:87–100

Ellis TP, Helfenbein KG, Tzagoloff A, Dieckmann CL (2004) Aep3p stabilizes the mitochondrial bicistronic mRNA encoding subunits 6 and 8 of the H$^+$-translocating ATP synthase of *Saccharomyces cerevisiae*. J Biol Chem 279:15728–15733

Ellis TP, Lukins HB, Nagley P, Corner BE (1999) Suppression of a nuclear *aep2* mutation in *Saccharomyces cerevisiae* by a base substitution in the 5′-untranslated region of the mitochondrial *oli1* gene encoding subunit 9 of ATP synthase. Genetics 151:1353–1363

Faye G, Simon M (1983) Analysis of a yeast nuclear gene involved in the maturation of mitochondrial pre-messenger RNA of the cytochrome oxidase subunit I. Cell 32:77–87

Fiori A, Mason TL, Fox TD (2003) Evidence that synthesis of the *Saccharomyces cerevisiae* mitochondrially encoded ribosomal protein Var1p may be membrane localized. Eukaryot Cell 2:651–653

Folley LS, Fox TD (1991) Site-directed mutagenesis of a *Saccharomyces cerevisiae* mitochondrial translation initiation codon. Genetics 129:659–668

Fontanesi F, Clemente P, Barrientos A (2011) Cox25 teams up with Mss51, Ssc1, and Cox14 to regulate mitochondrial cytochrome *c* oxidase subunit 1 expression and assembly in *Saccharomyces cerevisiae*. J Biol Chem 286:555–566

Fontanesi F, Soto IC, Barrientos A (2008) Cytochrome *c* oxidase biogenesis: new levels of regulation. IUBMB Life 60:557–568

Fontanesi F, Soto IC, Horn D, Barrientos A (2006) Assembly of mitochondrial cytochrome *c* oxidase, a complicated and highly regulated cellular process. Am J Physiol Cell Physiol 291:C1129–C1147

Fontanesi F, Soto IC, Horn D, Barrientos A (2010) Mss51 and Ssc1 facilitate translational regulation of cytochrome *c* oxidase biogenesis. Mol Cell Biol 30:245–259

Forsburg SL, Guarente L (1989) Communication between mitochondria and the nucleus in regulation of cytochrome genes in the yeast *Saccharomyces cerevisiae*. Annu Rev Cell Biol 5:153–180

Fox TD (1979) Five TGA "stop" codons occur within the translated sequence of the yeast mitochondrial gene for cytochrome *c* oxidase subunit II. Proc. Natl. Acad. Sci. U S A 76:6534–6538

Fox TD (1996) Translational control of endogenous and recoded nuclear genes in yeast mitochondria: regulation and membrane targeting. Experientia 52:1130–1135

Gray MW (1989) The evolutionary origins of organelles. Trends Genet 5:294–299

Groudinsky O, Bousquet I, Wallis MG, Slonimski PP, Dujardin G (1993) The *NAM1/MTF2* nuclear gene product is selectively required for the stability and/or processing of mitochondrial transcripts of the atp6 and of the mosaic, cox1 and cytb genes in *Saccharomyces cerevisiae*. Mol Gen Genet 240:419–427

Gruschke S, Kehrein K, Römpler K, Gröne K, Israel L, Imhof A et al (2011) Cbp3-Cbp6 interacts with the yeast mitochondrial ribosomal tunnel exit and promotes cytochrome *b* synthesis and assembly. J Cell Biol 193:1101–1114

Gruschke S, Ott M (2010) The polypeptide tunnel exit of the mitochondrial ribosome is tailored to meet the specific requirements of the organelle. BioEssays 32:1050–1057

Gruschke S, Rompler K, Hildenbeutel M, Kehrein K, Kuhl I, Bonnefoy N et al (2012) The Cbp3-Cbp6 complex coordinates cytochrome *b* synthesis with *bc₁* complex assembly in yeast mitochondria. J Cell Biol 199:137–150

Guarente L, Mason T (1983) Heme regulates transcription of the *CYC1* gene of *S. cerevisiae* via an upstream activation site. Cell 32:1279–1286

Haffter P, Fox TD (1992) Suppression of carboxy-terminal truncations of the yeast mitochondrial mRNA-specific translational activator *PET122* by mutations in two new genes, *MRP17* and *PET127*. Mol Gen Genet 235:64–73

Haffter P, McMullin TW, Fox TD (1991) Functional interactions among two yeast mitochondrial ribosomal proteins and an mRNA-specific translational activator. Genetics 127:319–326

Hell K, Neupert W, Stuart RA (2001) Oxa1p acts as a general membrane insertion machinery for proteins encoded by mitochondrial DNA. EMBO J 20:1281–1288

Hollingsworth MJ, Martin NC (1986) RNase P activity in the mitochondria of *Saccharomyces cerevisiae* depends on both mitochondrion and nucleus-encoded components. Mol Cell Biol 6:1058–1064

Islas-Osuna MA, Ellis TP, Marnell LL, Mittelmeier TM, Dieckmann CL (2002) Cbp1 is required for translation of the mitochondrial cytochrome *b* mRNA of *Saccharomyces cerevisiae*. J Biol Chem 277:37987–37990

Jia L, Dienhart M, Schramp M, McCauley M, Hell K, Stuart RA (2003) Yeast Oxa1 interacts with mitochondrial ribosomes: The importance of the C-terminal hydrophilic region of Oxa1. EMBO J 22:6438–6447

Jia L, Dienhart MK, Stuart RA (2007) Oxa1 directly interacts with Atp9 and mediates its assembly into the mitochondrial F_1F_0-ATP synthase complex. Mol Biol Cell 18:1897–1908

Kehrein K, Bonnefoy N, Ott M (2013) Mitochondrial Protein Synthesis: Efficiency and Accuracy. Antioxid Redox Signal. doi:10.1089/ars.2012.4896

Keng T, Guarente L (1987) Constitutive expression of the yeast *HEM1* gene is actually a composite of activation and repression. Proc. Natl. Acad. Sci. U S A 84:9113–9117

Khalimonchuk O, Bestwick M, Meunier B, Watts TC, Winge DR (2010) Formation of the redox cofactor centers during Cox1 maturation in yeast cytochrome oxidase. Mol Cell Biol 30:1004–1017

Khalimonchuk O, Bird A, Winge DR (2007) Evidence for a pro-oxidant intermediate in the assembly of cytochrome oxidase. J Biol Chem 282:17442–17449

Krause K, Lopes de Souza R, Roberts DG, Dieckmann CL (2004) The mitochondrial message-specific mRNA protectors Cbp1 and Pet309 are associated in a high-molecular weight complex. Mol Biol Cell 15:2674–2683

Kühl I, Fox TD, Bonnefoy N (2012) *Schizosaccharomyces pombe* homologs of the *Saccharomyces cerevisiae* mitochondrial proteins Cbp6 and Mss51 function at a post-translational step of respiratory complex biogenesis. Mitochondrion 12:381–390

Lazowska J, Jacq C, Slonimski PP (1980) Sequence of introns and flanking exons in wild-type and *box3* mutants of cytochrome *b* reveals an interlaced splicing protein coded by an intron. Cell 22:333–348

Li M, Tzagoloff A, Underbrink-Lyon K, Martin NC (1982) Identification of the paromomycin-resistance mutation in the 15 S rRNA gene of yeast mitochondria. J Biol Chem 257:5921–5928

Lipinski KA, Puchta O, Surendranath V, Kudla M, Golik P (2011) Revisiting the yeast PPR proteins–application of an Iterative Hidden Markov Model algorithm reveals new members of the rapidly evolving family. Mol Biol Evol 28:2935–2948

Lowry CV, Weiss JL, Walthall DA, Zitomer RS (1983) Modulator sequences mediate oxygen regulation of *CYC1* and a neighboring gene in yeast. Proc. Natl. Acad. Sci. USA 80:151–155

Manthey GM, McEwen JE (1995) The product of the nuclear gene *PET309* is required for translation of mature mRNA and stability or production of intron-containing RNAs derived from the mitochondrial *COX1* locus of *Saccharomyces cerevisiae*. EMBO J 14:4031–4043

Marykwas DL, Fox TD (1989) Control of the *Saccharomyces cerevisiae* regulatory gene *PET494*: transcriptional repression by glucose and translational induction by oxygen. Mol Cell Biol 9:484–491

McMullin TW, Fox TD (1993) *COX3* mRNA-specific translational activator proteins are associated with the inner mitochondrial membrane in *Saccharomyces cerevisiae*. J Biol Chem 268:11737–11741

McMullin TW, Haffter P, Fox TD (1990) A novel small-subunit ribosomal protein of yeast mitochondria that interacts functionally with an mRNA-specific translational activator. Mol Cell Biol 10:4590–4595

McStay GP, Su CH, and Tzagoloff A (2012) Modular assembly of yeast cytochrome oxidase. *Mol. Biol. Cell*

Mears JA, Sharma MR, Gutell RR, McCook AS, Richardson PE, Caulfield TR et al (2006) A structural model for the large subunit of the mammalian mitochondrial ribosome. J Mol Biol 358:193–212

Mick DU, Fox TD, Rehling P (2011) Inventory control: cytochrome *c* oxidase assembly regulates mitochondrial translation. Nat Rev Mol Cell Biol 12:14–20

Mick DU, Vukotic M, Piechura H, Meyer HE, Warscheid B, Deckers M et al (2010) Coa3 and Cox14 are essential for negative feedback regulation of *COX1* translation in mitochondria. J Cell Biol 191:141–154

Mittelmeier TM, Dieckmann CL (1995) In vivo analysis of sequences required for translation of cytochrome *b* transcripts in yeast mitochondria. Mol Cell Biol 15:780–789

Mulero JJ, Fox TD (1993a) Alteration of the *Saccharomyces cerevisiae COX2* mRNA 5′-untranslated leader by mitochondrial gene replacement and functional interaction with the translational activator protein PET111. Mol Biol Cell 4:1327–1335

Mulero JJ, Fox TD (1993b) *PET111* acts in the 5′-leader of the *Saccharomyces cerevisiae* mitochondrial *COX2* mRNA to promote its translation. Genetics 133:509–516

Mulero JJ, Fox TD (1994) Reduced but accurate translation from a mutant AUA initiation codon in the mitochondrial *COX2* mRNA of *Saccharomyces cerevisiae*. Mol Gen Genet 242:383–390

Muller PP, Reif MK, Zonghou S, Sengstag C, Mason TL, Fox TD (1984) A nuclear mutation that post-transcriptionally blocks accumulation of a yeast mitochondrial gene product can be suppressed by a mitochondrial gene rearrangement. J Mol Biol 175:431–452

Muroff I, Tzagoloff A (1990) *CBP7* codes for a co-factor required in conjunction with a mitochondrial maturase for splicing of its cognate intervening sequence. EMBO J 9:2765–2773

Myers AM, Crivellone MD, Koerner TJ, Tzagoloff A (1987) Characterization of the yeast *HEM2* gene and transcriptional regulation of *COX5* and *COR1* by heme. J Biol Chem 262:16822–16829

Naithani S, Saracco SA, Butler CA, Fox TD (2003) Interactions among *COX1*, *COX2*, and *COX3* mRNA-specific translational activator proteins on the inner surface of the mitochondrial inner membrane of *Saccharomyces cerevisiae*. Mol Biol Cell 14:324–333

Nobrega FG, Tzagoloff A (1980) Assembly of the mitochondrial membrane system. DNA sequence and organization of the cytochrome b gene in Saccharomyces cerevisiae D273–10B. J Biol Chem 255:9828–9837

Ooi BG, Lukins HB, Linnane AW, Nagley P (1987) Biogenesis of mitochondria: a mutation in the 5′-untranslated region of yeast mitochondrial oli1 mRNA leading to impairment in translation of subunit 9 of the mitochondrial ATPase complex. Nucleic Acids Res 15:1965–1977

Ott M, Herrmann JM (2010) Co-translational membrane insertion of mitochondrially encoded proteins. Biochim Biophys Acta 1803:767–775

Ott M, Prestele M, Baucrschmitt H, Funes S, Bonnefoy N, Herrmann JM (2006) Mba1, a membrane-associated ribosome receptor in mitochondria. EMBO J 25:1603–1610

Payne MJ, Finnegan PM, Smooker PM, Lukins HB (1993) Characterization of a second nuclear gene, *AEP1*, required for expression of the mitochondrial *OLI1* gene in *Saccharomyces cerevisiae*. Curr Genet 24:126–135

Payne MJ, Schweizer E, Lukins HB (1991) Properties of two nuclear pet mutants affecting expression of the mitochondrial *oli1* gene of *Saccharomyces cerevisiae*. Curr Genet 19:343–351

Pelissier P, Camougrand N, Velours G, Guerin M (1995) *NCA3*, a nuclear gene involved in the mitochondrial expression of subunits 6 and 8 of the F_0-F_1 ATP synthase of *S. cerevisiae*. Curr Genet 27:409–416

Perez-Martinez X, Broadley SA, Fox TD (2003) Mss51p promotes mitochondrial Cox1p synthesis and interacts with newly synthesized Cox1p. EMBO J 22:5951–5961

Perez-Martinez X, Butler CA, Shingu-Vazquez M, Fox TD (2009) Dual functions of Mss51 couple synthesis of Cox1 to assembly of cytochrome *c* oxidase in *Saccharomyces cerevisiae* mitochondria. Mol Biol Cell 20:4371–4380

Pierrel F, Bestwick ML, Cobine PA, Khalimonchuk O, Cricco JA, Winge DR (2007) Coa1 links the Mss51 post-translational function to Cox1 cofactor insertion in cytochrome *c* oxidase assembly. EMBO J 26:4335–4346

Pierrel F, Khalimonchuk O, Cobine PA, Bestwick M, Winge DR (2008) Coa2 is an assembly factor for yeast cytochrome *c* oxidase biogenesis that facilitates the maturation of Cox1. Mol Cell Biol 28:4927–4939

Poutre CG, Fox TD (1987) *PET111*, a Saccharomyces cerevisiae nuclear gene required for translation of the mitochondrial mRNA encoding cytochrome *c* oxidase subunit II. Genetics 115:637–647

Prestele M, Vogel F, Reichert AS, Herrmann JM, Ott M (2009) Mrpl36 is important for generation of assembly competent proteins during mitochondrial translation. Mol Biol Cell 20:2615–2625

Rak M, Gokova S, Tzagoloff A (2011) Modular assembly of yeast mitochondrial ATP synthase. EMBO J 30:920–930

Rak M, Tzagoloff A (2009) F1-dependent translation of mitochondrially encoded Atp6p and Atp8p subunits of yeast ATP synthase. Proc Natl Acad Sci USA 106:18509–18514

Rak M, Zeng X, Briere JJ, Tzagoloff A (2009) Assembly of F_0 in *Saccharomyces cerevisiae*. Biochim Biophys Acta 1793:108–116

Rodeheffer MS, Boone BE, Bryan AC, Shadel GS (2001) Nam1p, a protein involved in RNA processing and translation, is coupled to transcription through an interaction with yeast mitochondrial RNA polymerase. J Biol Chem 276:8616–8622

Rödel G (1997) Translational activator proteins required for cytochrome *b* synthesis in *Saccharomyces cerevisiae*. Curr Genet 31:375–379

Rödel G (1986) Two yeast nuclear genes, *CBS1* and *CBS2*, are required for translation of mitochondrial transcripts bearing the 5′-untranslated *COB* leader. Curr Genet 11:41–45

Rödel G, Fox TD (1987) The yeast nuclear gene *CBS1* is required for translation of mitochondrial mRNAs bearing the cob 5′ untranslated leader. Mol Gen Genet 206:45–50

Rödel G, Korte A, Kaudewitz F (1985) Mitochondrial suppression of a yeast nuclear mutation which affects the translation of the mitochondrial apocytochrome *b* transcript. Curr Genet 9:641–648

Sagan L (1967) On the origin of mitosing cells. J Theor Biol 14:255–274

Sanchirico ME, Fox TD, Mason TL (1998) Accumulation of mitochondrially synthesized *Saccharomyces cerevisiae* Cox2p and Cox3p depends on targeting information in untranslated portions of their mRNAs. EMBO J 17:5796–5804

Schatz G (1968) Impaired binding of mitochondrial adenosine triphosphatase in the cytoplasmic "petite" mutant of *Saccharomyces cerevisiae*. J Biol Chem 243:2192–2199

Shingu-Vazquez M, Camacho-Villasana Y, Sandoval-Romero L, Butler CA, Fox TD, Perez-Martinez X (2010) The carboxyl-terminal end of Cox1 is required for feedback-assembly regulation of Cox1 synthesis in *Saccharomyces cerevisiae* mitochondria. J Biol Chem 285:34382–34389

Simon M, Faye G (1984) Organization and processing of the mitochondrial oxi3/oli2 multigenic transcript in yeast. Mol Gen Genet 196:266–274

Smith PM, Fox JL, Winge DR (2012) Biogenesis of the cytochrome bc_1 complex and role of assembly factors. Biochim Biophys Acta 1817:276–286

Soto IC, Fontanesi F, Myers RS, Hamel P, Barrientos A (2012) A heme-sensing mechanism in the translational regulation of mitochondrial cytochrome *c* oxidase biogenesis. Cell Metab 16:801–813

Steele DF, Butler CA, Fox TD (1996) Expression of a recoded nuclear gene inserted into yeast mitochondrial DNA is limited by mRNA-specific translational activation. Proc Natl Acad Sci USA 93:5253–5257

Stock D, Gibbons C, Arechaga I, Leslie AG, Walker JE (2000) The rotary mechanism of ATP synthase. Curr Opin Struct Biol 10:672–679

Szyrach G, Ott M, Bonnefoy N, Neupert W, Herrmann JM (2003) Ribosome binding to the Oxa1 complex facilitates cotranslational protein insertion in mitochondria. EMBO J 22:6448–6457

Tavares-Carreon F, Camacho-Villasana Y, Zamudio-Ochoa A, Shingu-Vazquez M, Torres-Larios A, Perez-Martinez X (2008) The pentatricopeptide repeats present in Pet309 are necessary for translation but not for stability of the mitochondrial *COX1* mRNA in yeast. J Biol Chem 283:1472–1479

Tzagoloff A, Barrientos A, Neupert W, Herrmann JM (2004) Atp10p assists assembly of Atp6p into the F_0 unit of the yeast mitochondrial ATPase. J Biol Chem 279:19775–19780

Tzagoloff A, Crivellone MD, Gampel A, Muroff I, Nishikimi M, Wu M (1988) Mutational analysis of the yeast coenzyme QH2-cytochrome *c* reductase complex. Philos Trans R Soc Lond B Biol Sci 319:107–120

Valencik ML, McEwen JE (1991) Genetic evidence that different functional domains of the *PET54* gene product facilitate expression of the mitochondrial genes *COX1* and *COX3* in *Saccharomyces cerevisiae*. Mol Cell Biol 11:2399–2405

Weber ER, Dieckmann CL (1990) Identification of the *CBP1* polypeptide in mitochondrial extracts from *Saccharomyces cerevisiae*. J Biol Chem 265:1594–1600

Wollman FA, Minai L, Nechushtai R (1999) The biogenesis and assembly of photosynthetic proteins in thylakoid membranes. Biochim Biophys Acta 1411:21–85

Zambrano A, Fontanesi F, Solans A, de Oliveira RL, Fox TD, Tzagoloff A et al (2007) Aberrant translation of cytochrome c oxidase subunit 1 mRNA species in the absence of Mss51p in the yeast *Saccharomyces cerevisiae*. Mol Biol Cell 18:523–535

Zara V, Conte L, Trumpower BL (2009a) Biogenesis of the yeast cytochrome bc_1 complex. Biochim Biophys Acta 1793:89–96

Zara V, Conte L, Trumpower BL (2009b) Evidence that the assembly of the yeast cytochrome bc_1 complex involves the formation of a large core structure in the inner mitochondrial membrane. FEBS J 276:1900–1914

Zara V, Conte L, Trumpower BL (2007) Identification and characterization of cytochrome bc_1 subcomplexes in mitochondria from yeast with single and double deletions of genes encoding cytochrome bc_1 subunits. FEBS J 274:4526–4539

Zassenhaus HP, Martin NC, Butow RA (1984) Origins of transcripts of the yeast mitochondrial var1 gene. J Biol Chem 259:6019–6027

Zeng X, Barros MH, Shulman T, Tzagoloff A (2008) *ATP25*, a new nuclear gene of *Saccharomyces cerevisiae* required for expression and assembly of the Atp9p subunit of mitochondrial ATPase. Mol Biol Cell 19:1366–1377

Zeng X, Hourset A, Tzagoloff A (2007) The *Saccharomyces cerevisiae ATP22* gene codes for the mitochondrial ATPase subunit 6-specific translation factor. Genetics 175:55–63

Chapter 6
Kinetoplast-Mitochondrial Translation System in Trypanosomatids

Dmitri A. Maslov and Rajendra K. Agrawal

Abstract The mitochondrial translation system in trypanosomatids is unique in many ways. There is extensive post-transcriptional editing of its mRNAs, and its ribosomes are among those that contain the smallest sized RNAs but the largest number of proteins. Obvious questions in this field are: (i) how and if these ribosomes distinguish between the pre- and post-edited mRNAs?; (ii) do the mRNA editing machinery, also referred to as the editosome, and the ribosome interact during early stages of translation initiation to facilitate the mRNA recruitment to the ribosomal small subunit?; and (iii) how are these ribosomes structured and assembled? This review article touches each of the above questions, by providing some historical accounts, current understanding, and challenges associated with studying this system.

6.1 Introduction: General Features of the Kinetoplast-Mitochondrial Gene Expression System

Trypanosomatids represent a group of parasitic protists often found in the alimentary tract of insects (e.g., *Crithidia fasciculata* isolated from mosquito) or various tissues of vertebrates (e.g., *Leishmania tarentolae* from gecko) in which they are inoculated by a bite of an infected insect vector (see the following references for

D. A. Maslov (✉)
Department of Biology, University of California – Riverside, Riverside, CA 92521, USA
e-mail: maslov@ucr.edu

R. K. Agrawal
Division of Translational Medicine, Wadsworth Center,
New York State Department of Health, Albany, NY 12201, USA

R. K. Agrawal
Department of Biomedical Sciences, School of Public Health,
State University of New York at Albany, Albany, NY, USA

A.-M. Duchêne (ed.), *Translation in Mitochondria and Other Organelles*,
DOI: 10.1007/978-3-642-39426-3_6, © Springer-Verlag Berlin Heidelberg 2013

reviews: Maslov et al. 2013; Vickerman 1976). Some trypanosomatids are important agents of human diseases including African Trypanosomiasis, also known as
'sleeping sickness' (the case of tsetse fly-transmitted *Trypanosoma brucei*), South
American Chagas disease (reduviid bug-transmitted *Trypanosoma cruzi*) or various forms of leishmaniasis (several species of the genus *Leishmania* transmitted
by sand flies). The group is distinguished from other taxa by a unique arrangement
of mitochondrial DNA, which is found in a highly condensed disk-shape form in
a particular area of a single reticulated mitochondrion of a cell (Simpson et al.
2002; Vickerman and Preston 1976). This region is called a kinetoplast, a historical name reflecting the close proximity of this organelle to the basal body of the
cell's flagellum, an observation interpreted by early researchers as evidence for
involvement of this structure in locomotion. A physical connection between the
kinetoplast and the flagellar apparatus does exist, and it is related to a coordination
of the flagellar and mitochondrial replication processes during the cell division
(Liu et al. 2005). The diameter of the condensed kinetoplast DNA (kDNA) disk
is usually less than 500 nm; however, in a purified form kDNA spreads out into
a large (10–15 μm) round network-like structure composed of several thousand
interlocked circular DNA molecules (minicircles) (Chen et al. 1995). The minicircle size is constant within a network and usually varies from 1.0 to 2.5 kb among
trypanosomatid species (Simpson 1986). Minicircle sequence analyses revealed
that they are composed of a conserved region, shared by all minicircles of a network and containing the origins of DNA replication, and a variable region, defining multiple sequence classes of the network's minicircles (Ray 1989). However,
no minicircle coding function was immediately apparent, and their role, if any,
remained unclear. By contrast, the larger molecules ("maxicircles") present in
~50 copies per network were found to contain homologs of mitochondrial genes
found in other systems (Borst and Hoeijmakers 1979). These included genes for
two ribosomal RNAs (12S and 9S rRNAs), three subunits of cytochrome *c* oxidase
(COI, COII, COIII), apocytochrome *b* (Cyb), a subunit (A6) of F_1F_0 ATPase, several subunits of NADH dehydrogenase, as well as few unassigned reading frames,
but no tRNA genes (de la Cruz et al. 1984; Simpson et al. 1987). Several of the
maxicircle genes had encoded defects, such as frame-shifts (e.g. COII in several
species), missing initiation codons (e.g. Cyb and COIII in *L. tarentolae*), or even
larger regions or the entire genes being unrecognizable (e.g. A6 in *L. tarentolae*,
the entire COIII in *T. brucei*). Some other genes, including COI, ND1, and ND5,
appear to be normal. The universal genetic code is used with the exception of the
UGA codon which is decoded as tryptophan. This is achieved by the cytoplasmic Trp-tRNA with anticodon CCA which is converted into anticodon UCA by
enzymatic modification which takes place upon importation of this tRNA in the
mitochondrion (Alfonzo et al. 1999). The remaining tRNAs operating in the mitochondrion represent imported cytoplasmic tRNA of the same specificity (Alfonzo
and Soll 2009). The exception is the initiator tRNA which is derived from the
cytoplasmic elongation Met-tRNA$_m$ (Tan et al. 2002). It is converted into the initiator fMet-tRNA$_i$ in the mitochondrion by formylation. The fraction of the original
imported tRNA which is left unformylated serves for elongation. The nature of the

tRNA importation machinery remains unclear (see Alfonzo and Soll 2009; Spears et al. 2012 for recent reviews).

6.2 U-Insertion/Deletion RNA Editing

The kDNA mysteries began to unravel with the discovery of RNA editing in 1984 when it was demonstrated that a −1 frameshift on the COII gene is corrected at the post-transcriptional level by insertion of four uridylate residues (Benne et al. 1986). This observation was followed by demonstration of additional editing events including extreme cases of pan-editing, which represents extensive editing occurring throughout the entire length of an mRNA and involves insertions of hundreds of U-residues, as well as multiple deletions of encoded U's (Bhat et al. 1990; Feagin et al. 1988a; Koslowsky et al. 1990). Thus, six G-rich regions earlier found in certain positions in maxicircles (G1 to G6) turned out to represent pan-edited cryptogenes encoding subunits of NADH dehydrogenase and ribosomal protein S12 (RPS12) (Corell et al. 1994; Maslov et al. 1992; Read et al. 1992). The minicircle function was unraveled when it was found that their variable regions encode small (~40 nt) transcripts, termed "guide RNAs" (or gRNAs), which mediate interactions between the editing machinery and pre-edited mRNAs (Blum et al. 1990; Pollard et al. 1990; Sturm and Simpson 1990b). This process included multiple U-insertions and deletions until a perfect match would be achieved between the guide and the corresponding segment of the mRNA (Koslowsky et al. 1991; Yu and Koslowsky 2006). The specificity of a guide (finding the region to be edited, pre-edited region or PER) is defined by a small (~10 bp) sequence match between the 5′ end of a gRNA and a sequence immediately 3′ from PER ("anchor region duplex") (Blum and Simpson 1990). While only one guide is required for small PERs covering only a few nucleotides, multiple guides would be needed for editing of cryptogene transcripts (Corell et al. 1993; Riley et al. 1994). In such cases, the editing starts at the 3′ end of a pre-edited mRNA and extends in the 5′ direction (Decker and Sollner-Webb 1990). This orderly progression of editing is due to sequential creation of the anchor region for each subsequent gRNA by a preceding round of editing until the last gRNA performs the editing near the 5′ end of the message (Maslov and Simpson 1992). As discussed later (Sect. 6.3), the initiation codons are often (but not always) created during this concluding round of editing (Feagin et al. 1988b; Shaw et al. 1988). Although, as a rule, editing extends a few nucleotides upstream from the initiation codon, the 5′ end untranslated region (UTR) sequence of about 20–40 nt in length remains unedited in all studied mRNA. Similarly, the termination codon on the 3′ end of pan-edited templates is often created during the first round of editing, while most of the 3′ end UTR sequence remains unedited. The fact that anchor sequences are relatively short may be responsible for 'mis-editing by mis-guiding,' a situation where annealing of a spurious gRNA to the pre-edited region would result in production of an aberrant editing pattern (Decker and Sollner-Webb 1990; Sturm et al. 1992). Such 'mis-edited' or incompletely edited sequences are often observed at the boundary between the downstream (fully edited) and upstream (pre-edited) sections of a pan-edited mRNA. These junction regions may also represent intermediates of editing by a cognate gRNA (Maslov et al. 1994; Sturm et al. 1992; Sturm and Simpson 1990a). In any case, these aberrant sequences are eventually corrected. When numerous cDNA clones are aligned according to the $3' \rightarrow 5'$ polarity of editing, the consensus edited sequence becomes apparent. This consensus sequence is commonly regarded as 'completely' edited sequence, especially if it contains an ORF that generates an in silico protein product homologous to mitochondrial encoded proteins in other organisms. Such was the case for ribosomal protein S12 (RPS12) (Maslov et al. 1992; Maslov and Simpson 1994; Read et al. 1992), subunits 8 and 9 of NADH dehydrogenase (ND8, ND9), and several other cryptogenes (Corell et al. 1994; Read et al. 1994; Souza et al. 1992, 1993). However, with G3 and G4 cryptogenes the consensus edited sequences encoded polypeptides with no detectable homology outside of kinetoplastids (Thiemann et al. 1994). Moreover, in some cases more than one ORF can be found within the same edited consensus sequence (Corell et al. 1994; Maslov et al. 1999; Read et al. 1992). The conservation among trypanosomatids would then serve as a criterion for the functionality of the deduced edited sequence. A few cases were also described with suggested existence of alternative editing patterns within the stretch of an edited mRNA sequence (Maslov 2010; Read et al. 1994). A partially edited COIII mRNA in *T. brucei* has been proposed to encode a functionally important protein (Ochsenreiter et al. 2008; Ochsenreiter and Hajduk 2006). The mRNAs, such as COI, that contain encoded translatable reading frames, remain unedited.

It was hypothesized early on that the editing is performed by the 'enzyme cascade' mechanism (Blum et al. 1990). First, an endonuclease would cleave a pre-edited mRNA at the site of a mismatch with a gRNA. This would be followed by U-insertions performed by a terminal uridylyl transferase and U-removals by an exonuclease. These sequence alterations at the editing site would result in a match between the mRNA and the gRNA. A ligase activity would then seal the edited mRNA. This model has been supported by direct analyses performed over the last 10–15 years (see Aphasizhev and Aphasizheva 2011a, b for comprehensive reviews). The editing activities are assembled into two major RNA editing core complexes, RECC1 and RECC2, performing U-deletions and U-insertions, respectively. The RECCs, also known as 20S 'editosomes,' share the same structural scaffold made up of six proteins, but also contain different sets of enzymatic and auxiliary components. RECC1 includes REN1 endonuclease, REX1 and REX2 exonucleases, and REL1 ligase, while RECC2 contains REN2 endonuclease, RET2 uridylyl transferase, and REL2 ligase. These editing complexes interact with several additional protein complexes, including the MRP1/2 complex, which may promote gRNA annealing to (at least some) mRNA and is also involved in regulation of mRNA stability; GRBC, also termed MRB1, which stabilizes gRNAs and mediates higher order interactions among various complexes; KPAP1, which performs the mRNA polyadenylation; and a score of additional components that promote various non-catalytic functions related to editing. These complexes are engaged in highly dynamic interactions (Hashimi et al. 2013; Koslowsky 2009), including their interactions with the translation machinery as discussed below.

6.3 Problems for Translation Imposed by Existence of Editing

First discovered in trypanosomatids (Benne et al. 1986), the RNA editing was later found to be widespread among eukaryotes, including mitochondrial and chloroplast systems (Koslowsky 2004). Variable in type (insertional or substitutional) and mechanism, the process of RNA editing apparently involves multiple independent origins, however, the trypanosomatid U-insertion/deletion type has been found only in kinetoplastids (Gray 2012a; Simpson et al. 2006; Simpson et al. 2000), where the extent of sequence rearrangements, such as those occurring in the pan-edited mRNAs, is also unprecedented. Perhaps, because of that, as well as excessive complexity of the process itself, the trypanosomatid type of editing seems rather inefficient. For example, for most pan-edited mRNAs, the steady state includes an abundant population of the pre-edited transcripts and a variety of their partially edited and mis-edited forms (see e.g. Maslov et al. 1992; Neboháčová et al. 2009). In the absence of selection, these immature transcripts lacking a meaningful reading frame would produce nonfunctional translation products. The initiation of translation is therefore expected to involve specific recognition of translationally competent templates to exclude the immature forms.

An attractive hypothesis is that the extensive mRNA editing in trypanosomatids not only poses a problem, but also contributes to the solution, for the edited mRNA recognition. It is conceivable that the necessary activator *cis*-elements are created during the final rounds of editing at the 5′-end. Since the editing starts at the 3′-end and ends at the 5′-end of mRNA, the completion of the last editing round would attest that the rest of mRNA has been edited. Such a signal could be analogous to the presence of bacterial Shine-Dalgarno (SD) sequence, which participates in recruitment of the small ribosomal subunit (SSU) by interacting with the anti-SD sequence present at the 3′ end of the SSU rRNA. However, It should be noted that there is no single-stranded segment at the 5′ end of the 9S rRNA that could even be considered as a candidate to serve a function analogous to bacterial anti-SD sequence. Besides, no conserved sequence elements that would potentially interact with the SSU 9S rRNA, or with any other component of the SSU, have been found in the vicinity of the 5′ end in trypanosomatid mitochondrial mRNAs. It has been proposed that the translation is initiated at the 5′-most AuG created by editing (or AuA, as well as Auu, in some cases), as can be suggested by considering the pan-edited COIII mRNA in *T. brucei*, or 5′-edited COIII and Cyb mRNAs in *L. tarentolae* (Feagin et al. 1988b; Shaw et al. 1988). However, this is generally not the case, which is exemplified by the 5′-edited Cyb mRNA in *T. brucei*, with an out-of-frame AUG found in the 5′-untranslated (and unedited) region

(a)

```
DNA      GTTAAGAATAATGGTTATAAATTTTATATAAA A  G    CG  G AGA     A  A   ...
RNA-ed   GUUAAGAAUAAUGGUUAUAAAUUUUAUAUAAAuAuGuuuCGuuGuAGAuuuuuAuuAuuu...
Protein                                      M  F  R  C  R  F  L  L  F   ...
```

(b)

```
DNA      TAAACATATATAATGTATTAGATTAAAAGTAA G   G   A GA     G TTTTC   ...
RNA-ed   UAAACAUAUAUAAUGUAUUAGAUUAAAAG*AAuGuuuGuuuAuGAuuuuuGuUUUUCuuu...
Protein                                      M  F  V  Y  D  F  C  F  S  F ...
```

Fig. 6.1 The initiation codon selection problem. The 5′-terminal gene sequences (*DNA*) and edited RNA sequences (*RNA-ed*) and the deduced *N*-terminal amino acid (*protein*) sequences of *T. brucei* Cyb (**a**) and *L. tarentolae* ND8 (**b**) are shown. Encoded nucleotides are shown with *upper case*, uridylates inserted by the editing process with lower case U's, the deleted U-residue with an *asterisk*. The inferred methionine initiation codon AuG is shown in *bold*. The upstream out-of frame AUG (**a**) and the in-frame AUG (**b**) are highlighted in *yellow*

(Fig. 6.1a). Obviously, the initiation at such codons should be avoided. Even when an upstream codon is in-frame, as in the pan-edited ND8 mRNA of *L. tarento-lae* (Fig. 6.1b), it should not be used for initiation because the same codon is also present in the pre-edited and partially edited molecules that do not encode a functional product.

Another difficulty is that, in some cases the proper initiation codon is not created by editing but is encoded. An example is the COII mRNA, in which the editing is limited to small internal region (Benne et al. 1986; Shaw et al. 1989). Thus, the same encoded 5′ AUG codon should be used for initiation in the edited mRNA but avoided in the pre-edited mRNA. Thus, the problem of recognition of the fully edited mRNA seems inseparable from recognition of a proper initiation codon. However, both problems can be solved at once with the help of mRNA-specific translation activators, similar to those which operate in yeast mitochondria (Fox 2012; Herrmann et al. 2013). In this system, protein Pet309 activates translation of cytochrome c oxidase subunit 1 (*COX1*), Pet 111-*COX2* mRNA, and Cbp1-apocytochrome b (*COB*) mRNA. These activators are assembled into a large (0.9 MDa) complex interacting with the inner membrane-bound mitochondrial ribosomes (henceforth referred to as mitoribosomes) (Krause et al. 2004; Naithani et al. 2003). Evidence is accumulating that a somewhat similar system operates also in trypanosomatids.

6.4 Mitochondrial Translation Products

Early attempts to detect translation in kinetoplast-mitochondria were based primarily on the presumption that this system, as many other eukaryotic mitochondrial systems, would be sensitive to chloramphenicol; however, the experimental results obtained were controversial and inconclusive (see Maslov and Agrawal 2012 for a recent review). On the hindsight, after the realization that the mitochondrial system of trypanosomatids is one of the most highly diverged from the mammalian or fungal systems (Gray 2012a; Gray 2012b), the presumptions made on chloramphenicol sensitivity were rather naïve. Moreover, as the maxicircle DNA sequence data became available, resistance to this inhibitor was predicted by the analysis of the large ribosomal subunit (LSU) 12S rRNA region that participates in formation of the peptidyl transferase center in the mitoribosome (de la Cruz et al. 1985b; Eperon et al. 1983).

The very existence of extensively edited mRNAs, encoding seemingly functional and evolutionarily conserved polypeptides, has usually been regarded as the strong circumstantial evidence for the translation of such templates. Yet, a direct proof in the form of identified mitochondrially encoded proteins had been missing. Although the purified respiratory complexes have been shown to be enzymatically active, all identifiable components were found to be nuclear encoded subunits (Priest and Hajduk 1992; Speijer et al. 1996, 1997). That was puzzling because mitochondrial translation products represent indispensible subunits of cytochrome c oxidase (COI–COIII), cytochrome bc_1 (Cyb) and ATP synthase (subunit 6 or A6). The unifying property of the kinetoplast-mitochondrially encoded polypeptides is their extreme hydrophobicity which severely impedes analysis of such products due to poor efficiency of electrophoretic separation, staining, and proteolytic digestion procedures, as well as by losses due to aggregation (Breek et al. 1997). This hydrophobicity feature was utilized in a special 2D gel system to separate the mitochondrial components of cytochrome c oxidase and cytochrome bc_1 complexes from the nuclear encoded subunits. In this gel system, the hydrophobic mitochondrial polypeptides occupy a position off the main diagonal of the gel, while the polypeptides of the nuclear origin align on the main diagonal (Marres and Slater 1977) (Fig. 6.2).

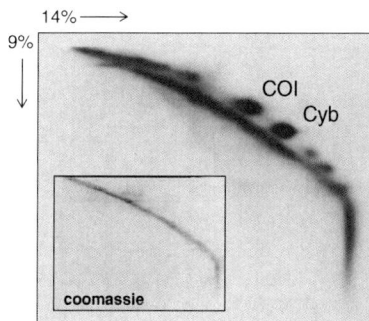

Fig. 6.2 The two-dimensional gel system employed to detect hydrophobic products of mitochondrial translation in trypanosomatids. The gel shown represents analysis of the ^{35}S-labeled products of protein biosynthesis in *T. brucei*, when cytosolic translation was inhibited with cycloheximide. Each gel dimension represents separation in a standard Laemmli-type Tris–glycine-SDS with the polyacrylamide concentration as shown. The two major ^{35}S-labeled spots represent COI and Cyb polypeptides. The identity of other labeled off-diagonal spots is only tentatively known: the faster migrating polypeptides likely represent COII and COIII, while A6 is expected to be found in an off-diagonal position closer to the start. The *inset* shows the corresponding Coomassie-stained gel

In the analysis of purified cytochrome *c* oxidase from *L. tarentolae*, a series of poorly stainable diffused spots were observed off the main diagonal (Horváth et al. 2000b). Several of these spots contained material reactive with antibodies against COI, suggesting the presence of various aggregated forms of this hydrophobic polypeptide. A partial *N*-terminal sequence derived from two most abundant spots confirmed that both represented the COI polypeptide. Short sequences matching the predicted COII and COIII amino acid sequences were derived from other spots. Since all three mitochondrially encoded subunits are supposed to be present in equimolar quantities within the intact cytochrome *c* oxidase complex, the higher abundance of the COI spots suggests that the other two polypeptides are prone to variable degrees of aggregation. By a similar analysis of cytochrome bc_1 complex from the same organism, monomeric and dimeric forms of Cyb were identified (Horváth et al. 2000a). The *N*-terminal amino acid sequence of this polypeptide is translated from a 5′-edited region of the mRNA, representing the first direct evidence that an edited mRNA can produce a functionally translatable template. The sequencing required an *N*-terminal deblocking, indicating the presence of formylmethionine at the *N*-terminus. However, mass spectrometry was largely proved to be inefficient for obtaining additional protein sequence data, as only a single peptide (FAFYCER) was detected by two independent mass spectrometry analyses of the off-diagonal Cyb material obtained from *T. brucei* (Nebohácová et al. 2004); (Aphasizhev, Škodová, Maslov, unpublished observations). It is conceivable that the extreme hydrophobicity of these polypeptides interferes with tryptic digestion and peptide recovery.

In vivo synthesis of COI, COII, and Cyb in *L. tarentolae* was also studied, using 2D gels by incorporation of ^{35}S-labeled amino acids in the presence of cycloheximide to inhibit cytosolic translation (Horváth et al. 2002). The synthesis was found to be insensitive to 100 ug/ml chloramphenicol, gentamycin, paromomycin, lincomycin, hygromycin, and tetracycline, but was sensitive to puromycin, a mimic of the peptidyl tRNA's CCA-end and a universal inhibitor of translation. The synthesis in isolated mitochondria was linear for almost 2 h and, surprisingly, did not depend on exogenous energy or amino acids. By using Blue Native (BN) gel electrophoresis it was shown that the synthesized products were concurrently incorporating into the respective respiratory complexes. A similar pattern of (in) sensitivity to antibiotics has been observed for the in vivo COI and Cyb synthesis in *T. brucei* (Nebohácová et al. 2004). RNAi-induced ablation of RET1 was concomitant with the inhibition of the COI and Cyb synthesis (Nebohácová et al. 2004). RET1 is the enzyme required for biogenesis of gRNAs and its down-regulation entails a decrease in gRNA abundance, which in turn affects editing (Aphasizhev et al. 2002, 2004). The gRNA-mediated mechanism could not, however, explain the puzzling effect on the COI synthesis that utilizes an unedited mRNA template. Moreover, the steady-state abundance of COI mRNA is not affected by RET1 RNAi. More recently, it has been shown that effect of RET1 down-regulation on the COI synthesis is largely caused by a collapse of the COI mRNA's 3′-end poly(A/U) tail, which is required for translation in this system (Aphasizheva and Aphasizhev 2010) (see also below).

The presence of a labeled material in the main diagonal of the 2D gels and its migration close to the running front in BN gels indicated the presence of a heterogeneous population of nascent non-hydrophobic polypeptides (Fig. 6.2). Other polypeptide synthesis properties, such as the time course, resistance to antibiotics, and in *T. brucei* sensitivity to RNAi, have paralleled that of COI and Cyb (Horváth et al. 2002; Neboháčová et al. 2004), suggesting that the diagonal material also represents products of mitochondrial translation. It still remains to be demonstrated whether these polypeptides are derived from pre-edited or partially edited mRNAs and whether they represent any functional product(s). The potential implication of such a product would be the lack of precision during translation initiation, involving initial selection of a functional translation template or an initiation codon, or frequent frame-shiftings during translation elongation.

The remaining polypeptides, including RPS12, A6, and subunits of NADH dehydrogenase complex, are yet to be found. Among these, A6 has recently been putatively identified by tracing the incorporation of a labeled mitochondrial translation product in the ATPase complex (Škodová, Maslov, unpublished results). There is little doubt that all or most of the maxicircle genes are expressed, at least during some stages of the natural life cycle. Mitochondrial translation is indispensible even in the bloodstream form of *T. brucei*, a stage without the electron transport chain (Cristodero et al. 2010). However, the lack of selective pressure in culture may entail the loss of functionality for certain genes. This is exemplified by the *L. tarentolae* UC and *Phytomonas serpens* 1G, the strains, which had lost the productive editing of at least few NADH dehydrogenase subunits (ND8, ND9 and others), and by *Leishmania donovani* 1S LdBob, where insertional inactivation of the maxicircle genes for ND1 subunit has been found in addition to the loss of ND3, ND8, and ND9 editing (Maslov et al. 1999; Neboháčová et al. 2009; Thiemann et al. 1994). However, this scenario does not apply to RPS12 mRNA, which is productively edited in all species examined, and therefore must be indispensably and consistently present in the mitoribosome (Maslov 2010; Maslov et al. 1992, 1999; Maslov and Simpson 1994; Nebohácová et al. 2009; Read et al. 1992). Paradoxically, RPS12 has been notoriously missing among components of the mitoribosomes in the recent proteomics analyses (Aphasizheva et al. 2011; Maslov et al. 2007; Zíková et al. 2008). RPS12 is a relatively small (10 kDa), basic (pI = 9.0) and not particularly hydrophobic polypeptide, but seems to carry an unexpected feature that has eluded its detection by the standard mass spectrometry procedures.

6.5 Mitochondrial Ribosomes

The early attempts to isolate kinetoplast ribosomes by analysis of A260/A280 absorbance profiles or by cosedimentation with nascent polypeptides, were not productive, apparently due to higher abundance of other high molecular weight complexes, including free minicircles (reviewed in Maslov and Agrawal 2012). Although the 9S and 12S rRNAs could be found in the large (>30S) heterogeneous complexes and the ribosome-like particles were detected by transmission electron microscopy (Scheinman et al. 1993), the exact compositional and structural nature of the mitoribosome remained elusive. A subsequent analysis performed using isolated mitochondria from *L. tarentolae* suggested that one of the major obstacles in isolation of a pure population of the monosomes (a monomeric ribosome including both its subunits) was that it represented only a minor fraction that was obscured by much more abundant other ribosomal ribonucleoprotein (rRNP) complexes (Maslov et al. 2006). One of them, termed SSU* complex with a sedimentation coefficient value of 45S, was among the most abundant rRNPs in the mitochondrial lysates (Maslov et al. 2007). This complex was found to be prone to dimerization, but was rather stable otherwise, withstanding mechanical impact by pelleting and resuspension in the presence of 0.5 M KCl. The SSU* complex contained only the 9S rRNA and a score of SSU ribosomal proteins, but none of the LSU rRNA or proteins. Several of the components represented pentatricopeptide repeat (PPR) proteins, which are known to participate in various aspects of RNA biogenesis in cellular organelles (Delannoy et al. 2007; Schmitz-Linneweber and Small 2008). There were also proteins with no discernible functions, including several unusually large proteins (~200 kDa) that have not been encountered in mitoribosomes from other organisms. The combined molecular weight of nearly 40 stably associated proteins was in excess of 2 MDa. Transmission electron microscopy (TEM) showed that the 45S SSU* complexes possess an unusual morphology showing two large lobes. The overall morphology of one of these lobes resembled that of a typical SSU. The current hypothesis is that these complexes represent a heterodimer composed of the 25–30S SSUs (also detectable as individual SSUs present in a low amount) and a large protein complex, which contained most of the unusual proteins identified with the 45S SSU* complex. All components found in the *L. tarentolae* SSU* complex have also been found in the S17-tagged complexes in *T. brucei,* suggesting a strong similarity in the overall organization of the translational apparatus in these two species. Direct analyses confirmed that in *T. brucei*, most of the SSUs of the mitoribosomes are tied within the abundant SSU* complexes (Maslov, Ridlon, Škodová, unpublished). Free LSU are also abundant, while assembled ribosomes (SSU-LSU monosomes) represent a relatively minor population of particles. However, the presence of such an unusual complex has not been reported in other systems. A further structural and functional characterization of the SSU* complex will shed light on its role in the kinetoplast mitochondria.

Individual LSU was detected as a 40S complex, which was readily affected by the ionic conditions during mitochondrial lysis, yielding products that varied in their sedimentation coefficient values in the 35–40S range (or even larger). This property apparently reflected the existence of some loosely associated components within the LSU or its interactions with other proteins, but this aspect has not been investigated further. Overall, the *L. tarentolae* LSU complexes were less stable, as compared to the 45S SSU* complexes, and were also prone to dimerization.

The monosomes were initially identified as the 50S shoulder of the more abundant 45S SSU* peak (Maslov et al. 2006). The resedimentation of this material allowed for revealing a distinct class of particles containing both SSU and LSU rRNAs, as expected. The conventional TEM showed that most particles in this fraction display the morphology of a typical bacterial ribosome. However, a more definite proof that the discovered 50S rRNP represents a monosome of the *Leishmania* mitoribosome came from a three-dimensional cryo-electron microscopy (cryo-EM) analysis of these particles (Sharma et al. 2009).

6.6 Three-Dimensional Cryo-EM Structure of the *Leishmania* mitoribosome

The rRNAs in *Leishmania* mitoribosomes are among the smallest. As described in the previous section, it has been a daunting task to purify these mitoribosomes from *L. tarentolae*, as there are several species of the mitoribosome complexes, in various combinations and stoichiometry of its two subunits. Furthermore, the 50S monosomes, i.e., the minimal functional units made of one each of its two subunits (SSU and LSU), are present in relatively low abundance, and are unusually protein-rich as compared to their bacterial or cytoplasmic counterparts. The *Leishmania* mitoribosomeSSU contains a 610 nucleotide-long 9S rRNA, while its LSU contains a 1,173 nucleotide-long 12S rRNA (de la Cruz et al. 1985a, b). When compared to bacterial rRNAs, these numbers account for ~60 % less rRNA content (Sharma et al. 2009). Because of the small size of rRNA and large number (>100) of ribosomal proteins (r-proteins), all r-proteins are less likely to interact directly with the rRNA-containing central core of the molecule. This compositional disproportionality suggests that there must be a great deal of quaternary protein–protein interactions. This situation apparently makes the isolation of homogeneous 50S monosomes in sufficient quantities that could be suitable for structural studies very challenging. These mitoribosomes are expected to be present in tight association with the mitochondrial membrane that makes them even more susceptible to isolating conditions, thereby further contributing to compositional heterogeneity (Maslov et al. 2006, 2007).

The first and only cryo-EM structure determined for the *Leishmania* 50S monosome at ~14 Å resolution was obtained after applying extensive classification schemes on a relatively heterogeneous single-particle dataset (Sharma et al. 2009). The structure (Fig. 6.3) revealed that, like any other known ribosome, the *Leishmania* mitoribosome is composed of two unequally sized subunits, the ~28S SSU and a ~40S LSU (Maslov et al. 2006, 2007), with intact functionally important and conserved regions of the rRNAs that are known to be involved in mRNA decoding on the SSU (Ogle et al. 2001) and peptide-bond formation on the LSU (Nissen et al. 2000) in the bacterial ribosome. The cryo-EM map also showed that the 50S monosome possesses most of the structural features that are typical for bacterial ribosomes, but the overall structure was found to be highly porous, with an overall diameter (~245 Å), slightly smaller than the bacterial 70S ribosome (~260 Å) and significantly smaller than the mammalian 55S mitoribosome

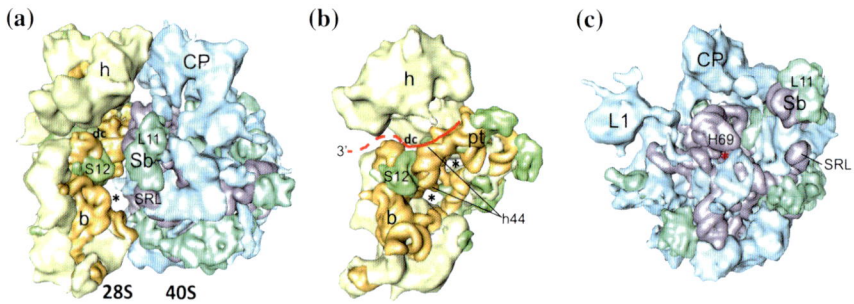

Fig. 6.3 Cryo-EM structure of the *Leishmania tarentolae* 50S mitoribosome. **a** RNA–protein segmented map of the mitoribosome displayed in a side view, with the 28S SSU on the *left* and the 40S LSU on the *right* side. **b** Structure of the SSU shown from the SSU-LSU interface side. **c** Structure of the LSU shown from the LSU-SSU interface side. Mito-specific MRPs of SSU and LSU are colored *yellow* and *blue*, respectively; conserved ribosomal proteins [here "conserved" refers to bacterial homologs present in the *L. tarentolae* mitoribosome] of SSU and LSU are colored *green* and *aquamarine*, respectively; and rRNAs of SSU and LSU are colored *orange* and *purple*, respectively. Landmarks of SSU: *b* body, *dc* mRNA decoding site, *h* head, *h44* SSU rRNA helix 44, *pt* platform, *S12* protein S12. The area marked with *asterisks* (*) in **a** show the absence of density due to truncation of SSU rRNA helix 44 in the mitoribosome as compared to that in the bacterial SSU rRNA. *Two asterisks* in **b** point to the areas where tunnel-like features are formed due to absence of bacterial SSU rRNA segments (see Sharma et al. 2009; Sharma et al., this volume for comparison with structures of other ribosomes). The putative mRNA path is indicated in *red*. Landmarks of LSU: *CP* central protuberance, *L1* protein L1 stalk, *Sb* Stalk base or L11 region, *SRL* α-sarcin-ricin stem loop (or helix 95 of the LSU rRNA, also see Fig. 6.4), *H69* LSU rRNA helix 69. The *red asterisk* below and behind H69 in **c** indicates the general location of the peptidyltransferase center

(~320 Å). Similar to the situation with mammalian mitoribosomes, the rRNAs of the *Leishmania* mitoribosome are largely shielded by proteins. The cryo-EM structure also allowed partial delineation of the secondary and tertiary structures of the *Leishmania* mito-rRNAs, which are subject to further refinement as the higher resolution structure becomes available (for recent reviews, see, Agrawal et al. 2011, Agrawal and Sharma 2012). In the next two paragraphs we briefly describe the unique features of the *Leishmania* mito-SSU and LSU, respectively.

Even with dramatically reduced size of the 9S rRNA, the SSU portion of the cryo-EM map displays all the typical features of an SSU structure, relating closely to its bacterial counterpart and highlighting the importance of the overall architecture of SSU in the ribosome function. Molecular analysis of the cryo-EM map revealed that some of the missing bacterial rRNA segments are replaced in part by the *Leishmania* mito-specific proteins (LMSP) in three-dimensional space. The level of compensation of missing bacterial rRNA segments by LMSPs is much more extensive in *Leishmania* mito-SSU compared to that in the mammalian mito-SSU (Sharma et al. 2003). Despite an extensive compensation by LMSPs the absence of several rRNA segments leads to formation of a large number of tunnel-like gaps, both in the head and main body regions of the SSU (Fig. 6.3b;

Sharma et al. 2009). The critical nucleotides, known to be directly involved in the decoding process, are retained in SSU rRNA helix 44 (all rRNA helices are cited according bacterial rRNA numbering), despite a dramatic reduction in size of the helix 44 that creates a large gap in the lower inter-subunit bridge forming region (Fig. 6.3a; Sharma et al. 2009). The *Leishmania* mito-SSU possesses more than 50 proteins. This estimate is based on an exhaustive mass spectrometric analysis of mitoribosomes from both *Leishmania* and its close relative *Trypanosoma* (Aphasizheva et al. 2011; Zíková et al. 2008). Of these proteins only 10 have been found to be homologous to bacterial SSU proteins, including proteins S5, S6, S8, S9, S11, S12, and S15–S18. In addition to occupying the majority of solvent exposed side of the subunit, proteins occupy most of the interface side of the SSU head, thereby significantly altering the composition of mRNA and tRNA paths in the *Leishmania* mito-SSU (Sharma et al. 2009) as compared to those in bacterial and other organellar ribosome structures (Manuell et al. 2007; Sharma et al. 2003, 2007) that have been studied so far (also see Agrawal et al. 2011).

As is the case with SSU, the LSU portion of the cryo-EM structure of the *Leishmania* 50S monosome carries structural features of a typical eubacterial LSU, i.e., including a central protuberance, with two stalk-like features on either side of the central protuberance (Fig. 6.3c). The greatly reduced rRNA content in the mito-LSU is primarily due to the absence of the segments that constitute the peripheral regions of the bacterial LSU (Schuwirth et al. 2005; Selmer et al. 2006) and segments that form the inner lining of the nascent polypeptide chain exit tunnel (Nissen et al. 2000), suggesting that the topology of the nascent polypeptide exit tunnel in the *Leishmania* mito-LSU is dramatically remodeled to facilitate perhaps the co-translational insertion of the mitochondrially encoded polypeptides into the mitochondrial inner membrane. Among the structurally most conserved regions of the LSU rRNA are the peptidyl transferase center (PTC) and the α-sarcin/ricin stem-loop (SRL) regions. However, the SLR region shows a relative spatial shift toward the PTC, possibly due to the truncation of two LSU rRNA helices, helix 89 and helix 91, in the *Leishmania* 12S rRNA (Fig. 6.4).

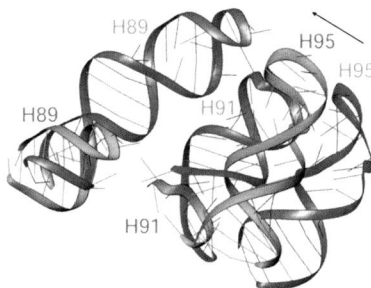

Fig. 6.4 Depiction of the spatial shift of LSU rRNA helix 95 (H95, SRL) in the *L. tarentolae* mitoribosome. Most of the bacterial 23S rRNA (*gray*) helix 91 (H91) and a major portion of helix 89 (H89) are absent in the 12S mito-rRNA (*pink*), which might have caused the spatial shift of the SRL closer to the peptidyl-transferase center (PTC) (see Fig. 6.3c for positions of SRL and PTC on the mito-LSU). Modified from Supplementary Information in Sharma et al. (2009)

Thus, the SRL and the PTC regions in the *Leishmania* LSU are closer by ~25 Å, as compared to that in the bacterial ribosome. The rRNA helices 89 and 91 seem to provide direct communication between the PTC and SRL regions during the translation elongation cycle. Relative closeness of PTC and SRL in *Leishmania* mito-LSU suggests that there would be substantial alterations in size and structural organizations of *Leishmania* mitochondrial translation elongation factors. Like its SSU, the *Leishmania* mito-LSU is also estimated to possess a large number, ~77, of ribosomal proteins (Aphasizheva et al. 2011; Zíková et al. 2008). Of these only 11 proteins are homologous to bacterial LSU proteins, including L2–L4, L9, L11–L17, L20–L24, L27–L30, and L33. Similar to its SSU, additional proteins in *Leishmania* mito-LSU occupy some of the functional regions on the interface side, including the tRNA corridor between the ribosomal aminoacyl (A site) and peptidyl (P) sites (Sharma et al. 2009), suggesting that the dynamics of tRNA movement on the *Leishmania* mitoribosome will be substantially different from that on the bacterial ribosome.

6.7 Pentatricopeptide Repeat Proteins and Their Role in Translation

Pentatricopeptide repeat (PPR) proteins were originally discovered in plants where they are present in highest abundance among all living systems. So far, more than 400 family members of PPR proteins have been discovered by genome analyses (Lurin et al. 2004; Schmitz-Linneweber and Small 2008). The PPR domain represents a rather diverse 35 amino acid sequence organized as a set of tandem repeats. PPR proteins have been ascribed multiple roles centered on various functions of RNAs in the cellular organelles (transcription, splicing, editing, translation of mRNA), involving some specific interactions with a stretch of the RNA sequence (Lightowlers and Chrzanowska-Lightowlers 2008; Rackham and Filipovska 2012). Although found in many eukaryotes, these proteins are considerably less numerous outside of plants. For instance, only seven PPR proteins, including a component of the mitochondrial RNA polymerase and mitoribosomal protein MRPS27, have been found in mammals (Davies et al. 2009). Yet, this is probably an underestimation since some of these proteins might have escaped their detection due to the degeneracy of the PPR motifs.

By any measure, trypanosomatids stand out among other eukaryotes as they carry about 40 members of the PPR family, and represent the second largest group after plants (Aphasizheva et al. 2011; Mingler et al. 2006; Pusnik et al. 2007; Pusnik and Schneider 2012). Functions of several of these PPR proteins have been investigated. Six PPR proteins are associated with the previously described (Sect. 6.5) SSU* complex in *L. tarentolae* (Table 6.1). Six different PPR proteins, labeled TbPPR2–TbPPR7, are required for stability of the mitochondrial rRNAs in *T. brucei*, while the other two PPR proteins in the same series (TbPPR1 and TbPP8) are yet to be assigned a function (Pusnik et al.

Table 6.1 Mitochondrial ribosomal PPR (and putative PPR) proteins in *T. brucei*

Gene ID	Name	Size (kDa)	Probability PPR (%)	Number of PPR units	Orthologs found in *Leishmania* SSU*	Inferred putative localization	Identified by Pusnik et al. (2007)	Identified by Zíková et al. (2008)
Tb927.11.5500 (Tb11.02.3180)	KRIPP1	95.0	28.3	3	LmjF.24.0830	SSU*, monosomes	+	+
Tb927.1.2990	KRIPP2	116.0	100.0	8		SSU in monosomes	TbPPR2	+
Tb927.1.1160	KRIPP3	58.5	100.0	7		Monosomes	TbPPR3	+
Tb927.10.13200 (Tb10.389.0260)	KRIPP4	102.2	100.0	6		SSU*	TbPPR4	
Tb927.10.380 (Tb10.70.7960)	KRIPP5	39.7	98.4	2		LSU in monosomes	TbPPR5	+
Tb927.3.4550	KRIPP6	74.8	100.0	8		LSU	TbPPR6	
Tb927.4.4720	KRIPP7	59.1	100.0	9		LSU	TbPPR7	
Tb927.3.5240	KRIPP8	66.7	48.1	3	LmjF.29.0430	SSU*, monosomes	+	+
Tb927.8.3170	KRIPP9	90.5	63.0	3		LSU in monosomes	+	+
Tb927.8.4860	KRIPP10	76.6	100.0	10		SSU in monosomes	+	+
Tb927.8.6040	KRIPP11	26.9	100.0	4		SSU*	+	
Tb927.4.2790	KRIPP12	62.9	94.2	8		LSU in monosomes	+	
Tb927.11.11470 (Tb11.01.3300)	KRIPP14	32.1	75.2	5	LmjF.28.2180	SSU*, monosomes		
Tb927.7.2490	KRIPP15	80.0	100.0	11		SSU*		
Tb927.9.5280	KRIPP16	32.0	0.5	2	LmjF.15.1400	SSU*, monosomes		
Tb927.10.11820 (Tb09.160.3800)	KRIPP17	39.2	75.1	5		SSU*		+
Tb927.6.610 (Tb10.26.0050)	KRIPP18	65.1	25.1	3		SSU*		
Tb927.7.5280	KRIPP19	220.2	1.8	7		LSU in monosomes		
Tb927.3.3050	KRIPP20	112.1	4.2	14		LSU		

(continued)

Table 6.1 (continued)

Gene ID	Name	Size (kDa)	Probability PPR (%)	Number of PPR units	Orthologs found in *Leishmania* SSU*	Inferred putative localization	Identified by Pusnik et al. (2007)	Identified by Zíková et al. (2008)
Tb927.2.4460	KRIPP21	67.9	73.4	8		LSU		
Tb927.6.2080	KRIPP22	46.0	3.2	4	LmjF.30.0650	SSU*		
Tb927.11.4650 (Tb11.02.2250)	KRIPP13	167.9	N/D	N/D		Monosomes		
Tb927.5.1790	PPR29	72.1	N/D	N/D	LmjF.15.0410	SSU*, monosomes		

The table is largely based on the proteomics analysis of mitochondrial ribosomal RNP complexes by Aphasizheva et al. (2011). Complexes were isolated by affinity chromatography, after TAP-tagging of ribosomal protein S17 (which is part of the free SSU, monosomes and SSU* complexes) and L3 (which is part of the free LSU and monosomes). Preferential localization of ribosomal proteins was tentatively inferred from its occurrence and a relative frequency of the peptide hits in these two fractions. Thus, a protein is attributed to SSU* if it occurred only in the S17-TAP fraction; to free LSU—if only in L3-TAP fraction; to free LSU—if only in L3-TAP fraction; to SSU in monosomes—in both fractions but more frequently in L3-TAP. When a protein was also found in *L. tarentolae* SSU* complexes (Maslov et al. 2007), as well as S17-TAP and L3-TAP fractions, it was attributed to SSU* and monosomes. Search for PPR motifs was performed using TPRpred. It should be noted that the low probability PPR scores may at least be partially due to the intrinsic degeneracy of the PPR units, and the evolutionary divergence of the trypanosomatid lineage. In the work of Pusnik and Schneider (2012), the described PPR protein (TbPPR9) had a low probability and is no longer recognizable as PPR by the current version of TPRpred. The same is the case with KRIPP13 and PPR29 in this table (*N/D* not determined)

2007). Several PPR proteins (including TbPPR2, 3 and 5) were later identified as components of *T. brucei* mitoribosomes isolated by co-fractionation with affinity-tagged ribosomal proteins S17 and L3 (Zíková et al. 2008). A recent comprehensive proteomics analysis of mitoribosomes obtained by a similar approach resulted in expanding this list to more than 20 PPR proteins, termed KRIPPs for *k*inetoplast *ri*bosomal *P*PR-containing *p*roteins (Aphasizheva et al. 2011).

Participation of PPR proteins in the formation and functioning of the mitoribosome are not the only roles demonstrated for these proteins. For example, KRIPP1 (a component of the SSU* complex, Table 6.1), was also discovered within multiprotein complexes implicated in gRNA binding (GRBC or MRB1 complex), edited mRNA stability (MERS1) and polyadenylation (KPAP1) (Ammerman et al. 2012; Weng et al. 2008). Interestingly, two non-PPR members of the 45S SSU* complexes, LmjF.27.0630 and LmjF.33.2510, have also been found in association with these three complexes (Weng et al. 2008). Additional PPR components of the *T. brucei* ribosomes with a putative localization in the SSU (KRIPP2, 5, 8, and 9) have also been found in the KPAP1 poly-A polymerase complex and other complexes (Aphasizheva et al. 2011). This property of individual PPR proteins to serve as versatile building blocks in the formation of multiple macromolecular complexes may allow them to facilitate interactions among mRNA processing, editing, and translation machineries (Aphasizhev and Aphasizheva 2011a, b).

Among non-ribosomal PPR proteins, KPAF1 (TbPPR1, Tb927.2.3180) and KPAF2 (Tb927.11.14380) were found to induce formation of the long A/U heteropolymer tails by the KPAP1 poly-A polymerase and RET1 terminal uridylyl transferase which occurs upon completion of mRNA editing (Aphasizheva et al. 2011; Etheridge et al. 2008). These and other observations, as well as the available information from other mitochondrial systems, led the authors (Aphasizhev and Aphasizheva 2011a) to hypothesize that the processing of primary polycistronic transcript is coupled with a binding of a specific PPR protein to the mRNA's 3′ end. This would also serve to stabilize the mRNA 3′ end prior to polyadenylation. Indeed, TbPPR9 (Tb927.11.16250) was shown to stabilize the COI and COII mitochondrial mRNAs, but not the other mRNAs that were investigated (Pusnik and Schneider 2012). This 3′-end binding *trans*-factor may dissociate following the first editing event. Another proposed role for PPR proteins was to mediate the processes of editing and a long poly(A/U)-tail formation (Aphasizhev and Aphasizheva 2011a). Within the frame of this hypothesis, the 5′-end sequence created after the completion of editing would be recognizable by a specific PPR protein, which would in turn facilitate the recruitment of the KPAF1/KPAF2/KPAP1/RET1 complex to the 3′ end. The resulting acquisition of a long poly(A/U) tail by the mRNA is required for the recognition of the latter by the mitoribosome (Aphasizheva et al. 2011). While the putative 3′ and 5′ binding factors are yet to be identified, the aforementioned stabilization mechanism by TbPPR9 is also unknown. In any case, the translational activation of mRNA would involve a 3′ end ↔5′ end cross-talk mediated by PPR proteins.

6.8 Proposed Involvement of Various RNP Complexes in Translation

The available data indicate that translation in trypanosomatid mitochondria is a highly dynamic process. First of all, the assembled 50S monosomes constitute only a minor fraction of all ribosomal particles. The majority of the LSUs are found as individual particles, seemingly associated with the inner membrane of mitochondria. Most of the SSUs are sequestered within the 45S SSU* complexes which are localized in the mitochondrial matrix. Therefore, it is conceivable that the translating mitoribosomes assemble around the template and disassemble upon completion of translation. In-frame of this hypothesis, the SSU* complex might be involved in recognition of long polyA/U-tailed translation-competent mRNAs. This may be facilitated by the PPR proteins of the SSU* complex or/and by its other components in alliance with unknown factors/complexes. The mRNA recognition function probably resides within the non-SSU component of the SSU* complex, i.e., the large protein complex (described under Sect. 6.5), which would dissociate from the actual SSU portion of the complex and disassemble once its putative function to recruit the fully edited mRNA to the SSU has been accomplished. The outcome of these interactions would be the formation of an early-stage initiation complex of the SSU with the long-tailed mRNA. Indeed, by application of a rapid cryogenic cell rupture combined with affinity pull-down of TAP-tagged ribosomal complexes, it was demonstrated that the long-tailed edited RPS12 mRNA preferentially associated with the S17-TAP tagged material (Aphasizheva et al. 2011). Since the same fraction was also enriched with the SSU's 9S rRNA as compared to the LSU's 12S rRNA, it should mainly represent the SSU (as individual subunit or in the form of SSU*) as opposed to the whole mitoribosome. This result strongly implicates SSU in a specific interaction with the long-tailed mRNA species. However, a detailed analysis of the pull-down fraction, to determine the relative content of various SSU-containing complexes (individual SSU, SSU* and monosomes), is required to further elucidate this interaction, as well as involvement of any canonical translation factor(s) (see e.g., Sharma et al., this volume, for the list of mammalian mitochondrial translation factors), which are yet to be identified and characterized for the trypanosomatid mitochondrial systems.

As the next step of translation, the SSU initiation complex has been hypothesized to be recruited to the membrane-bound LSU. The mitoribosome is thus assembled on the inner mitochondrial membrane, where the translation elongation (see e.g., (Agrawal et al. 2000; Schmeing and Ramakrishnan 2009)) and the co-translational membrane insertion of the nascent polypeptide would occur (see e.g., Agrawal et al. 2011; Agrawal and Sharma 2012; Sharma et al., this volume). In this context, a candidate 80S elongation complex has been also identified, which included, in addition to rRNAs, the long-tailed edited RPS12 mRNA, and tRNA^{Cys} with a repaired 3'-end CCA. Therefore, it is likely that the 80S complex represents the fraction of actively translating mitoribosomes (Aphasizheva et al. 2011). However, since only a relatively small fraction of the total mitoribosomal RNAs and the long-tailed mRNA were found in this complex, in conjunction with the fact that the individual 50S monosomes are present in relatively low abundance, it is conceivable that the detected 50S monosome particles represent a breakdown product of the translating 80S complex. As shown more recently, not only the RPS12 mRNAs but also the long-tailed COI and Cyb mRNAs are associated with the 80S complex, as expected (Maslov, unpublished). Further analyses are required to determine if the 80S complex also contains elongation factors, membrane tethering subunits (e.g. ortholog of Mdm38/LetM1), and protein-folding machinery (e.g. orthologs of Oxa1 and Cox11), as expected for a membrane-bound active translation complex. Indeed, interactions of Mdm38/LetM1 (Tb927.3.4920) with the mitoribosome, as well as with the KPAF/KPAP1 complex, have been detected by proteomics analyses of the affinity-tagged complexes (Aphasizheva et al. 2011).

In addition to the interaction between the mitoribosomes and polyadenylation complexes, most likely mediated by an SSU (either within SSU* or monosomes or both), a growing body of evidence suggests that mitoribosomes also interact directly with the RNA editing machinery. While the rapid affinity pull-down experiments (as mentioned above) has shown the binding of SSU to the long-tailed mRNA, the LSU has been also shown to associate directly with the editing substrates (pre-edited RNA, guide RNA) and enzymatic complexes (RET1, RECC, and GRBC) (Aphasizheva et al. 2011). Additional data in support of these interactions come from the immunofluorescence microscopy studies in which the RNA editing ligase (REL1, a marker for RECC) and gRNA displacement helicase (REH1) co-localized with ribosomal proteins S17 and L3 in the kinetoplast region of *L. tarentolae* (Li et al. 2011). Thus, a close physical proximity of RNA editing and translation machinery strongly supports the existence of functional interactions between these highly complex cellular machineries.

6.9 Concluding Remarks

The peculiarities of the mitochondrial translation described in this article can be only partially explained by the divergence of the trypanosomatid lineage. Probably the most important factor defining the idiosyncrasy of this system is the existence of the U-insertion/deletion RNA editing, which appears to have imposed a requirement for the selection of a properly edited mRNA for the translation initiation. The enormity of this task has probably prevented it from being resolved only with the help of mRNA-specific translational activators, similar to those identified in the yeast mitochondrial system. Instead, the trypanosomatid mitochondrial system seems to have evolved through a tighter coordination during various stages of mRNA processing and dynamic interactions between the mRNA processing and translational machineries. Although many details of these processes still need to be elucidated at the molecular level, the emerging view is that the gene expression in trypanosomatid mitochondria is achieved through tightly regulated interactions of the mRNA processing, editing, and polyadenylation complexes with each other and with the mitoribosome. Interactions among these apparently tailor-made complexes seem to be mediated by their shared components including the PPR proteins, which perform their functions via specific recognition of the mRNA sequences. The properties of the 45S SSU* complex suggest its role in the recognition and recruitment of fully edited mRNAs suitable for translation. However, further high-resolution structural analyses and additional functional characterizations of the 45S SSU* and the 50S mitoribosome, as well as related RNP complexes, will be needed to understand the intricacies of this highly unusual translation system.

Acknowledgments The work in the authors' laboratories has been supported by National Institutes of Health grants AI088292 (to DAM) and GM61576 (to RKA). We thank Manjuli Sharma for providing Fig. 6.3 and for the critical comments on the manuscript, and Ruslan Aphasizhev and Larry Simpson for discussions.

References

Agrawal RK, Sharma MR (2012) Structural aspects of mitochondrial translational apparatus. Curr Opin Struct Biol 22:797–803

Agrawal RK, Spahn CM, Penczek P, Grassucci RA, Nierhaus KH, Frank J (2000) Visualization of tRNA movements on the *Escherichia coli* 70S ribosome during the elongation cycle. J Cell Biol 150:447–460

Agrawal RK, Sharma MR, Yassin AS, Lahiri I, Spremulli L (2011) Structure and function of organellar ribosomes as revealed by cryo-EM. In: Rodnina M, Wintermeyer W, Green R (eds) Ribosomes: structure, function, and dynamics. Springer, Wien, pp 83–96

Alfonzo JD, Soll D (2009) Mitochondrial tRNA import—the challenge to understand has just begun. Biol Chem 390:717–722

Alfonzo JD, Blanc V, Estevez AM, Rubio MA, Simpson L (1999) C to U editing of the anticodon of imported mitochondrial tRNA(Trp) allows decoding of the UGA stop codon in *Leishmania tarentolae*. EMBO J 18:7056–7062

Ammerman ML, Downey KM, Hashimi H, Fisk JC, Tomasello DL, Faktorová D, Kafková L, King T, Lukeš J, Read LK (2012) Architecture of the trypanosome RNA editing accessory complex, MRB1. Nucleic Acids Res 40:5637–5650

Aphasizhev R, Aphasizheva I (2011a) Mitochondrial RNA processing in trypanosomes. Res Microbiol 162:655–663

Aphasizhev R, Aphasizheva I (2011b) Uridine insertion/deletion editing in trypanosomes: a playground for RNA-guided information transfer. Wiley Interdiscip Rev RNA 2:669–685

Aphasizhev R, Sbicego S, Peris M, Jang SH, Aphasizheva I, Simpson AM, Rivlin A, Simpson L (2002) Trypanosome mitochondrial 3′ terminal uridylyl transferase (TUTase): the key enzyme in U-insertion/deletion RNA editing. Cell 108:637–648

Aphasizheva I, Aphasizhev R (2010) RET1-catalyzed uridylylation shapes the mitochondrial transcriptome in *Trypanosoma brucei*. Mol Cell Biol 30:1555–1567

Aphasizheva I, Aphasizhev R, Simpson L (2004) RNA-editing terminal uridylyl transferase 1: identification of functional domains by mutational analysis. J Biol Chem 279:24123–24130

Aphasizheva I, Maslov D, Wang X, Huang L, Aphasizhev R (2011) Pentatricopeptide repeat proteins stimulate mRNA adenylation/uridylation to activate mitochondrial translation in trypanosomes. Mol Cell 42:106–117

Benne R, Van den Burg J, Brakenhoff J, Sloof P, Van Boom J, Tromp M (1986) Major transcript of the frameshifted coxII gene from trypanosome mitochondria contains four nucleotides that are not encoded in the DNA. Cell 46:819–826

Bhat GJ, Koslowsky DJ, Feagin JE, Smiley BL, Stuart K (1990) An extensively edited mitochondrial transcript in kinetoplastids encodes a protein homologous to ATPase subunit 6. Cell 61:885–894

Blum B, Simpson L (1990) Guide RNAs in kinetoplastid mitochondria have a nonencoded $3'$ oligo-(U) tail involved in recognition of the pre-edited region. Cell 62:391–397

Blum B, Bakalara N, Simpson L (1990) A model for RNA editing in kinetoplastid mitochondria: "Guide" RNA molecules transcribed from maxicircle DNA provide the edited information. Cell 60:189–198

Borst P, Hoeijmakers JHJ (1979) Kinetoplast DNA. Plasmid 2:20–40

Breek CK, Speijer D, Dekker H, Muijsers AO, Benne R (1997) Further evidence for the presence of mitochondrially encoded subunits in cytochrome *c* oxidase of the trypanosomatid *Crithidia fasciculata*. Biol Chem Hoppe Seyler 378:837–841

Chen J, Rauch CA, White JH, Englund PT, Cozzarelli NR (1995) The topology of the kinetoplast DNA network. Cell 80:61–69

Corell RA, Feagin JE, Riley GR, Strickland T, Guderian JA, Myler PJ, Stuart K (1993) *Trypanosoma brucei* minicircles encode multiple guide RNAs which can direct editing of extensively overlapping sequences. Nucleic Acids Res 21:4313–4320

Corell RA, Myler P, Stuart K (1994) *Trypanosoma brucei* mitochondrial CR4 gene encodes an extensively edited mRNA with completely edited sequence only in bloodstream forms. Mol Biochem Parasitol 64:65–74

Cristodero M, Seebeck T, Schneider A (2010) Mitochondrial translation is essential in bloodstream forms of *Trypanosoma brucei*. Mol Microbiol 78:757–769

Davies SM, Rackham O, Shearwood AM, Hamilton KL, Narsai R, Whelan J, Filipovska A (2009) Pentatricopeptide repeat domain protein 3 associates with the mitochondrial small ribosomal subunit and regulates translation. FEBS Lett 583:1853–1858

de la Cruz V, Neckelmann N, Simpson L (1984) Sequences of six structural genes and several open reading frames in the kinetoplast maxicircle DNA of *Leishmania tarentolae*. J Biol Chem 259:15136–15147

de la Cruz V, Lake JA, Simpson AM, Simpson L (1985a) A minimal ribosomal RNA: sequence and secondary structure of the 9S kinetoplast ribosomal RNA from *Leishmania tarentolae*. Proc Natl Acad Sci U S A 82:1401–1405

de la Cruz V, Simpson A, Lake J, Simpson L (1985b) Primary sequence and partial secondary structure of the 12S kinetoplast (mitochondrial) ribosomal RNA from *Leishmania tarentolae*: conservation of peptidyl-transferase structural elements. Nucl Acids Res 13:2337–2356

Decker CJ, Sollner-Webb B (1990) RNA editing involves indiscriminate U changes throughout precisely defined editing domains. Cell 61:1001–1011

Delannoy E, Stanley WA, Bond CS, Small ID (2007) Pentatricopeptide repeat (PPR) proteins as sequence-specificity factors in post-transcriptional processes in organelles. Biochem Soc Trans 35:1643–1647

Eperon I, Janssen J, Hoeijmakers J, Borst P (1983) The major transcripts of the kinetoplast DNA of *T. brucei* are very small ribosomal RNAs. Nucl Acids Res 11:105–125

Etheridge RD, Aphasizheva I, Gershon PD, Aphasizhev R (2008) $3'$ adenylation determines mRNA abundance and monitors completion of RNA editing in *T. brucei* mitochondria. EMBO J 27:1596–1608

Feagin JE, Abraham J, Stuart K (1988a) Extensive editing of the cytochrome *c* oxidase III transcript in *Trypanosoma brucei*. Cell 53:413–422

Feagin JE, Shaw JM, Simpson L, Stuart K (1988b) Creation of AUG initiation codons by addition of uridines within cytochrome *b* transcripts of kinetoplastids. PNAS 85:539–543

Fox TD (2012) Mitochondrial protein synthesis, import, and assembly. Genetics 192:1203–1234

Gray MW (2012a) Evolutionary origin of RNA editing. Biochemistry 51:5235–5242

Gray MW (2012b) Mitochondrial evolution. Cold Spring Harb Perspect Biol 4:a011403

Hashimi H, Zimmer SL, Ammerman ML, Read LK, Lukeš J (2013) Dual core processing: MRB1 is an emerging kinetoplast RNA editing complex. Trends Parasitol 29:91–99

Herrmann JM, Woellhaf MW, Bonnefoy N (2013) Control of protein synthesis in yeast mitochondria: the concept of translational activators. Biochim Biophys Acta 1833:286–294

Horváth A, Berry EA, Maslov DA (2000a) Translation of the edited mRNA for cytochrome *b* in trypanosome mitochondria. Science 287:1639–1640

Horváth A, Kingan TG, Maslov DA (2000b) Detection of the mitochondrially encoded cytochrome *c* oxidase subunit I in the trypanosomatid protozoan *Leishmania tarentolae*. J Biol Chem 275:17160–17165

Horváth A, Neboháčová M, Lukeš J, Maslov DA (2002) Unusual polypeptide synthesis in the kinetoplast-mitochondria from *Leishmania tarentolae*. Identification of individual *de novo* translation products. J Biol Chem 277:7222–7230

Koslowsky DJ (2004) A historical perspective on RNA editing: how the peculiar and bizarre became mainstream. Methods Mol Biol 265:161–197

Koslowsky DJ (2009) Complex interactions in the regulation of trypanosome mitochondrial gene expression. Trends Parasitol 25:252–255

Koslowsky DJ, Bhat GJ, Perrollaz AL, Feagin JE, Stuart K (1990) The MURF3 gene of *T. brucei* contains multiple domains of extensive editing and is homologous to a subunit of NADH dehydrogenase. Cell 62:901–911

Koslowsky DJ, Bhat GJ, Read LK, Stuart K (1991) Cycles of progressive realignment of gRNA with mRNA in RNA editing. Cell 67:537–546

Krause K, de Lopes SR, Roberts DG, Dieckmann CL (2004) The mitochondrial message-specific mRNA protectors Cbp1 and Pet309 are associated in a high-molecular weight complex. Mol Biol Cell 15:2674–2683

Li F, Herrera J, Zhou S, Maslov DA, Simpson L (2011) Trypanosome REH1 is an RNA helicase involved with the 3′-5′ polarity of multiple gRNA-guided uridine insertion/deletion RNA editing. Proc Natl Acad Sci U S A 108:3542–3547

Lightowlers RN, Chrzanowska-Lightowlers ZM (2008) PPR (pentatricopeptide repeat) proteins in mammals: important aids to mitochondrial gene expression. Biochem J 416:e5–e6

Liu B, Liu Y, Motyka SA, Agbo EE, Englund PT (2005) Fellowship of the rings: the replication of kinetoplast DNA. Trends Parasitol 21:363–369

Lurin C, Andres C, Aubourg S, Bellaoui M, Bitton F, Bruyere C, Caboche M, Debast C, Gualberto J, Hoffmann B, Lecharny A, Le RM, Martin-Magniette ML, Mireau H, Peeters N, Renou JP, Szurek B, Taconnat L, Small I (2004) Genome-wide analysis of *Arabidopsis* pentatricopeptide repeat proteins reveals their essential role in organelle biogenesis. Plant Cell 16:2089–2103

Manuell AL, Quispe J, Mayfield SP (2007) Structure of the chloroplast ribosome: novel domains for translation regulation. PLoS Biol 5:e209

Marres CAM, Slater EC (1977) Polypeptide composition of purified QH2: cytochrome *c* oxidoreductase from beef-heart mitochondria. Biochim Biophys Acta 462:531–548

Maslov DA (2010) Complete set of mitochondrial pan-edited mRNAs in *Leishmania mexicana amazonensis* LV78. Mol Biochem Parasitol 173:107–114

Maslov DA, Agrawal RK (2012) Mitochondrial translation in trypanosomatids. In: Bindereif A (ed) RNA metabolism in trypanosomes. Springer, Berlin, pp 215–236

Maslov DA, Simpson L (1992) The polarity of editing within a multiple gRNA-mediated domain is due to formation of anchors for upstream gRNAs by downstream editing. Cell 70:459–467

Maslov DA, Simpson L (1994) RNA editing and mitochondrial genomic organization in the cryptobiid kinetoplastid protozoan, *Trypanoplasma borreli*. Mol Cell Biol 14:8174–8182

Maslov DA, Sturm NR, Niner BM, Gruszynski ES, Peris M, Simpson L (1992) An intergenic G-rich region in *Leishmania tarentolae* kinetoplast maxicircle DNA is a pan-edited cryptogene encoding ribosomal protein S12. Mol Cell Biol 12:56–67

Maslov DA, Thiemann O, Simpson L (1994) Editing and misediting of transcripts of the kinetoplast maxicircle G5 (ND3) cryptogene in an old laboratory strain of *Leishmania tarentolae*. Mol Biochem Parasitol 68:155–159

Maslov DA, Nawathean P, Scheel J (1999) Partial kinetoplast-mitochondrial gene organization and expression in the respiratory deficient plant trypanosomatid *Phytomonas serpens*. Mol Biochem Parasitol 99:207–221

Maslov DA, Sharma MR, Butler E, Falick AM, Gingery M, Agrawal RK, Spremulli LL, Simpson L (2006) Isolation and characterization of mitochondrial ribosomes and ribosomal subunits from *Leishmania tarentolae*. Mol Biochem Parasitol 148:69–78

Maslov DA, Spremulli LL, Sharma MR, Bhargava K, Grasso D, Falick AM, Agrawal RK, Parker CE, Simpson L (2007) Proteomics and electron microscopic characterization of the unusual mitochondrial ribosome-related 45S complex in *Leishmania tarentolae*. Mol Biochem Parasitol 152:203–212

Maslov DA, Votýpka J, Yurchenko V, Lukeš J (2013) Diversity and phylogeny of insect trypanosomatids: all that is hidden shall be revealed. Trends Parasitol 29:43–52

Mingler MK, Hingst AM, Clement SL, Yu LE, Reifur L, Koslowsky DJ (2006) Identification of pentatricopeptide repeat proteins in *Trypanosoma brucei*. Mol Biochem Parasitol 150:37–45

Naithani S, Saracco SA, Butler CA, Fox TD (2003) Interactions among COX1, COX2, and COX3 mRNA-specific translational activator proteins on the inner surface of the mitochondrial inner membrane of *Saccharomyces cerevisiae*. Mol Biol Cell 14:324–333

Neboháčová M, Maslov DA, Falick AM, Simpson L (2004) The effect of RNA interference down-regulation of RNA editing 3′-terminal uridylyl transferase (TUTase) 1 on mitochondrial *de novo* protein synthesis and stability of respiratory complexes in *Trypanosoma brucei*. J Biol Chem 279:7819–7825

Neboháčová M, Kim CE, Simpson L, Maslov DA (2009) RNA editing and mitochondrial activity in promastigotes and amastigotes of *Leishmania donovani*. Int J Parasitol 39:635–644

Nissen P, Hansen J, Ban N, Moore PB, Steitz TA (2000) The structural basis of ribosome activity in peptide bond synthesis. Science 289:920–930

Ochsenreiter T, Hajduk SL (2006) Alternative editing of cytochrome c oxidase III mRNA in trypanosome mitochondria generates protein diversity. EMBO Rep 7:1128–1133

Ochsenreiter T, Cipriano M, Hajduk SL (2008) Alternative mRNA editing in trypanosomes is extensive and may contribute to mitochondrial protein diversity. PLoS ONE 3:e1566

Ogle JM, Brodersen DE, Clemons WM Jr, Tarry MJ, Carter AP, Ramakrishnan V (2001) Recognition of cognate transfer RNA by the 30S ribosomal subunit. Science 292:897–902

Pollard VW, Rohrer SP, Michelotti EF, Hancock K, Hajduk SL (1990) Organization of minicircle genes for guide RNAs in *Trypanosoma brucei*. Cell 63:783–790

Priest JW, Hajduk SL (1992) Cytochrome *c* reductase purified from *Crithidia fasciculata* contains an atypical cytochrome c_1. J Biol Chem 267:20188–20195

Pusnik M, Schneider A (2012) A trypanosomal pentatricopeptide repeat protein stabilizes the mitochondrial mRNAs of cytochrome oxidase subunits 1 and 2. Eukaryot Cell 11:79–87

Pusnik M, Small I, Read LK, Fabbro T, Schneider A (2007) Pentatricopeptide repeat proteins in *Trypanosoma brucei* function in mitochondrial ribosomes. Mol Cell Biol 27:6876–6888

Rackham O, Filipovska A (2012) The role of mammalian PPR domain proteins in the regulation of mitochondrial gene expression. Biochim Biophys Acta 1819:1008–1016

Ray D (1989) Conserved sequence blocks in kinetoplast DNA minicircles from diverse species of trypanosomes. Mol Cell Biol 9:1365–1367

Read LK, Myler PJ, Stuart K (1992) Extensive editing of both processed and preprocessed maxicircle CR6 transcripts in *Trypanosoma brucei*. J Biol Chem 267:1123–1128

Read LK, Wilson KD, Myler PJ, Stuart K (1994) Editing of *Trypanosoma brucei* maxicircle CR5 mRNA generates variable carboxy terminal predicted protein sequences. Nucleic Acids Res 22:1489–1495

Riley GR, Corell RA, Stuart K (1994) Multiple guide RNAs for identical editing of *Trypanosoma brucei* apocytochrome *b* mRNA have an unusual minicircle location and are developmentally regulated. J Biol Chem 269:6101–6108

Scheinman A, Aguinaldo A-M, Simpson AM, Peris M, Shankweiler G, Simpson L, Lake JA (1993) Reconstitution of a minimal small ribosomal subunit. In: Nierhaus K (ed) The translation apparatus. Plenum Press, New York, pp 719–726

Schmeing TM, Ramakrishnan V (2009) What recent ribosome structures have revealed about the mechanism of translation. Nature 461:1234–1242

Schmitz-Linneweber C, Small I (2008) Pentatricopeptide repeat proteins: a socket set for organelle gene expression. Trends Plant Sci 13:663–670

Schuwirth BS, Borovinskaya MA, Hau CW, Zhang W, Vila-Sanjurjo A, Holton JM, Cate JH (2005) Structures of the bacterial ribosome at 3.5 Å resolution. Science 310:827–834

Selmer M, Dunham CM, Murphy FV, Weixlbaumer A, Petry S, Kelley AC, Weir JR, Ramakrishnan V (2006) Structure of the 70S ribosome complexed with mRNA and tRNA. Science 313:1935–1942

Sharma MR, Koc EC, Datta PP, Booth TM, Spremulli LL, Agrawal RK (2003) Structure of the mammalian mitochondrial ribosome reveals an expanded functional role for its component proteins. Cell 115:97–108

Sharma MR, Wilson DN, Datta PP, Barat C, Schluenzen F, Fucini P, Agrawal RK (2007) Cryo-EM study of the spinach chloroplast ribosome reveals the structural and functional roles of plastid-specific ribosomal proteins. Proc Natl Acad Sci U S A 104:19315–19320

Sharma MR, Booth TM, Simpson L, Maslov DA, Agrawal RK (2009) Structure of a mitochondrial ribosome with minimal RNA. Proc Natl Acad Sci U S A 106:9637–9642

Shaw J, Feagin JE, Stuart K, Simpson L (1988) Editing of mitochondrial mRNAs by uridine addition and deletion generates conserved amino acid sequences and AUG initiation codons. Cell 53:401–411

Shaw J, Campbell D, Simpson L (1989) Internal frameshifts within the mitochondrial genes for cytochrome oxidase subunit II and maxicircle unidentified reading frame 3 in *Leishmania tarentolae* are corrected by RNA editing: evidence for translation of the edited cytochrome oxidase subunit II mRNA. Proc Natl Acad Sci 86:6220–6224

Simpson L (1986) Kinetoplast DNA in trypanosomid flagellates. Int Rev Cytol 99:119–179

Simpson L, Neckelmann N, de la Cruz V, Simpson A, Feagin J, Jasmer D, Stuart K (1987) Comparison of the maxicircle (mitochondrial) genomes of *Leishmania tarentolae* and *Trypanosoma brucei* at the level of nucleotide sequence. J Biol Chem 262:6182–6196

Simpson L, Thiemann OH, Savill NJ, Alfonzo JD, Maslov DA (2000) Evolution of RNA editing in trypanosome mitochondria. Proc Natl Acad Sci U S A 97:6986–6993

Simpson AGB, Lukeš J, Roger AJ (2002) The evolutionary history of kinetoplastids and their kinetoplasts. Mol Biol Evol 19:2071–2083

Simpson AG, Stevens JR, Lukeš J (2006) The evolution and diversity of kinetoplastid flagellates. Trends Parasitol 22:168–174

Souza AE, Myler PJ, Stuart K (1992) Maxicircle CR1 transcripts of *Trypanosoma brucei* are edited, developmentally regulated, and encode a putative iron-sulfur protein homologous to an NADH dehydrogenase subunit. Mol Cell Biol 12:2100–2107

Souza AE, Shu H-H, Read LK, Myler PJ, Stuart KD (1993) Extensive editing of CR2 maxicircle transcripts of *Trypanosoma brucei* predicts a protein with homology to a subunit of NADH dehydrogenase. Mol Cell Biol 13:6832–6840

Spears JL, Rubio MAT, Sample P, Alfonzo JD (2012) tRNA biogenesis and processing. In: Bindereif A (ed) RNA metabolism in trypanosomes. Springer, Heidelberg, pp 99–121

Speijer D, Muijsers AO, Dekker H, De Haan A, Breek CKD, Albracht SPJ, Benne R (1996) Purification and characterization of cytochrome *c* oxidase from the insect trypanosomatid *Crithidia fasciculata*. Mol Biochem Parasitol 79:47–59

Speijer D, Breek CKD, Muijsers AO, Hartog AF, Berden JA, Albracht SPJ, Samyn B, Van Beeumen J, Benne R (1997) Characterization of the respiratory chain from cultured *Crithidia fasciculata*. Mol Biochem Parasitol 85:171–186

Sturm NR, Simpson L (1990a) Partially edited mRNAs for cytochrome *b* and subunit III of cytochrome oxidase from *Leishmania tarentolae* mitochondria: RNA editing intermediates. Cell 61:871–878

Sturm NR, Simpson L (1990b) Kinetoplast DNA minicircles encode guide RNAs for editing of cytochrome oxidase subunit III mRNA. Cell 61:879–884

Sturm NR, Maslov DA, Blum B, Simpson L (1992) Generation of unexpected editing patterns in Leishmania tarentolae mitochondrial mRNAs: misediting produced by misguiding. Cell 70:469–476

Tan TH, Bochud-Allemann N, Horn EK, Schneider A (2002) Eukaryotic-type elongator tRNAMet of *Trypanosoma brucei* becomes formylated after import into mitochondria. Proc Natl Acad Sci U S A 99:1152–1157

Thiemann OH, Maslov DA, Simpson L (1994) Disruption of RNA editing in *Leishmania tarentolae* by the loss of minicircle-encoded guide RNA genes. EMBO J 13:5689–5700

Vickerman K (1976) The diversity of the kinetoplastid flagellates. In: Lumsden WHR, Evans DA (eds) Biology of the kinetoplastida. Academic Press, London, pp 1–34

Vickerman K, Preston TM (1976) Comparative cell biology of the kinetoplastid flagellates. In: Lumsden WHR, Evans DA (eds) Biology of the kinetoplastida. Academic Press, London, pp 35–130

Weng J, Aphasizheva I, Etheridge RD, Huang L, Wang X, Falick AM, Aphasizhev R (2008) Guide RNA-binding complex from mitochondria of trypanosomatids. Mol Cell 32:198–209

Yu LE, Koslowsky DJ (2006) Interactions of mRNAs and gRNAs involved in trypanosome mitochondrial RNA editing: structure probing of a gRNA bound to its cognate mRNA. RNA 12:1050–1060

Zíková A, Panigrahi AK, Dalley RA, Acestor N, Anupama A, Ogata Y, Myler PJ, Stuart KD (2008) *Trypanosoma brucei* mitochondrial ribosomes: affinity purification and component identification by mass spectrometry. Mol Cell Proteomics 7:1286–1296

Chapter 7
Translation in Mitochondria and Apicoplasts of Apicomplexan Parasites

Ankit Gupta, Afreen Haider, Suniti Vaishya and Saman Habib

Abstract Protozoan parasites of the phylum apicomplexa are the causal organisms for several important infectious diseases. They contain three cellular compartments - the cytoplasm, mitochondrion, and the relic plastid called the apicoplast, where active translation takes place. Early investigations of the components and activities of the translation machinery in parasite organelles have provided evidence for intriguing variations from other systems. Understanding these mechanisms also offers novel possibilities for drug intervention. In this article, we discuss the translation machinery in the apicoplast and mitochondrion of the apicomplexan cell, with emphasis on current knowledge from *Plasmodium* and *Toxoplasma*.

7.1 Introduction

Apicomplexa is a large group of protists comprising parasites that are causative agents of many diseases. The most prominent of the apicomplexan parasites of humans are *Plasmodium* spp., the causal organism for malaria. This infection was

A. Gupta · A. Haider · S. Vaishya · S. Habib (✉)
Division of Molecular and Structural Biology, CSIR-Central Drug Research Institute,
10-Jankipuram Extension, Sitapur Road, Lucknow 226021, Uttar Pradesh, India
e-mail: saman.habib@gmail.com; saman_habib@cdri.res.in

A. Gupta
e-mail: guptaankitbiochem07@yahoo.co.in

A. Haider
e-mail: afreen.haider@gmail.com

S. Vaishya
e-mail: suniti.vaish08@gmail.com

A.-M. Duchêne (ed.), *Translation in Mitochondria and Other Organelles*,
DOI: 10.1007/978-3-642-39426-3_7, © Springer-Verlag Berlin Heidelberg 2013

responsible for an estimated 0.65 million deaths in 2011 in 104 countries that are endemic for the disease (WHO 2012). Among the other major apicomplexan parasites are AIDS-associated opportunistic pathogens (*Toxoplasma, Cryptosporidium*) and economically important veterinary pathogens of livestock and poultry (*Eimeria, Theileria, Babesia*). Apicomplexans belong to the Infrakingdom Alveolata that includes two other divergent major eukaryotic groups: dinoflagellates, and ciliates (Adl et al. 2005). Apicomplexa is closely related to the dinoflagellate-ciliate clade (Wolters 1991) with the sister apicomplexa and dinoflagellate lineages having diverged about 800–900 million years ago (Baldauf et al. 2000).

In addition to the essential organelle—the mitochondrion, most apicomplexan parasites possess a relic plastid called the apicoplast (Wilson and Williamson 1997). It is believed that mitochondria originated from an alpha-proteobacterium by means of endosymbiosis approximately two billion years ago (Gray et al. 2001, 2004). Extant mitochondrial (mt) genomes exhibit significant variations in structure and size (Gray et al. 2004). The largest mt genome is found in land plants with a size range of 180–2,400 kb (Palmer et al. 1992; Ward et al. 1981) while the smallest is the 6 kb mt genome of the apicomplexan *Plasmodium* (Wilson and Williamson 1997; Preiser et al. 1996). The mitochondrion of *Cryptosporidium parvum* is degenerate and has completely lost its mt genome (Abrahamsen et al. 2004). Within apicomplexa, *Babesia* and *Theileria* have monomeric linear 6.6–7.1 kb mt genomes with the gene order and direction of transcription completely distinct from the tandemly repeated linear 6 kb mt genome of *Plasmodium* (Hikosaka et al. 2010; Brayton et al. 2007; Kairo et al. 1994). The mt genome of apicomplexan parasites encodes only three proteins, cytochrome oxidase subunit I (CoxI), subunit III (CoxIII), and apocytochrome b (CytB) with highly fragmented large subunit (LSU) and small subunit (SSU) rRNA (Feagin et al. 1997). None of the tRNAs and essential translation factors are encoded by the apicomplexan mt genome and would have to be imported by the orgenelle (Rusconi and Cech 1996; Esseiva et al. 2004).

The apicoplast of apicomplexan parasites is a four-membrane bound organelle that evolved through secondary endosymbiosis in which one eukaryote engulfed and retained another eukaryote of red algal lineage that contained a plastid resulted from primary endosymbiosis (Lim and McFadden 2010). However, unlike most apicomplexans, *Cryptosporidium* lacks an apicoplast (Zhu et al. 2000; Abrahamsen et al. 2004). The apicoplast genome is homologous to plant and algal plastid genomes but is highly reduced and has completely lost the genes involved in photosynthesis (Wilson et al. 1996; Kohler et al. 1997; McFadden and Waller 1997; Wilson and Williamson 1997), similar to the plastid genome of a nonphotosynthetic parasitic flowering plant *Epifagus virginiana* (Wolfe et al. 1992). Although apicoplasts are nonphotosynthetic, they have retained other plastid functions such as fatty acid, isoprenoid, and heme biosynthesis with products of these pathways being required for important and often critical functions at different stages of the parasite infection cycle (Jomaa et al. 1999; Tarun et al. 2009; Waller et al. 2000; Sato et al. 2004). Many plastid genes have been transferred to the nuclear genome during evolution and their protein products are targeted back to the apicoplast (Martin et al. 1998; Waller et al. 1998).

Nuclear-encoded apicoplast-targeted (NEAT) proteins possess a bipartite N-terminal sequence which mediates their transfer from the cytosol into the apicoplast via the secretory pathway (Waller et al. 1998; Foth et al. 2003; Harb et al. 2004). The first 20–30 amino acids of these proteins is a classical secretory signal sequence (SS) which is cleaved by a signal peptidase within the ER. This exposes a 50- to 200-aa long transit peptide (TP) that is critical for targeting the protein to the apicoplast. TPs in *P. falciparum* contain no consensus sequences or conserved secondary structures but are rich in positively charged amino acids. Targeting prediction algorithms, PlamoAP (Foth et al. 2003), and PATS (Zuegge et al. 2001), have been developed to predict apicoplast import in *Plasmodium* and a Python-based program ApicoAP developed recently can be used to identify NEAT proteins in other apicomplexans (Cilingir et al. 2012). Translocation of NEAT proteins across the four apicoplast membranes is believed to involve an ERAD-like complex and homologs of the chloroplast translocons TOC and TIC (Tonkin et al. 2008; van Dooren et al. 2001).

7.2 Components of the Apicoplast/Mitochondrial Translation Machinery are Encoded by Organellar or Nuclear Genomes

Apicomplexan parasites such as *Plasmodium* and *Toxoplasma* have an A + T-rich (86.9 % and 78.4 % A + T, respectively), 35 kb circular apicoplast genome that is actively translated (Chaubey et al. 2005). Although the universal genetic code is followed in the cytosol and mitochondrion, there is evidence that the stop codon UGA is recognized as tryptophan within specific ORFs in the apicoplast of *Neospora*, *Toxoplasma* and *Plasmodium* (Lang-Unnasch and Aiello 1999; Wilson 2002). Suppression of termination at the UGA codon has been suggested to be a result of the sequence of the SSU rRNA gene that corresponds to helix 34 of its *E. coli* counterpart (Lang-Unnasch and Aiello 1999).

Most of the essential components for protein translation are encoded by the apicoplast itself; these include elongation factor Tu (EF-Tu), 18 ribosomal proteins, large and small subunit ribosomal RNAs (16S and 23S rRNA), and a minimal set of 26 tRNA isoacceptors encoded by 35 genes (Wilson et al. 1996; Wilson and Williamson 1997) (Fig. 7.1). While the tRNA genes are sufficient to support translation in the apicoplast, the remaining apicoplast ribosomal proteins, aminoacyl tRNA synthetases (aaRS), and crucial translation factors are imported by the organelle (Biswas et al. 2011; Jackson et al. 2011; Johnson et al. 2011). Apart from components required for translation, the apicoplast genome also encodes RNA polymerase subunits, a ClpC-like protease and the SufB protein of the [Fe–S] complexation pathway (Wilson et al. 1996; Wilson and Williamson 1997).

The presence of mitochondria-targeted EF-Tu and cytb mRNA in the mitoplast fraction of *T. gondii* has indicated the presence of an active translation machinery

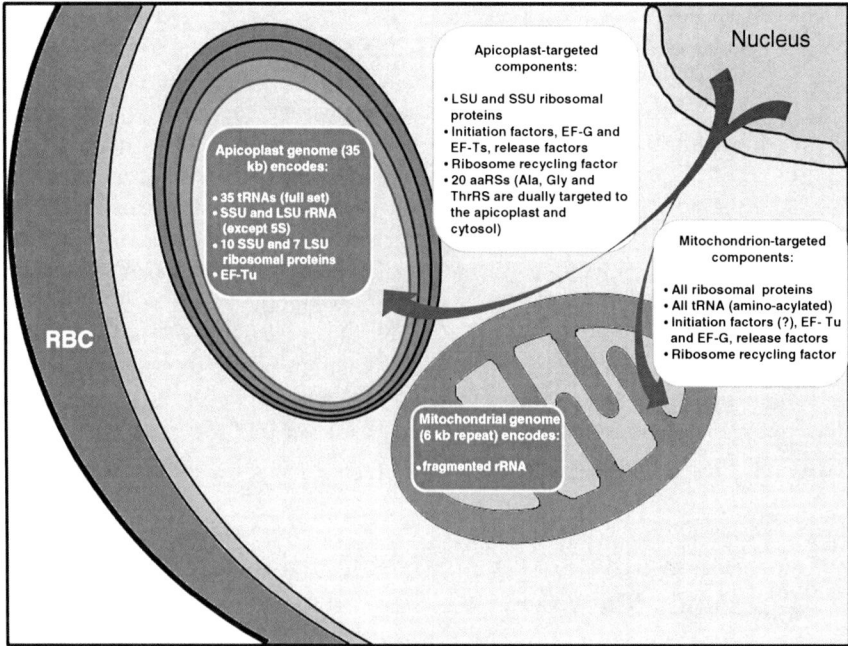

Fig. 7.1 Translation components of *Plasmodium* organelles

in the organelle (Pino et al. 2010). None of the tRNAs and essential translation factors are encoded by the apicomplexan mt genome and would have to be
imported from the cytosol (Rusconi and Cech 1996; Esseiva et al. 2004). The first
evidence of mitochondrial tRNA import was found in *T. gondii*, where a percentage of nuclear-encoded tRNAs- tRNA^{Ala}, tRNA^{Ile}, tRNA^{Ser}, tRNA^{Trp}, tRNA^{Gln},
and tRNA^{Met-e}—were localized in the mitochondrial fraction in addition to the
cytosol (Esseiva et al. 2004). Although mitochondria lack any resident aaRS gene
and there is no evidence for transport of these proteins from the cytosol (Bhatt
et al. 2009), 80 % of tRNA^{Trp} and tRNA^{Ile} recovered from the mitochondrial fraction were found to be aminoacylated (Pino et al. 2010), indicating that the *T. gondii*
mitochondria import charged tRNAs from the cytosol. Similar mitochondrial
import of a charged tRNA has been reported for yeast tRNA^{Lys} (Kamenski et al.
2007). The lack of accumulation of uncharged tRNAs in *T. gondii* mitochondria is
suggestive of either degradation after use or retrograde transport to the cytosol for
recycling by cytosolic aaRSs (Pino et al. 2010).

Aminoacyl-tRNA synthetases (aaRS) that charge tRNAs with their cognate amino acid by esterification have been completely lost from the genomes
of both organellar compartments. There is a reduced set of 37 genes for aaRSs
in *Plasmodium*, but these are sufficient to translate genomes of the nuclear, apicoplast, and mitochondrial compartments. AaRSs are not targeted to the mitochondrion but those needed for charging tRNA in the apicoplast would have to

be imported from the cytosol. Targeting of IleRS encoded by two nuclear genes to the cytosol and apicoplast, respectively has been demonstrated (Istvan et al. 2011). There is also evidence of sharing of some aaRSs between the cytosolic and apicoplast compartment by dual targeting in *Plasmodium* spp. AlaRS, GlyRS, and ThrRS encoded by single copy genes in the nucleus are dually targeted to the apicoplast and cytosol, possibly by using alternate start sites for initiation (Jackson et al. 2011).

7.3 Apicoplast and Mitochondrial Ribosomes

Bioinformatic predictions suggest that the composition and size of apicomplexan organellar ribosomes is different from bacterial, or plastid and mitochondrial ribosomes. For instance, a total of only 16 SSU and 21 LSU ribosomal proteins (Brehelin et al. 2010) would constitute the *Plasmodium falciparum* apicoplast ribosome when both apicoplast-encoded as well as targeted proteins are taken into account; this is in contrast to *E. coli*, which has 22 SSU and 34 LSU proteins (Arnold and Reilly 1999). All apicomplexan mitoribosome proteins must be imported from the cytoplasm but a complete identification of these has not been made thus far. A 5S rRNA gene is not identifiable on apicoplast DNA, raising the possibility of its import from the cytosol (Sato 2011). Moreover, rRNAs encoded by the mitochondrial genome are highly fragmented and range in size from 23 to 190 nt, but potentially form secondary structures similar to those proposed for continuous rRNAs (Feagin et al. 1997; Feagin et al. 2012). A total of 27 rRNA sequences that would comprise fragmented SSU and LSU rRNA (12 SSU rRNAs and 15 LSU rRNAs) have been identified in *Plasmodium* mitochondria (Feagin et al. 1997; Hikosaka et al. 2011). Only six fragmented LSU rRNA sequences have been identified in *Babesia* and *Theileria* mt genomes and the pattern of fragmentation differs from that predicted in *Plasmodium* (Hikosaka et al. 2010).

Comparisons with corresponding *E. coli* and apicoplast sequences showed 50–60 % conservation in core sequences of *P. falciparum* mitochondrial rRNA regions (Feagin et al. 1992). However, some core sequences and other regions including the sarcin/ricin loop of the LSU rRNA and portions of the 5' end and the 790 loop of the SSU rRNA are missing in mitochondrial rRNA (Feagin et al. 1997; Feagin et al. 2012). It might be possible that these missing regions are either nonfunctional or have not yet been identified due to very small size or lack of sequence conservation (Feagin et al. 1992). Interestingly, nonencoded short A-tails have been shown to be present at the 3' end of most, but not all the small mitochondrial rRNAs of *P. falciparum* and *Theileria parva* (Nene et al. 1998; Gillespie et al. 1999). All mt rRNA are predicted to base pair with at least one other rRNA, possibly aiding in the correct location and orientation of these molecules on the ribosome (Feagin et al. 2012). Superimposition on the 3D ribosome structural model has shown that the fragmented rRNAs cluster on the interfaces of the two ribosomal subunits (Feagin et al. 2012).

7.4 Translation Initiation and the Involvement of fMet-tRNA^{fMet} in Parasite Organelles

Translation initiation comprises the early steps involved in formation of the initiation complex which precedes the formation of the very first peptide bond in a polypeptide chain. In *E. coli*, initiation requires the ribosomes, mRNA, fMet-tRNA^{fMet}, GTP and three protein initiation factors—IF1, IF2, and IF3 (Gualerzi and Pon 1990) and takes place via sequential formation of intermediary complexes—the 30S pre-initiation complex (30S PIC), the 30S Initiation complex (30S IC) and finally, the 70S initiation complex (70S IC), which marks the completion of the initiation phase of translation. Briefly, IF3 which is responsible for keeping the ribosome in a split state, binds to the small 30S ribosomal subunit together with the mRNA and completes the formation of the 30S PIC. Subsequently, IF1 binds to the 30S subunit at the A-site and facilitates the binding of IF2. The fMet-tRNA^{fMet}.IF2.GTP ternary complex recognizes and binds to the start codon to form the complete 30S-IC (Hartz et al. 1989; Gualerzi and Pon 1990). Finally, IF2-mediated association of the 50S subunit results in the 70S-IC (Lockwood et al. 1971; Antoun et al. 2003, 2004) with the simultaneous release of IF-1 and IF-3. Upon subunit association, the GTPase activity of IF2 is also activated causing hydrolysis of GTP to release GDP and P$_i$ and enabling escape of IF-2 from the complex.

The initiation machinery differs in cytosolic and organellar protein synthesis in *Plasmodium/Toxoplasma*; being of prokaryotic descent, the latter bears resemblance to bacterial systems. Translation is initiated on the AUG codon in apicoplast-encoded mRNAs. On the other hand, AUG is used only for the *cytb* gene in *Plasmodium* mitochondrial translation while AUA and AUU are the initiation codons for *coxI* and *coxIII* genes (Kairo et al. 1994; Feagin 1992; Rehkopf et al. 2000). The assembly of the mRNA and 30S ribosomal subunit is not promoted through complementary base pairing of the purine-rich Shine-Dalgarno (SD) sequence (Shine and Dalgarno 1975) with the 16S rRNA, as the SD sequence is not conserved in organellar mRNAs of apicomplexan parasites. Apicoplast genes are transcribed as mono- or polycistrons and have short intergenic spaces that are generally A/T rich. It is likely that these interact with the 16S rRNA similar to a class of chloroplast mRNAs of the green alga *Euglena gracilis* that have an A/T-rich 5′ UTR instead of the SD sequence (Betts and Spremulli 1994). The pyrimidine-rich sequence in the bacterial 16S rRNA that pairs with the SD sequence is also replaced by an A/T rich sequence in the *Plasmodium* apicoplast-encoded 16S rRNA. As in mammalian mitochondria, the three mitochondrial mRNAs in *Plasmodium* possess short 5′ UTRs and may thus employ some other mechanism for precise pairing with the fragmented organellar rRNA (Jackson et al. 2011).

None of the three initiation factors are encoded by the organellar genomes of apicomplexa (Fig. 7.1). Nuclear-encoded IF1 candidates with putative N-terminal bipartite apicoplast targeting sequence (Foth et al. 2003) are found

Table 7.1 Translation factors putatively targeted to the mitochondrion and apicoplast of apicomplexan parasites

	Plasmodium falciparum		Toxoplasma gondii		Eimeria tenella		Babesia bovis		Theileria parva	
	Apicoplast	Mitochondria	Apicoplast	Mitochondria	Apicoplast	Mitochondria	Apicoplast	Mitochondria	Apicoplast	Mitochondria
IF1	PF14_0658#	–	TGME49_217620	–	NOT FOUND	–	BBOV_III011550	–	TP02_0767	–
IF2	PFE0830c	PF08_0018 PF13_0069	TGGT1_124680	TGVEG_031600	Multiple homologs, organellar distribution not clearly predictable				TP01_0597	–
IF3	MAL8P1.27	NOT FOUND	TGME49_224235	NOT FOUND	ETH_00024530	NOT FOUND	BBOV_II002530	NOT FOUND	TP01_0849 TA14970	NOT FOUND
EF-Tu	PFC10_api0028[a]	PF3D7_1330600	TGME49_302050	TGME49_262380	AAO40237.1	ETH_00018510	BBOV_V00020	BBOV_III00500	TP05_0019	TP05_0698
EF-Ts	PF3D7_0305000[b]	NOT FOUND	TGME49_209010	TGME49_307610	ETH_00022805	ETH_00028485	BBOV_II003360	NOT FOUND	TP04_0253	NOT FOUND
EF-G	PFF0115c[c]	PFL1590C[c]	TGME49_223970	TGME49_260170	ETH_00034460 (prediction for mitochondrial targeting)	NOT FOUND	BBOV_IV004710	BBOV_III000360	TP01_0278	TP03_0816
RFs	RF1: PF14_0265; RF2:MAL7P1.20	C12orf65: PFD0480w; ICT1: PF11_0182; RF1a: PF11575c	RF1 TGME49_007630	RF1a: TGME49_114270; C12orf65: TGVEG_043320	NOT FOUND	NOT FOUND	RF1: BBOV_IV006160: RF2: BBOV_II003430	C12orf65: BBOV_IV005260	RF1: TP 01_0422: RF2: TP04_0242	C12orf65: TP01_0213: ICT1: TP03_0303
RRF	PFB0390 W#[d]	PFD0990 W#[d]	TGME49_257980	TGME49_216530	ETH_00014315	NOT FOUND	BBOV_I004040	BBOV_II000580	TP02_0711	NOT FOUND

[a]Chaubey et al. 2005, [b]Biswas et al. 2011; [c]Johnson et al. 2011; [d]Gupta et al. 2013
#unpublished data, our laboratory

across apicomplexans except *Eimeria* (Table 7.1). Similar to mitochondria in other organisms, an IF-1 targeted to mitochondria of apicomplexans also cannot be identified (Koc and Spremulli 2002; Towpik 2005). It has been proposed that an insertion in the mitochondrial IF2 (mtIF2) sequence of *Bos taurus* compensates for the functional absence of IF1 (Gaur et al. 2008). It is possible that the mtIF2 in *Plasmodium* may contain an insert similar to one found in *Bos taurus* mtIF2; alternatively, translation in these mitochondria may not require IF1. The latter hypothesis is supported by phylogenetic analyses of IF2 which demonstrates poor conservation of the insertion amongst invertebrates (Atkinson et al. 2012). There is evidence in support of red algal descent of the apicoplast (Lim and McFadden 2010) and an IF2 is encoded by the plastid genomes of red alga *Porphyra Purpurea* (Reith and Munholland 1995), *Guillardia theta* (Douglas and Penny 1999), and *Cyanidioschyzon merolae* (Ohta et al. 2003). However, all apicomplexan apicoplast genomes have lost an IF2. Three putative IF2 encoding sequences are found on the nuclear genome in *Plasmodium falciparum* (Jackson et al. 2011) although targeting of their protein products to the apicoplast/mitochondria remains to be confirmed. Putative apicoplast-targeted orthologs of IF3 are predicted but a mitochondria-targeted IF3 is not identified across apicomplexan genomes (Table 7.1). This is unlike most mitochondrial translation systems, which are known to involve an IF3 (Koc and Spremulli 2002; Atkinson et al. 2012).

The use of formyl-methionine (fMet) as the initiator amino acid and fMet-tRNA^{fMet} as the initiator tRNA for the recognition of the start codon is presumed for translation of all apicoplast genes as the plastid DNA encodes its own initiator tRNA^{fMet}. In addition, methionyl formyltransferase (MFT) and peptide deformylase (PDF), needed for formylation and de-formylation of initiator Met-tRNA, have N-terminal apicoplast targeting sequences. The apicoplast localization of MFT and PDF has been confirmed in *T. gondii*, pointing to clear use of formylated initiator methionine in plastid translation (Pino et al. 2010). No such prediction can be made regarding apicomplexan mitochondria because they lack resident tRNAs as well as imported enzymatic machinery for aminoacylation and subsequent formylation or deformylation. It has been suggested that these mitochondria may sidestep the use of a prokaryotic initiator fMet-tRNA^{fMet} or just use the eukaryotic elongator Met-tRNA^{Met} imported from the cytosol (Pino et al. 2010). The absence of genes encoding MFT and PDF in the genomes of *Babesia* and *Theileria* strengthens the view that organellar translation can proceed in the absence of tRNA^{Met} formylation; the use of fMet-tRNA^{fMet} in both apicoplasts and mitochondria might be completely bypassed in *Babesia* and *Theileria*. The possibility of transport of the fMet-tRNA^{fMet} from the apicoplast to the mitochondrion for initiation from the only mitochondrial gene (*cytb*) containing an AUG start codon has been proposed (Howe and Purton 2007). This scenario is unlikely considering the four apicoplast and two mitochondrial membranes that the charged tRNA would have to traverse and yet maintain its charged status. Overall, it seems that apicomplexan mitochondria support translation without the requirement of formylated initiator methionine.

7.5   Elongation Factors and the Absence of Mitochondrial EF-Ts in *Plasmodium* spp.

The translation elongation phase involves addition of amino acids to the C-termini of growing nascent polypeptide chains (Mitra et al. 2005). The elongation cycle in prokaryotes requires the involvement of three factors; elongation factor Tu (EF-Tu), elongation factor Ts (EF-Ts), and elongation factor G (EF-G). Elongation initiates with the entry of the initiator tRNA (fMet-tRNA) to the P-site that leads to a conformational change which opens the A-site for binding a new aminoacyl-tRNA (aa-tRNA). This binding is assisted by EF-Tu bound with GTP in a ternary complex (aa-tRNA.EF-Tu.GTP). The codon-anticodon interaction at the P-site activates GTP hydrolysis by EF-Tu on the ribosome (Pape et al. 1998; Rodnina et al. 1995; Vorstenbosch et al. 1996) and the conformation of EF-Tu changes from the GTP- to the GDP-bound form (Abel et al. 1996; Polekhina et al. 1996). The GDP form has lower affinity for tRNA (Dell et al. 1990) and aa-tRNA is released from EF-Tu.GDP and accommodated at the A-site while EF-Tu.GDP is released from the ribosome. The release of the EF-Tu.GDP complex is facilitated by the guanine nucleotide exchange factor, EF-Ts, through the formation of an intermediate EF-Tu.GDP.EF-Ts complex; GDP is released from the complex followed by replacement of EF-Ts by GTP and regeneration of the active GTP form of EF-Tu (Dahl et al. 2006b). EF-G binds the ribosome after EF-Tu release and activates translocation, the movement of deacylated-tRNA from the ribosome P-site to the E-site and concurrent transfer of the peptidyl-tRNA from the A-site to the P-site. The A-site is thus made vacant for entry of another aa-tRNA.EF-Tu.GTP ternary complex (Yu et al. 2009).

The *tufA* gene encoding EF-Tu is found in the apicoplast of all apicomplexan parasites (Chaubey et al. 2005) while mitochondrial EF-Tu is nuclear-encoded (Pino et al. 2010). Nuclear-encoded nucleotide exchange factor EF-Ts has been shown to be targeted to the apicoplast in *P. falciparum* and its ability to mediate GDP-GTP exchange on apicoplast EF-Tu has been demonstrated (Biswas et al. 2011). The picture in the mitochondrion seems to differ as an obvious mitochondrial EF-Ts is lacking in the parasite genome. The related apicomplexans *Toxoplasma* and *Eimeria* each possess two bacterial EF-Ts genes that may function in both the apicoplast and mitochondrion. However, only a single EF-Ts is apparent in *Theileria* and *Plasmodium* species. It seems that the extremely reduced mitochondrial translation in *Plasmodium* may proceed without an EF-Ts and that slow recycling of EF-Tu.GDP to EF-Tu.GTP may suffice for elongation in the organelle. The mitochondria of budding yeast *Saccharomyces cerevisiae* also lack EF-Ts (Chiron et al. 2005) and it has been proposed that mitochondrial EF-Tu might compensate for the absence of EF-Ts by acquiring the ability of self exchange through decreased affinity for GDP or increased affinity for GTP (Chiron et al. 2005). It remains to be established whether such a mechanism exists in the mitochondria of apicomplexan parasites.

The presence of two nuclear-encoded EF-Gs targeted separately to the apicoplast and mitochondria has been shown in *P. falciparum* (Johnson et al. 2011).

Experiments from our laboratory have established that these organellar EF-Gs exhibit GTPase activity and participate in ribosome recycling in the presence of the corresponding organellar recycling factor (Gupta et al. 2013).

7.6 Translation Termination and Putative Release Factors in Organelles of Apicomplexan Parasites

Termination of translation is brought about by three stop codons which, unlike amino-acid specifying sense codons, are not recognized by tRNA but by proteins referred to as class I release factors (RF). Bacterial class I RFs mimic aminoacylated elongator-tRNAs in the ribosomal A-site and promote peptide chain release from the last elongator tRNA positioned at the P-site by hydrolysis of the ester bond (Capecchi 1967; Scolnick et al. 1968; Caskey et al. 1968). A GGQ motif which mediates hydrolysis is conserved throughout all ribosome-dependent RFs (Frolova 1999). Prokaryotic termination is supported by two class I release factors, RF1 and RF2, that recognize the stop codons UAA/UAG and UAA/UGA through their codon recognition motifs PxT and SPF, respectively (Scolnick et al. 1968); in eukaryotes recognition of all termination codons is by a single release factor, eRF1 (Dontsova 2000; Ito 2002).

Despite the fact that RF1 and RF2 are sufficient for effective translation termination, there has been evolution of an expanded RF protein family in eukaryotic organelles. It comprises of mtRF1a, mtRF2a, pRF1, and pRF2 that are mitochondrial and plastid versions of RF1 and RF2. Apart from these mtRF1, mtRF2b, mtRF2c, ICT1 (immature colon carcinoma transcript-1), and C12orf65 are noncanonical RFs that form a part of this extended RF family in organellar termination in mitochondria of some eukrayotes (Raczynska et al. 2006; Chrzanowska-Lightowlers et al. 2011). *P. falciparum* shows evidence of the existence of a putative noncanonical RF, mtRF1 that lacks the characteristic PxT motif and instead carries a PExGxS motif (Duarte et al. 2012). This mtRF1 may have evolved from an alpha-proteobacterial ancestor as inferred from phylogenetic analyses (Duarte et al. 2012). Apicomplexans *P. falciparum*, *T. gondii* and *B. bovis* are predicted to have a C12orf65 ortholog, which has been lost in green algae and land plants during evolution. *P. falciparum* and *T. parva* also contain a probable ICT1 which carries a GGQ motif but lacks a distinctive codon recognition motif. *P. falciparum*, *B. bovis* and *T. parva* nuclear genomes are predicted to encode putative pRF1 and pRF2 homologs for the apicoplast while *T. gondii* seems to lack a pRF2 (Duarte et al. 2012). During the course of evolution, both pRF2 and the TGA stop-codon seem to have been lost in *T. gondii* (Denny et al. 1998) as none of its 26 plastid ORFs are predicted to end with a TGA codon (Duarte et al. 2012). UAA is the predominant stop codon in the *P. falciparum* apicoplast with the remaining ORFs terminating with TGA. The targeting of both RF1 and RF2 to the *P. falciparum* apicoplast remains to be confirmed as does the possible redundancy in termination codon recognition by apicoplast-targeted RFs. The class II release factor RF3 is a GTPase which essentially mediates the recycling of class I RFs and helps in their removal from the ribosomal A site after GTP-dependent hydrolysis of the peptidyl-tRNA bond (Freistroffer et al. 1997). However, like many bacteria (Margus et al. 2007), RF3 seems to be missing for both apicoplast and mitochondria in apicomplexa (Jackson et al. 2011).

7.7 Recycling of Organellar Ribosomes

After the termination of protein synthesis and release of the nascent polypeptide chain, the post-termination complex is dissembled by ribosome recycling factor (RRF) (Hirashima and Kaji 1972) with concerted action of EF-G to start the next round of translation (Hirashima and Kaji 1973). Eukaryotes usually carry a homolog of RRF only in their organelles but not in the cytoplasm. The recycling of the 80S ribosome in eukaryotic cytoplasmic translation was a mystery for a long time till it was shown that eukaryotic translocase eEF2 (eukaryotic elongation factor 2) exhibits a novel activity that splits 80S ribosome into subunits in the presence of ATP (Demeshkina et al. 2007). A few other reports showed that the ribosome recycling step is catalyzed by a ribosome-dependent ATPase, eukaryotic elongation factor 3 (eEF3) in the cytoplasm of yeast (Kurata et al. 2010) and rabbit reticulocyte system (Pisarev et al. 2007).

RRF is universally conserved in eubacteria and its inactivation is deleterious for bacterial growth (Janosi et al. 1994). *Mycoplasma genitalium* which has one of the smallest genomes reported till date retains RRF rather than RF2 and RF3, further supporting the key role of this protein in the prokaryotic translation (Janosi et al. 1996). RRF is also reported to play a key role in both plastid (Rolland et al. 1999) as well as mitochondrial translation (Rorbach et al. 2008). Deletion of mitochondrial RRF is lethal with gross mitochondrial dysfunction and dysmorphology (Rorbach et al. 2008), while plastid RRF is essential for embryogenesis and chloroplast biogenesis as shown in *Arabidopsis thaliana* (Rolland et al. 1999). Targeting of a nuclear-encoded RRF homolog each to the apicoplast and mitochondrion of *P. falciparum* has been shown by N-terminal coding sequence-GFP fusions and their ability to split 70S ribosomes has been confirmed in our laboratory (Gupta et al. 2013).

7.8 Translation Components as Targets for Drug Intervention Against Apicomplexan Parasites

Since apicomplexan parasites are causative agents of several infectious diseases affecting humans and livestock there is need for the development of new effective drugs. This gains immense significance due to increasing drug resistance observed against drugs currently being used against parasites such as *Plasmodium*. Having surveyed the participants and the mechanisms governing the translation process and also having identified potential candidates from the apicomplexan proteome that are expected to function as organellar translation factors, we now briefly examine the current status of research addressing the possibility of using organellar translation components as targets for drug intervention against malaria.

The apicoplast was demonstrated to be essential for parasite survival in blood stages of *P. falciparum* by using antibiotics that inhibited apicoplast genome replication (Fichera and Roos 1997). Since a functional apicoplast is essential for a viable parasite cell, inhibition of housekeeping functions of the plastid leads to parasite death, although this often follows a 'delayed-death' phenotype wherein the blood-stage parasite dies in the cycle subsequent to the one in which it has been exposed to the drug. The apicoplast harbours biosynthetic pathways such as type II fatty acid synthesis (Waller et al. 2000; Surolia and Surolia 2001), the synthesis of heme (Sato et al. 2004; Dhanasekaran et al. 2004) as well as [Fe–S] complexation (Kumar et al. 2011). However, the production of isopentenyl pyrophosphate (IPP) via the nonmevalonate isoprenoid biosynthesis pathway seems to be the only essential function of the apicoplast (Yeh and DeRisi 2011). There is evidence that components of the apicoplast and/or mitochondrial translation apparatus are targets of antibiotics that exhibit anti-malarial effects. Translation inhibitory antibiotics that are detrimental to *Plasmodium* and/or *Toxoplasma* survival are listed in Table 7.2.

The ribosome is an important target for drug intervention as a variety of translation inhibitory antibiotics bind to ribosomes and block protein synthesis thereby inhibiting cell growth. One of the critical steps in developing new translation inhibitors is the characterization of ribosomal components and their interaction with these molecules. However, there is limited understanding of organellar ribosome composition in apicomplexan parasites. Mutational studies with apicomplexan rRNAs confirm the mode of inhibition of several antibiotics such as clindamycin (Camps et al. 2002) and thiostrepton (Clough et al. 1997; Rogers et al. 1997) that inhibit parasite growth by interacting with apicoplast encoded rRNA. Another macrolide, azithromycin, inhibits not only human blood and liver stage parasites but also parasite development in the mosquito (Shimizu et al. 2010). In vitro studies with blood stage *P. falciparum* culture reveal that azithromycin occupies the 50S polypeptide exit tunnel by interacting with apicoplast-encoded RPL4, nuclear encoded RPL22 and large subunit rRNA (Sidhu et al. 2007; Poehlsgaard and Douthwaite 2005). A point mutation in the LSU rRNA gene of the *T. gondii* apicoplast has been shown to confer resistance to clindamycin in vitro (Camps et al. 2002). Tetracycline inhibits translation by preventing the binding of peptidyl t-RNA to the acceptor site on ribosomal small subunit (Brodersen et al. 2000) and apicoplast-specific effects such as disruption of protein import into the organelle by clindamycin and tetracycline has been observed (Goodman et al. 2007). The translation inhibitor doxycycline is in clinical use as a malaria prophylactic agent for travelers to malaria-endemic areas. Doxycycline-mediated block in expression of the apicoplast genome resulting in the distribution of nonfunctional apicoplasts during erythrocytic schizogony has been implicated

Table 7.2 Translation inhibitory antibiotics affecting growth of *Plasmodium falciparum* and/or *Toxoplasma gondii*

Drug target	Drug	IC$_{50}$	Reference
Apicoplast 23S rRNA	Clindamycin	*Pf* 20nM	Fichera et al. (1995), Fichera and Roos (1997), Pfefferkorn and Borotz (1994)
	Azithromycin	*Pf* 2 μM, *Tg* 2 μM	Fichera et al. (1995), Pfefferkorn and Borotz (1994), Sidhu et al. (2007)
	Thiostrepton	*Pf* 2 μM	Clough et al. (1997), McConkey et al. 1997); Rogers et al. (1997), Sullivan et al. 92000), Chaubey et al. (2005); Goodman et al. (2007)
	Micrococcin	*Pf* 35nM	Rogers et al. (1997), Beckers et al. (1995), Budimulja et al. (1997), Fichera and Roos (1997)
	Chloramphenicol	*Pf* 10 μM, *Tg* 5 μM	Goodman et al. (2007), Beckers et al. (1995), Budimulja et al. (1997), Fichera and Roos (1997)
Apicoplast 16S rRNA	Doxycycline	*Pf* 11.3 μM	Budimulja et al. (1997), Pradines et al. (2000)
	Tetracycline	*Pf* 10 μM, *Tg* 20 μM	Budimulja et al. (1997), Dahl et al. (2006a), Pradines et al. (2000)
Apicoplast elongation factor-Tu	Amythiamycin	*Pf* 10nM	Clough et al. (1999)
	Kirromycin	*Pf* 50 μM	
	GE2270A	*Pf* 0.3 μM	
Apicoplast elongation factor-G	Fusidic Acid	*Pf* 60 μM	Black et al. (1985)
Apicoplast Ile-RS	Mupirocin	*Pf* 50nM	Istvan et al. (2011), Jackson et al. (2011)
Dual-targeted Thr-RS	Borrelidin	*Pf* 1.4nM	Istvan et al. (2011), Jackson et al. (2011)
Cytosolic Lys-RS	Cladosporin	*Pf* 45.4 nM	Hoepfner et al. (2012)

in the anti-malarial action of the drug (Dahl et al. 2006a). Live microscopy on *P. berghei* has shown that microbial translation inhibitors also block the development of the apicoplast during liver stage schizogony and lead to impaired parasite maturation (Stanway et al. 2009).

The identification of additional targets in the organellar protein translation apparatus may offer new sites for drug intervention and contribute to enhancing the drug repertoire against resistant strains. Recent studies on tRNA synthetases in *Plasmodium* have identified mupirocin as a potent inhibitor of apicoplast Ile-RS (Istvan et al. 2011; Jackson et al. 2011). The drug exhibits a classic delayed-death effect typical of apicoplast-specific action. This is in contrast to the Thr-RS inhibitor borrelidin, which kills parasites rapidly and whose target Thr-RS has now been demonstrated to be dually targeted to the cytoplasm and apicoplast of *P. falciparum* (Ishiyama et al. 2011; Jackson et al. 2011). EF-Tu is an important drug target as many antibiotics interact directly with the factor. A member of the thiazolyl peptide family, GE2270A, and kirromycin show antimalarial activity by inhibiting apico-plast translation in *P. falciparum* (Clough et al. 1999). GE2270A inhibits bacterial protein synthesis (Kettenring et al. 1991; Selva et al. 1991) by competing with aa-tRNA for binding on EF-Tu and therefore blocks the GDP to GTP conformational change (Heffron and Jurnak 2000). Other antibiotics such as pulvomycin and kir-romycin also act directly on EF-Tu and inhibit protein synthesis (Wolf et al. 1978; Wolf et al. 1974) by disturbing the allosteric changes required for the switch from EF-Tu.GTP to EF-Tu.GDP. Interaction of kirromycin and its effect on GDP release by *P. falciparum* apicoplast EF-Tu has been demonstrated (Biswas et al. 2011). Organellar EF-Gs are also candidate targets for drug intervention against apicompl-exan parasites. The steroid antibiotic fusidic acid shows an anti-parasiticidal activ-ity on *P. falciparum* blood stages (Johnson et al. 2011) as well as on veterinary parasites *Babesia* and *Theileria* (Salama et al. 2012). The drug acts by locking EF-G.GDP onto the ribosome after translocation thus inhibiting the next round of elon-gation (Willie et al. 1975; Ticu et al. 2009). Two nuclear-encoded EF-Gs targeted to the apicoplast and mitochondria have been identified in *P. falciparum* (Johnson et al. 2011). Experiments with recombinant apicoplast and mitochondrial EF-Gs in our laboratory suggest that FA has a more pronounced effect of apicoplast EF-G compared to the mitochondrial factor (unpublished data).

Recent interest in translation mechanisms of the apicoplast and mitochondrion of apicomplexan parasites might help identify potential targets for drug interven-tion and help in management of diseases like malaria. It is important that studies be taken up to confirm the localization and function of these factors in parasite organelles. In this context, the evaluation of the precise target-specific activities of lead compounds is a challenge, and the possibility of off-targets at other loca-tions in the parasite needs to be addressed. A recently developed procedure that will aid in confirming the apicoplast as the site of action of specific anti-malarials is the chemical rescue of apicoplast-minus *P. falciparum* parasites in asexual blood stages by IPP, the product of the essential nonmevalonate isoprenoid biosynthe-sis pathway in the apicoplast (Yeh and DeRisi 2011). The evaluation of the effect of potential translation inhibitors on liver stages of the parasite cycle would also require intensive exploration of the translation process in these exo-erythrocytic stages.

References

Abel K, Yoder MD, Hilgenfeld R, Jurnak F (1996) An alpha to beta conformational switch in EF-Tu. Structure 4(10):1153–1159

Abrahamsen MS, Templeton TJ, Enomoto S, Abrahante JE, Zhu G, Lancto CA, Deng M, Liu C, Widmer G, Tzipori S, Buck GA, Xu P, Bankier AT, Dear PH, Konfortov BA, Spriggs HF, Iyer L, Anantharaman V, Aravind L, Kapur V (2004) Complete genome sequence of the api-complexan Cryptosporidium parvum. Science 304(5669):441–445

Adl SM, Simpson AG, Farmer MA, Andersen RA, Anderson OR, Barta JR, Bowser SS, Brugerolle G, Fensome RA, Fredericq S, James TY, Karpov S, Kugrens P, Krug J, Lane CE,

Lewis LA, Lodge J, Lynn DH, Mann DG, McCourt RM, Mendoza L, Moestrup O, Mozley-Standridge SE, Nerad TA, Shearer CA, Smirnov AV, Spiegel FW, Taylor MF (2005) The new higher level classification of eukaryotes with emphasis on the taxonomy of protists. J Eukaryot Microbiol 52(5):399–451

Antoun A, Pavlov MY, Andersson K, Tenson T, Ehrenberg M (2003) The roles of initiation factor 2 and guanosine triphosphate in initiation of protein synthesis. EMBO J 22(20):5593–5601

Antoun A, Pavlov MY, Tenson T, Ehrenberg MM (2004) Ribosome formation from subunits studied by stopped-flow and rayleigh light scattering. Biol Proced Online 6:35–54

Arnold RJ, Reilly JP (1999) Observation of Escherichia coli ribosomal proteins and their posttranslational modifications by mass spectrometry. Anal Biochem 269(1):105–112

Atkinson GC, Kuzmenko A, Kamenski P, Vysokikh MY, Lakunina V, Tankov S, Smirnova E, Soosaar A, Tenson T, Hauryliuk V (2012) Evolutionary and genetic analyses of mitochondrial translation initiation factors identify the missing mitochondrial IF3 in S. cerevisiae. Nucleic Acids Res 40(13):6122–6134

Baldauf SL, Roger AJ, Wenk-Siefert I, Doolittle WF (2000) A kingdom-level phylogeny of eukaryotes based on combined protein data. Science 290(5493):972–977

Beckers CJ, Roos DS, Donald RG, Luft BJ, Schwab JC, Cao Y, Joiner KA (1995) Inhibition of cytoplasmic and organellar protein synthesis in Toxoplasma gondii. Implications for the target of macrolide antibiotics. J Clin Invest 95(1):367–376

Betts L, Spremulli LL (1994) Analysis of the role of the Shine-Dalgarno sequence and mRNA secondary structure on the efficiency of translational initiation in the Euglena gracilis chloroplast atpH mRNA. J Biol Chem 269(42):26456–26463

Bhatt TK, Kapil C, Khan S, Jairajpuri MA, Sharma V, Santoni D, Silvestrini F, Pizzi E, Sharma A (2009) A genomic glimpse of aminoacyl-tRNA synthetases in malaria parasite Plasmodium falciparum. BMC Genomics 10:644

Biswas S, Lim EE, Gupta A, Saqib U, Mir SS, Siddiqi MI, Ralph SA, Habib S (2011) Interaction of apicoplast-encoded elongation factor (EF) EF-Tu with nuclear-encoded EF-Ts mediates translation in the Plasmodium falciparum plastid. Int J Parasitol 41(3–4):417–427

Black FT, Wildfang IL, Borgbjerg K (1985) Activity of fusidic acid against Plasmodium falciparum in vitro. Lancet 1(8428):578–579

Brayton KA, Lau AO, Herndon DR, Hannick L, Kappmeyer LS, Berens SJ, Bidwell SL, Brown WC, Crabtree J, Fadrosh D, Feldblum T, Forberger HA, Haas BJ, Howell JM, Khouri H, Koo H, Mann DJ, Norimine J, Paulsen IT, Radune D, Ren Q, Smith RK Jr, Suarez CE, White O, Wortman JR, Knowles DP Jr, McElwain TF, Nene VM (2007) Genome sequence of Babesia bovis and comparative analysis of apicomplexan hemoprotozoa. PLoS Pathog 3(10):1401–1413

Brehelin L, Florent I, Gascuel O, Marechal E (2010) Assessing functional annotation transfers with inter-species conserved coexpression: application to Plasmodium falciparum. BMC Genomics 11:35

Brodersen DE, Clemons WM Jr, Carter AP, Morgan-Warren RJ, Wimberly BT, Ramakrishnan V (2000) The structural basis for the action of the antibiotics tetracycline, pactamycin, and hygromycin B on the 30S ribosomal subunit. Cell 103(7):1143–1154

Budimulja AS, Syafruddin Tapchaisri P, Wilairat P, Marzuki S (1997) The sensitivity of Plasmodium protein synthesis to prokaryotic ribosomal inhibitors. Mol Biochem Parasitol 84(1):137–141

Camps M, Arrizabalaga G, Boothroyd J (2002) An rRNA mutation identifies the apicoplast as the target for clindamycin in Toxoplasma gondii. Mol Microbiol 43(5):1309–1318

Capecchi MR (1967) Polypeptide chain termination in vitro: isolation of a release factor. Proc Natl Acad Sci U S A 58(3):1144–1151

Caskey CT, Tompkins R, Scolnick E, Caryk T, Nirenberg M (1968) Sequential translation of trinucleotide codons for the initiation and termination of protein synthesis. Science 162(3849):135–138

Chaubey S, Kumar A, Singh D, Habib S (2005) The apicoplast of Plasmodium falciparum is translationally active. Mol Microbiol 56(1):81–89

Chiron S, Suleau A, Bonnefoy N (2005) Mitochondrial translation: elongation factor tu is essential in fission yeast and depends on an exchange factor conserved in humans but not in budding yeast. Genetics 169(4):1891–1901

Chrzanowska-Lightowlers ZM, Pajak A, Lightowlers RN (2011) Termination of protein synthesis in mammalian mitochondria. J Biol Chem 286(40):34479–34485

Cilingir G, Broschat SL, Lau AO (2012) ApicoAP: the first computational model for identifying apicoplast-targeted proteins in multiple species of apicomplexa. PLoS ONE 7(5):e36598

Clough B, Rangachari K, Strath M, Preiser PR, Wilson RJ (1999) Antibiotic inhibitors of organellar protein synthesis in Plasmodium falciparum. Protist 150(2):189–195

Clough B, Strath M, Preiser P, Denny P, Wilson IR (1997) Thiostrepton binds to malarial plastid rRNA. FEBS Lett 406(1–2):123–125

Dahl EL, Shock JL, Shenai BR, Gut J, DeRisi JL, Rosenthal PJ (2006a) Tetracyclines specifically target the apicoplast of the malaria parasite Plasmodium falciparum. Antimicrob Agents Chemother 50(9):3124–3131

Dahl LD, Wieden HJ, Rodnina MV, Knudsen CR (2006b) The importance of P-loop and domain movements in EF-Tu for guanine nucleotide exchange. J Biol Chem 281(30):21139–21146

Dell VA, Miller DL, Johnson AE (1990) Effects of nucleotide- and aurodox-induced changes in elongation factor Tu conformation upon its interactions with aminoacyl transfer RNA. A fluorescence study. Biochemistry 29(7):1757–1763

Demeshkina N, Hirokawa G, Kaji A, Kaji H (2007) Novel activity of eukaryotic translocase, eEF2: dissociation of the 80S ribosome into subunits with ATP but not with GTP. Nucleic Acids Res 35(14):4597–4607

Denny P, Preiser P, Williamson D, Wilson I (1998) Evidence for a single origin of the 35 kb plastid DNA in apicomplexans. Protist 149(1):51–59

Dhanasekaran S, Chandra NR, Chandrasekhar Sagar BK, Rangarajan PN, Padmanaban G (2004) Delta-aminolevulinic acid dehydratase from Plasmodium falciparum: indigenous versus imported. J Biol Chem 279(8):6934–6942

Dontsova M, Frolova L, Vassilieva J, Piendl W, Kisselev L, Garber M (2000) Translation termination factor aRF1 from the archaeon Methanococcus jannaschii is active with eukaryotic ribosomes. FEBS Lett 472(2–3):213–216

Douglas SE, Penny SL (1999) The plastid genome of the cryptophyte alga, Guillardia theta: complete sequence and conserved synteny groups confirm its common ancestry with red algae. J Mol Evol 48(2):236–244

Duarte I, Nabuurs SB, Magno R, Huynen M (2012) Evolution and diversification of the organellar release factor family. Mol Biol Evol 29(11):3497–3512

Esseiva AC, Naguleswaran A, Hemphill A, Schneider A (2004) Mitochondrial tRNA import in Toxoplasma gondii. J Biol Chem 279(41):42363–42368

Feagin JE (1992) The 6-kb element of Plasmodium falciparum encodes mitochondrial cytochrome genes. Mol Biochem Parasitol 52(1):145–148

Feagin JE, Harrell MI, Lee JC, Coe KJ, Sands BH, Cannone JJ, Tami G, Schnare MN, Gutell RR (2012) The fragmented mitochondrial ribosomal RNAs of Plasmodium falciparum. PLoS ONE 7(6):e38320

Feagin JE, Mericle BL, Werner E, Morris M (1997) Identification of additional rRNA fragments encoded by the Plasmodium falciparum 6 kb element. Nucleic Acids Res 25(2):438–446

Feagin JE, Werner E, Gardner MJ, Williamson DH, Wilson RJ (1992) Homologies between the contiguous and fragmented rRNAs of the two Plasmodium falciparum extrachromosomal DNAs are limited to core sequences. Nucleic Acids Res 20(4):879–887

Fichera ME, Bhopale MK, Roos DS (1995) In vitro assays elucidate peculiar kinetics of clindamycin action against Toxoplasma gondii. Antimicrob Agents Chemother 39(7):1530–1537

Fichera ME, Roos DS (1997) A plastid organelle as a drug target in apicomplexan parasites. Nature 390(6658):407–409

Foth BJ, Ralph SA, Tonkin CJ, Struck NS, Fraunholz M, Roos DS, Cowman AF, McFadden GI (2003) Dissecting apicoplast targeting in the malaria parasite Plasmodium falciparum. Science 299(5607):705–708

Freistroffer DV, Pavlov MY, MacDougall J, Buckingham RH, Ehrenberg M (1997) Release factor RF3 in *E.coli* accelerates the dissociation of release factors RF1 and RF2 from the ribosome in a GTP-dependent manner. EMBO J 16(13):4126–4133

Frolova LY, Tsivkovskii RY, Sivolobova GF, Oparina NY, Serpinsky OI, Blinov VM, Tatkov SI, Kisselev LL (1999) Mutations in the highly conserved GGQ motif of class 1 polypeptide release factors abolish ability of human eRF1 to trigger peptidyl-tRNA hydrolysis. RNA 5(8):1014–1020

Gaur R, Grasso D, Datta PP, Krishna PD, Das G, Spencer A, Agrawal RK, Spremulli L, Varshney U (2008) A single mammalian mitochondrial translation initiation factor functionally replaces two bacterial factors. Mol Cell 29(2):180–190

Gillespie DE, Salazar NA, Rehkopf DH, Feagin JE (1999) The fragmented mitochondrial ribosomal RNAs of Plasmodium falciparum have short a tails. Nucleic Acids Res 27(11):2416–2422

Goodman CD, Su V, McFadden GI (2007) The effects of anti-bacterials on the malaria parasite Plasmodium falciparum. Mol Biochem Parasitol 152(2):181–191

Gray MW, Burger G, Lang BF (2001) The origin and early evolution of mitochondria. Genome Biol 2 (6):REVIEWS1018

Gray MW, Lang BF, Burger G (2004) Mitochondria of protists. Annu Rev Genet 38:477–524

Gualerzi CO, Pon CL (1990) Initiation of mRNA translation in prokaryotes. Biochemistry 29(25):5881–5889

Gupta A, Mir SS, Jackson KE, Lim EE, Shah P, Sinha A, Siddiqi MI, Ralph SA, Habib S (2013) Recycling factors for ribosome disassembly in the apicoplast and mitochondrion of Plasmodium falciparum. Mol Microbiol 88(5):891–905

Harb OS, Chatterjee B, Fraunholz MJ, Crawford MJ, Nishi M, Roos DS (2004) Multiple functionally redundant signals mediate targeting to the apicoplast in the apicomplexan parasite Toxoplasma gondii. Eukaryot Cell 3(3):663–674

Hartz D, McPheeters DS, Gold L (1989) Selection of the initiator tRNA by *E. coli* initiation factors. Genes Dev 3(12A):1899–1912

Heffron SE, Jurnak F (2000) Structure of an EF-Tu complex with a thiazolyl peptide antibiotic determined at 2.35 A resolution: atomic basis for GE2270A inhibition of EF-Tu. Biochemistry 39(1):37–45

Hikosaka K, Watanabe Y, Kobayashi F, Waki S, Kita K, Tanabe K (2011) Highly conserved gene arrangement of the mitochondrial genomes of 23 Plasmodium species. Parasitol Int 60(2):175–180

Hikosaka K, Watanabe Y, Tsuji N, Kita K, Kishine H, Arisue N, Palacpac NM, Kawazu S, Sawai H, Horii T, Igarashi I, Tanabe K (2010) Divergence of the mitochondrial genome structure in the apicomplexan parasites Babesia and Theileria. Mol Biol Evol 27(5):1107–1116

Hirashima A, Kaji A (1972) Purification and properties of ribosome-releasing factor. Biochemistry 11(22):4037–4044

Hirashima A, Kaji A (1973) Role of elongation factor G and a protein factor on the release of ribosomes from messenger ribonucleic acid. J Biol Chem 248(21):7580–7587

Hoepfner D, McNamara CW, Lim CS, Studer C, Riedl R, Aust T, McCormack SL, Plouffe DM, Meister S, Schuierer S, Plikat U, Hartmann N, Staedtler F, Cotesta S, Schmitt EK, Petersen F, Supek F, Glynne RJ, Tallarico JA, Porter JA, Fishman MC, Bodenreider C, Diagana TT, Movva NR, Winzeler EA (2012) Selective and specific inhibition of the Plasmodium falciparum lysyl-tRNA synthetase by the fungal secondary metabolite cladosporin. Cell Host Microbe 11(6):654–663

Howe CJ, Purton S (2007) The little genome of apicomplexan plastids: its raison d'etre and a possible explanation for the 'delayed death' phenomenon. Protist 158(2):121–133

Ishiyama A, Iwatsuki M, Namatame M, Nishihara-Tsukashima A, Sunazuka T, Takahashi Y, Omura S, Otoguro K (2011) Borrelidin, a potent antimalarial: stage-specific inhibition profile of synchronized cultures of Plasmodium falciparum. J Antibiot (Tokyo) 64(5):381–384

Istvan ES, Dharia NV, Bopp SE, Gluzman I, Winzeler EA, Goldberg DE (2011) Validation of isoleucine utilization targets in Plasmodium falciparum. Proc Natl Acad Sci U S A 108(4):1627–1632

Ito K, Frolova L, Seit-Nebi A, Karamyshev A, Kisselev L, Nakamura Y (2002) Omnipotent decoding potential resides in eukaryotic translation termination factor eRF1 of variant-code organisms and is modulated by the interactions of amino acid sequences within domain 1. Proc Natl Acad Sci U S A 99(13):8494–8499

Jackson KE, Habib S, Frugier M, Hoen R, Khan S, Pham JS, Ribas de Pouplana L, Royo M, Santos MA, Sharma A, Ralph SA (2011) Protein translation in Plasmodium parasites. Trends Parasitol 27(10):467–476

Janosi L, Ricker R, Kaji A (1996) Dual functions of ribosome recycling factor in protein bio-synthesis: disassembling the termination complex and preventing translational errors. Biochimie 78(11–12):959–969

Janosi L, Shimizu I, Kaji A (1994) Ribosome recycling factor (ribosome releasing factor) is essential for bacterial growth. Proc Natl Acad Sci U S A 91(10):4249–4253

Johnson RA, McFadden GI, Goodman CD (2011) Characterization of two malaria parasite orga-nelle translation elongation factor G proteins: the likely targets of the anti-malarial fusidic acid. PLoS ONE 6(6):e20633

Jomaa H, Wiesner J, Sanderbrand S, Altincicek B, Weidemeyer C, Hintz M, Turbachova I, Eberl M, Zeidler J, Lichtenthaler HK, Soldati D, Beck E (1999) Inhibitors of the nonmevalonate pathway of isoprenoid biosynthesis as antimalarial drugs. Science 285(5433):1573–1576

Kairo A, Fairlamb AH, Gobright E, Nene V (1994) A 7.1 kb linear DNA molecule of Theileria parva has scrambled rDNA sequences and open reading frames for mitochondrially encoded proteins. EMBO J 13(4):898–905

Kamenski P, Kolesnikova O, Jubenot V, Entelis N, Krasheninnikov IA, Martin RP, Tarassov I (2007) Evidence for an adaptation mechanism of mitochondrial translation via tRNA import from the cytosol. Mol Cell 26(5):625–637

Kettenring J, Colombo L, Ferrari P, Tavecchia P, Nebuloni M, Vekey K, Gallo GG, Selva E (1991) Antibiotic GE2270 a: a novel inhibitor of bacterial protein synthesis II. Structure elu-cidation. J Antibiot (Tokyo) 44(7):702–715

Koc EC, Spremulli LL (2002) Identification of mammalian mitochondrial translational initiation factor 3 and examination of its role in initiation complex formation with natural mRNAs. J Biol Chem 277(38):35541–35549

Kohler S, Delwiche CF, Denny PW, Tilney LG, Webster P, Wilson RJ, Palmer JD, Roos DS (1997) A plastid of probable green algal origin in apicomplexan parasites. Science 275(5305):1485–1489

Kumar B, Chaubey S, Shah P, Tanveer A, Charan M, Siddiqi MI, Habib S (2011) Interaction between sulphur mobilisation proteins SufB and SufC: evidence for an iron-sulphur cluster biogenesis pathway in the apicoplast of Plasmodium falciparum. Int J Parasitol 41(9):991–999

Kurata S, Nielsen KH, Mitchell SF, Lorsch JR, Kaji A, Kaji H (2010) Ribosome recycling step in yeast cytoplasmic protein synthesis is catalyzed by eEF3 and ATP. Proc Natl Acad Sci U S A 107(24):10854–10859

Lang-Unnasch N, Aiello DP (1999) Sequence evidence for an altered genetic code in the Neospora caninum plastid. Int J Parasitol 29(10):1557–1562

Lim L, McFadden GI (2010) The evolution, metabolism and functions of the apicoplast. Philos Trans R Soc Lond B Biol Sci 365(1541):749–763

Lockwood AH, Chakraborty PR, Maitra U (1971) A complex between initiation factor IF2, guanosine triphosphate, and fMet-tRNA: an intermediate in initiation complex formation. Proc Natl Acad Sci U S A 68(12):3122–3126

Margus T, Remm M, Tenson T (2007) Phylogenetic distribution of translational GTPases in bac-teria. BMC Genomics 8:15

Martin W, Stoebe B, Goremykin V, Hapsmann S, Hasegawa M, Kowallik KV (1998) Gene trans-fer to the nucleus and the evolution of chloroplasts. Nature 393(6681):162–165

McConkey GA, Rogers MJ, McCutchan TF (1997) Inhibition of plasmodium falciparum protein synthesis. Targeting the plastid-like organelle with thiostrepton. J Biol Chem 272(4):2046–2049

McFadden GI, Waller RF (1997) Plastids in parasites of humans. BioEssays 19(11):1033–1040

Mitra K, Schaffitzel C, Shaikh T, Tama F, Jenni S, Brooks CL, 3rd, Ban N, Frank J (2005) Structure of the E. coli protein-conducting channel bound to a translating ribosome. Nature 438 (7066):318–324

Nene V, Morzaria S, Bishop R (1998) Organisation and informational content of the Theileria parva genome. Mol Biochem Parasitol 95(1):1–8

Ohta N, Matsuzaki M, Misumi O, Miyagishima SY, Nozaki H, Tanaka K, Shin IT, Kohara Y, Kuroiwa T (2003) Complete sequence and analysis of the plastid genome of the unicellular red alga Cyanidioschyzon merolae. DNA Res 10(2):67–77

Palmer JD, Soltis D, Soltis P (1992) Large size and complex structure of mitochondrial DNA in two nonflowering land plants. Curr Genet 21(2):125–129

Pape T, Wintermeyer W, Rodnina MV (1998) Complete kinetic mechanism of elongation factor Tu-dependent binding of aminoacyl-tRNA to the A site of the E. coli ribosome. EMBO J 17(24):7490–7497

Pfefferkorn ER, Borotz SE (1994) Comparison of mutants of Toxoplasma gondii selected for resistance to azithromycin, spiramycin, or clindamycin. Antimicrob Agents Chemother 38(1):31–37

Pino P, Aeby E, Foth BJ, Sheiner L, Soldati T, Schneider A, Soldati-Favre D (2010) Mitochondrial translation in absence of local tRNA aminoacylation and methionyl tRNA Met formylation in Apicomplexa. Mol Microbiol 76(3):706–718

Pisarev AV, Hellen CU, Pestova TV (2007) Recycling of eukaryotic posttermination ribosomal complexes. Cell 131(2):286–299

Poehlsgaard J, Douthwaite S (2005) The bacterial ribosome as a target for antibiotics. Nat Rev Microbiol 3(11):870–881

Polekhina G, Thirup S, Kjeldgaard M, Nissen P, Lippmann C, Nyborg J (1996) Helix unwinding in the effector region of elongation factor EF-Tu-GDP. Structure 4(10):1141–1151

Pradines B, Spiegel A, Rogier C, Tall A, Mosnier J, Fusai T, Trape JF, Parzy D (2000) Antibiotics for prophylaxis of Plasmodium falciparum infections: in vitro activity of doxycycline against Senegalese isolates. Am J Trop Med Hyg 62(1):82–85

Preiser PR, Wilson RJ, Moore PW, McCready S, Hajibagheri MA, Blight KJ, Strath M, Williamson DH (1996) Recombination associated with replication of malarial mitochondrial DNA. EMBO J 15(3):684–693

Raczynska KD, Le Ret M, Rurek M, Bonnard G, Augustyniak H, Gualberto JM (2006) Plant mitochondrial genes can be expressed from mRNAs lacking stop codons. FEBS Lett 580(24):5641–5646

Rehkopf DH, Gillespie DE, Harrell MI, Feagin JE (2000) Transcriptional mapping and RNA processing of the Plasmodium falciparum mitochondrial mRNAs. Mol Biochem Parasitol 105(1):91–103

Reith M, Munholland J (1995) Complete nucleotide sequence of the Porphyra purpurea chloroplast genome. Plant Mol Biol Rep 13(4):333–335

Rodnina MV, Pape T, Fricke R, Wintermeyer W (1995) Elongation factor Tu, a GTPase triggered by codon recognition on the ribosome: mechanism and GTP consumption. Biochem Cell Biol 73(11–12):1221–1227

Rogers MJ, Bukhman YV, McCutchan TF, Draper DE (1997) Interaction of thiostrepton with an RNA fragment derived from the plastid-encoded ribosomal RNA of the malaria parasite. RNA (New York, NY 3 (8):815–820

Rolland N, Janosi L, Block MA, Shuda M, Teyssier E, Miege C, Cheniclet C, Carde JP, Kaji A, Joyard J (1999) Plant ribosome recycling factor homologue is a chloroplastic protein and is bactericidal in Escherichia coli carrying temperature-sensitive ribosome recycling factor. Proc Natl Acad Sci U S A 96(10):5464–5469

Rorbach J, Richter R, Wessels HJ, Wydro M, Pekalski M, Farhoud M, Kuhl I, Gaisne M, Bonnefoy N, Smeitink JA, Lightowlers RN, Chrzanowska-Lightowlers ZM (2008) The human mitochondrial ribosome recycling factor is essential for cell viability. Nucleic Acids Res 36(18):5787–5799

Rusconi CP, Cech TR (1996) Mitochondrial import of only one of three nuclear-encoded glutamine tRNAs in Tetrahymena thermophila. EMBO J 15(13):3286–3295

Salama AA, Aboulaila M, Moussa AA, Nayel MA, El-Sify A, Terkawi MA, Hassan HY, Yokoyama N, Igarashi I (2012) Evaluation of in vitro and in vivo inhibitory effects of fusidic acid on Babesia and Theileria parasites. Vet Parasitol

Sato S (2011) The apicomplexan plastid and its evolution. Cell Mol Life Sci 68(8):1285–1296

Sato S, Clough B, Coates L, Wilson RJ (2004) Enzymes for heme biosynthesis are found in both the mitochondrion and plastid of the malaria parasite Plasmodium falciparum. Protist 155(1):117–125

Scolnick E, Tompkins R, Caskey T, Nirenberg M (1968) Release factors differing in specificity for terminator codons. Proc Natl Acad Sci U S A 61(2):768–774

Selva E, Beretta G, Montanini N, Saddler GS, Gastaldo L, Ferrari P, Lorenzetti R, Landini P, Ripamonti F, Goldstein BP et al (1991) Antibiotic GE2270 a: a novel inhibitor of bacterial protein synthesis I. Isolation and characterization. J Antibiot (Tokyo) 44(7):693–701

Shimizu S, Osada Y, Kanazawa T, Tanaka Y, Arai M (2010) Suppressive effect of azithromycin on Plasmodium berghei mosquito stage development and apicoplast replication. Malar J 9:73

Shine J, Dalgarno L (1975) Determinant of cistron specificity in bacterial ribosomes. Nature 254(5495):34–38

Sidhu AB, Sun Q, Nkrumah LJ, Dunne MW, Sacchettini JC, Fidock DA (2007) In vitro efficacy, resistance selection, and structural modeling studies implicate the malarial parasite apicoplast as the target of azithromycin. J Biol Chem 282(4):2494–2504

Stanway RR, Witt T, Zobiak B, Aepfelbacher M, Heussler VT (2009) GFP-targeting allows visualization of the apicoplast throughout the life cycle of live malaria parasites. Biol Cell 101 (7):415–430, 415 p following 430

Sullivan M, Li J, Kumar S, Rogers MJ, McCutchan TF (2000) Effects of interruption of apicoplast function on malaria infection, development, and transmission. Mol Biochem Parasitol 109(1):17–23

Surolia N, Surolia A (2001) Triclosan offers protection against blood stages of malaria by inhibiting enoyl-ACP reductase of Plasmodium falciparum. Nat Med 7(2):167–173

Tarun AS, Vaughan AM, Kappe SH (2009) Redefining the role of de novo fatty acid synthesis in Plasmodium parasites. Trends Parasitol 25(12):545–550

Ticu C, Nechifor R, Nguyen B, Desrosiers M, Wilson KS (2009) Conformational changes in switch I of EF-G drive its directional cycling on and off the ribosome. EMBO J 28(14):2053–2065

Tonkin CJ, Kalanon M, McFadden GI (2008) Protein targeting to the malaria parasite plastid. Traffic 9(2):166–175

Towpik J (2005) Regulation of mitochondrial translation in yeast. Cell Mol Biol Lett 10(4):571–594

van Dooren GG, Schwartzbach SD, Osafune T, McFadden GI (2001) Translocation of proteins across the multiple membranes of complex plastids. Biochim Biophys Acta 1541(1–2):34–53

Vorstenbosch E, Pape T, Rodnina MV, Kraal B, Wintermeyer W (1996) The G222D mutation in elongation factor Tu inhibits the codon-induced conformational changes leading to GTPase activation on the ribosome. EMBO J 15(23):6766–6774

Waller RF, Keeling PJ, Donald RG, Striepen B, Handman E, Lang-Unnasch N, Cowman AF, Besra GS, Roos DS, McFadden GI (1998) Nuclear-encoded proteins target to the plastid in toxoplasma gondii and Plasmodium falciparum. Proc Natl Acad Sci U S A 95(21):12352–12357

Waller RF, Reed MB, Cowman AF, McFadden GI (2000) Protein trafficking to the plastid of Plasmodium falciparum is via the secretory pathway. EMBO J 19(8):1794–1802

Ward BL, Anderson RS, Bendich AJ (1981) The mitochondrial genome is large and variable in a family of plants (cucurbitaceae). Cell 25(3):793–803

WHO (2012) World malaria report

Willie GR, Richman N, Godtfredsen WP, Bodley JW (1975) Some characteristics of and structural requirements for the interaction of 24,25-dihydrofusidic acid with ribosome—elongation factor g complexes. Biochemistry 14(8):1713–1718

Wilson RJ (2002) Progress with parasite plastids. J Mol Biol 319(2):257–274

Wilson RJ, Denny PW, Preiser PR, Rangachari K, Roberts K, Roy A, Whyte A, Strath M, Moore DJ, Moore PW, Williamson DH (1996) Complete gene map of the plastid-like DNA of the malaria parasite Plasmodium falciparum. J Mol Biol 261(2):155–172

Wilson RJ, Williamson DH (1997) Extrachromosomal DNA in the apicomplexa. Microbiol Mol Biol Rev 61(1):1–16

Wolf H, Assmann D, Fischer E (1978) Pulvomycin, an inhibitor of protein biosynthesis preventing ternary complex formation between elongation factor Tu, GTP, and aminoacyl-tRNA. Proc Natl Acad Sci U S A 75(11):5324–5328

Wolf H, Chinali G, Parmeggiani A (1974) Kirromycin, an inhibitor of protein biosynthesis that acts on elongation factor Tu. Proc Natl Acad Sci U S A 71(12):4910–4914

Wolfe KH, Morden CW, Palmer JD (1992) Function and evolution of a minimal plastid genome from a nonphotosynthetic parasitic plant. Proc Natl Acad Sci U S A 89(22):10648–10652

Wolters J (1991) The troublesome parasites–molecular and morphological evidence that apicomplexa belong to the dinoflagellate-ciliate clade. Biosystems 25(1–2):75–83

Yeh E, DeRisi JL (2011) Chemical rescue of malaria parasites lacking an apicoplast defines organelle function in blood-stage Plasmodium falciparum. PLoS Biol 9(8):e1001138

Yu H, Chan YL, Wool IG (2009) The identification of the determinants of the cyclic, sequential binding of elongation factors tu and g to the ribosome. J Mol Biol 386(3):802–813

Zhu G, Marchewka MJ, Keithly JS (2000) Cryptosporidium parvum appears to lack a plastid genome. Microbiology 146(Pt 2):315–321

Zuegge J, Ralph S, Schmuker M, McFadden GI, Schneider G (2001) Deciphering apicoplast targeting signals–feature extraction from nuclear-encoded precursors of Plasmodium falciparum apicoplast proteins. Gene 280(1–2):19–26

Chapter 8
Mitochondrial Translation in Green Algae and Higher Plants

Thalia Salinas, Claire Remacle and Laurence Maréchal-Drouard

Abstract This review focuses on the mitochondrial translation system of higher plants and algae. A few mitochondrial mRNAs have to be expressed in the mitochondrion, and a functional translational machinery is required. With the exception of some ribosomal proteins, ribosomal RNAs and part of the transfer RNA population, all the other components are nucleus-encoded and depend on numerous macromolecular trafficking processes. The presence of a second endosymbiotic organelle within the plant cell, i.e., the chloroplast increases the complexity of the mitochondrial translation machinery by having several important repercussions on the origin as well as on the targeting of the mitochondrial translation components. As an illustration of this complexity, our present knowledge on the mitochondrial aminoacyl-tRNA synthetase (aaRS) population and the mitochondrial transfer RNA (tRNA) population in both higher plants and in green algae, are summarized. Concerning the translation process by itself little is known. The existence of *cis*- and *trans*-acting factors and the emergence of novel family proteins such as the PentatricoPeptide Repeat (PPR) proteins either as direct components of the ribosome or implicated in the regulation of mitochondrial translation is also tackled.

8.1 Introduction

In 1997, the first sequence of a mitochondrial genome of a higher plant has been published, that of *Arabidopsis thaliana*. This genome still encodes a set of about 30 proteins and similar numbers were found in mitochondrial genomes of higher plants

T. Salinas · L. Maréchal-Drouard (✉)
Institut de Biologie Moléculaire des Plantes, UPR 2357 Centre National de la Recherche Scientifique, University of Strasbourg, 12 rue du Général Zimmer, Strasbourg 67084, Cedex, France
e-mail: laurence.drouard@ibmp-cnrs.unistra.fr

C. Remacle
Génétique des Microorganismes, Department of Life Sciences, Institute of Botany, B22, University of Liège, 4000 Liège, Belgium

A.-M. Duchêne (ed.), *Translation in Mitochondria and Other Organelles*,
DOI: 10.1007/978-3-642-39426-3_8, © Springer-Verlag Berlin Heidelberg 2013

sequenced afterwards. The set of mitochondrial-encoded proteins slightly differs between different species but in all cases, these proteins correspond to components of the respiratory chain complexes, components involved in cytochrome c biogenesis or ribosomal proteins. Therefore, these proteins are essential elements for proper biogenesis and function of mitochondria and their synthesis necessitates an active translational machinery [i.e., ribosomes, translation factors, aminoacyl-tRNA synthetases (aaRSs), and transfer RNAs (tRNAs)]. Quite paradoxically, most components of this machinery are now expressed from nuclear genes, thus implying important macromolecular trafficking and regulatory cross-talk between the different compartments of the plant cell. Due to the endosymbiotic origin of mitochondria, the translation apparatus is expected to be very similar to bacterial translation. Indeed many ribosomal proteins, translation factors or aaRSs are of prokaryotic origins. However, striking differences also exist. As a matter of fact, a number of eukaryotic proteins have been recruited during evolution as for instance the PentatricoPeptide Repeat (PPR) proteins found to be associated to mitoribosomes or involved in translation regulation. Moreover, the presence of a second endosymbiotic organelle, the chloroplast, leads to an intricate share of several components between the two organellar translation machineries. Of particular interest is the mitochondrial tRNA population, which is complex and constituted by tRNAs of three different genetic origins (mitochondrial, plastidial, and nuclear). All these features present in higher plants and algae will be described. Our knowledge on the translation process and its regulation as well as the identification of *cis*- or *trans*-factors has been mainly prevented by technical limitations. Thereby only few data on this process are available and a brief overview is presented.

8.2 Components of the Mitochondrial Translation Machinery

8.2.1 Mitochondrial Ribosomes

Ribosomes are the central elements of the mitochondrial machinery. They comprise two subunits, the large subunit (LSU) and the small subunit (SSU) constituted of ribosomal RNAs (rRNA) and ribosomal proteins (MRP for Mitochondrial Ribosomal proteins). Mitoribosomes have a bacterial origin and contain many components that are conserved and found at structurally and functionally important sites. However, they evolved differently from their bacterial ancestor and between the different eukaryotic lineages (Kehrein et al. 2012). Plant mitoribosomes appear to be more bacteria-like than the other eukaryotic mitoribosomes, and harbor specific features.

Mitoribosomes from higher plants sediment at 77–78S and thus are larger compared to the bacterial ribosomes (Leaver and Harmey 1976) (Fig. 8.1). They are composed of three rRNAs molecules, the 26S and the 5S for the LSU and the 18S for the SSU. The protein composition of higher plant mitoribosomes has not been fully elucidated. Nevertheless, electrophoretic analysis of potato and broad bean mitoribosomes allowed determining that about 68–80 proteins constituted them

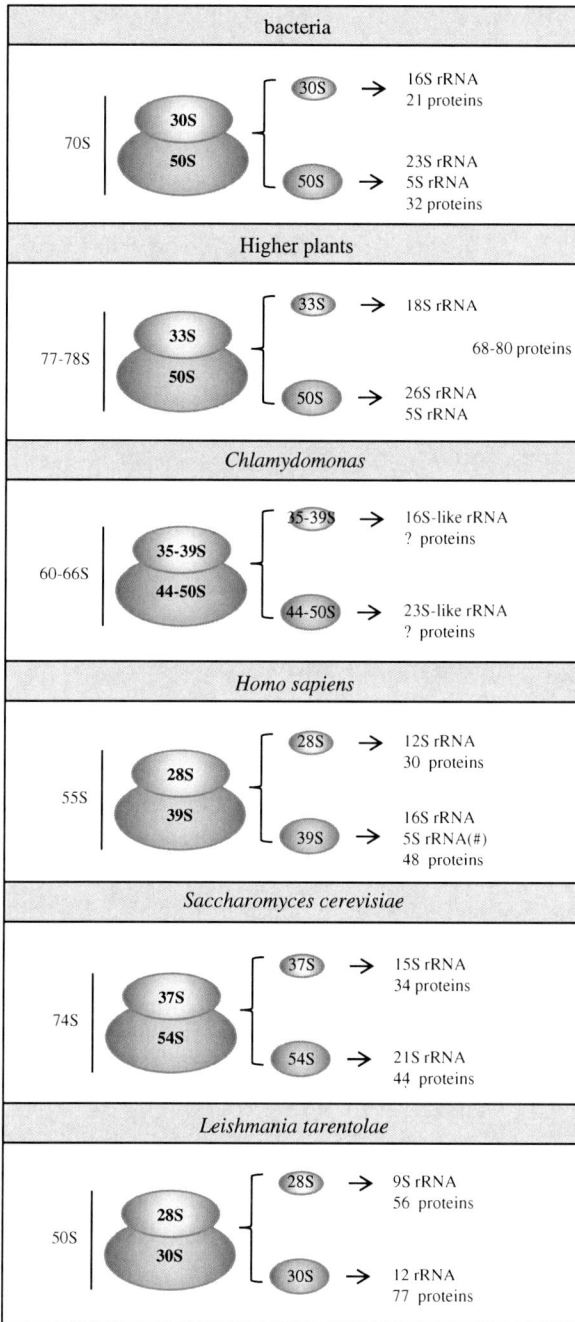

Fig. 8.1 Composition of bacterial and mitochondrial ribosomes from diverse eukaryotic lineages. The composition of mitochondrial ribosomes from *Homo sapiens* and *Saccharomyces cerevisiae* is according to (kehrein et al. 2012) and the one from *Leishmania tarentolae* is according to (Agrawal and Sharma 2012). (#) For the discussion concerning the presence or absence of the 5S rRNA in the mitochondrial ribosomes of mammals see Chap. 4

(Pinel et al. 1986; Maffey et al. 1997). What is important to point out here is that plant mitoribosomes contain more proteins than bacterial ribosomes. This is a feature shared with mammalian, yeast and protozoan mitoribosomes. In all these organisms, little is known about the role of the additional proteins acquired during eukaryotic evolution (Agrawal and Sharma 2012). Since a significant reduction in the size of rRNAs is observed in mitoribosomes from these organisms, it was thought that these extra proteins could compensate this reduction. However, the cryo-electron microscopy structures from mammalian mitoribosomes (*Bos taurus*) and from *Leishmania tarentolae* mitoribosomes have shown that missing rRNA fragments are only partially compensated by these extra proteins and that many of these proteins occupied new spatial positions in the ribosomal structure (Sharma et al. 2003, 2009). This probably reflects the introduction of novel functions in mitoribosomes as an adaptation of the translation process in mitochondria.

Our knowledge on mitoribosomes from green algae is still scarce. Analysis on *Chlamydomonas reinhardtii* and *Chlamydomonas eugametos* mitoribosomes indicates a sedimentation at approximately 60–66S (Denovan-Wright and Lee 1995), which is less than mitoribosomes from higher plants and even from bacteria ribosomes (70S) (Fig. 8.1). They are composed of rRNAs molecules for LSU and SSU that display particular features (see below) but nothing is known about their protein composition.

8.2.1.1 Ribosomal RNAs

Mitochondrial genomes from higher plants contain the 3 *rrn* genes encoding the LSU rRNAs 26S and 5S and the SSU rRNA 18S. The genes copy numbers vary between plant species, but in all cases the 26S rRNA is transcribed alone whereas the 18S and the 5S rRNAs are co-transcribed from a different locus (Bonen and Gray 1980).

Mitochondrial genomes from green algae all contain LSU 23S-like *rrn* and SSU 16-like *rrn* genes. In contrast, *Prototheca wickerhamii* and *Nephroselmis olivacea* mitochondrial genomes still code for a 5S *rrn* gene whereas the mitochondrial genomes from *Pedinomonas minor* and the chlorophyceae group lost their 5S *rrn* gene (Turmel et al. 1999). In addition, in the chlorophycean green-algal group, the two *rrn* genes are discontinuous and scrambled within the genome. For example, in *C. reinhardtii* the LSU and SSU rRNAs split into eight and four modules respectively those are interspersed with one another, with protein-coding genes and tRNA genes (Fig. 8.2). These modules are thought to hold together by specific intermolecular pairing to form conventional rRNA molecules (Bonen and Gray 1980; Nedelcu 1997; Fan et al. 2003).

8.2.1.2 Mitochondrial Ribosomal Proteins

The MRP proteins can be divided into two groups: the proteins that derived from the ancestral α-proteobacterial endosymbiont and the proteins that have been recruited during the course of evolution.

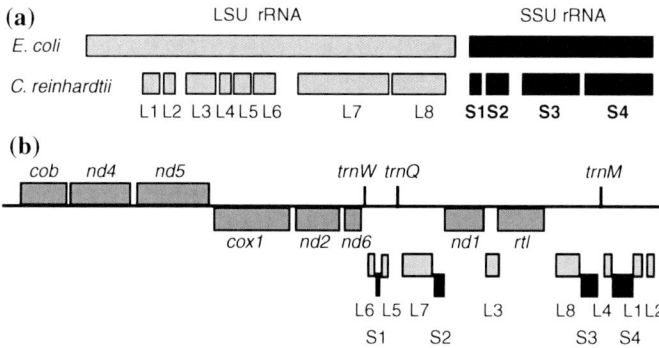

Fig. 8.2 Fragmented and scrambled mitochondrial rRNAs of *C. reinhardtii*. In (**a**) are represented the LSU and SSU rRNAs from *Escherichia coli* and the corresponding mitochondrial LSU (L1–L8) and SSU (S1–S4) RNAs fragments of *C. reinhardtii*. (**b**) Genetic organization of the rRNA gene modules in the *C. reinhardtii* mitochondrial genome. The LSU and SSU rRNA modules (light gray and black boxes respectively) of *C. reinhardtii* are rearranged and interspersed with one another, with protein-coding genes and tRNA genes. *Dark gray* boxes represent protein-coding genes: (*cob*) apocytochrome *b* of complex III; (*nd1, 2, 4, 5,* and *6*) subunits of complex I; (*cox1*) subunit 1 of complex IV; (*rtl*) reverse transcriptase-like protein. The 3 *trn* genes are indicated with the one letter code (W, Q and M)

MRPs with bacterial origin

The particularity of plants with the exception of the chlorophyceae group is that a set of MRPs is still encoded by the mitochondrial genome (Table 8.1). The number of these *mrp* genes varies from one plant to the other. The mitochondrial genome of the bryophyte *Marchantia polymorpha* encodes 16 *mrp* genes, which is the highest number observed in plants till now (Takemura et al. 1992). In Angiosperms, the number of mitochondrial-encoded MRPs genes goes to a maximum of 14, namely the *rpl2, rpl5,* and *rpl16* of the LSU and the *rps1, rps2, rps3, rps4, rps7, rps9, rps10, rps11, rps12, rps13,* and *rps19* of the SSU (Adams and Palmer 2003). It is worth mentioning that the study of ribosomal protein gene content in plant mitochondrial genomes has provided insights into the tempo of gene transfer events. Angiosperms are particularly interesting because gene transfer to the nucleus is still ongoing (Adams et al. 2002b). Indeed, the identification of *mrp* pseudogenes in various plant mitochondrial genomes have been correlated with recently relocated functional nuclear copies (Adams et al. 2002a). Interestingly, there is a case listed in wheat in which the *rpl5* gene is functionally present in both the mitochondrion and the nucleus (Sandoval et al. 2004). Furthermore, in some plant lineages as in *Arabidopsis*, the *rpl2* gene is fractured into two segments, with one encoded in the nucleus and the other in the mitochondrion thus showing the complex evolutionary histories of these MRPs (Adams et al. 2001).

All the other identified bacterial type MRPs genes that are not mitochondrion-encoded, are found in the nucleus. Only a few of these proteins have been characterized, namely the RPS10, RPS11, RPS12, RPS14, RPS19, RPL11, and RPL12 (Grohmann et al. 1992; Wischmann and Schuster 1995; Kadowaki et al. 1996;

Table 8.1 SSU and SSL plant MRPs candidates homologous to the bacterial ribosomal proteins

		A. thaliana	O. sativa	C. reinhardtii	M. polymorpha
SSU	S1		mt		mt
	S2	+	mt	+	mt
	S3	mt	mt		mt
	S4	mt	mt		mt
	S5	+	+	+	
	S6	+	+	+	
	S7	mt	mt	+	mt
	S8 [a]	cytosol-type		+	mt
	S9	+	+	+	
	S10 [b]	+	+	(MS/MS)	mt
	S11 [c]	+	+ (Ψmt)	+	mt
	S12 [d]	mt	mt	+ (MS/MS)	mt
	S13 [a]	chloroplast-type	mt	+	mt
	S14 [e]	+ (Ψmt)	+ (Ψmt)	+	mt
	S15	+	+	+	
	S16	+	+	+	
	S17	+	+	+	
	S18	+	+	+	
	S19 [f]	+ (Ψmt)	mt	+	mt
	S20				
	S21	+	+		
LSU	L1	+	+	+	
	L2	mt (5')	mt	+	mt
	L3	+	+	+	
	L4	+	+	+	
	L5	mt	mt		mt
	L6	+	+	+	mt
	L7	+ (MS/MS)	+		
	L9	+	+		
	L10	+	+		
	L11 [g]	+	+	+	
	L12 [h]	+ (MS/MS)	+	+	
	L13	+ (MS/MS)	+	+	
	L14	+	+	+	
	L15	+	+	+	
	L16	mt	mt	+	mt
	L17	+	+	+	
	L18	+	+		
	L19	+	+	+	
	L20	+	+	+	
	L21	+	+	+	
	L22	+	+	+	
	L23	+	+	+ (MS/MS)	
	L24	+	+	+ (MS/MS)	
	L25	(MS/MS)		(MS/MS)	
	L27	+	+	+ (MS/MS)	
	L28	+	+		
	L29	+	+	+	
	L30	+	+	+	
	L31				
	L32	+	+		
	L33	+	+	+	
	L34				
	L35				
	L36	+	+	+	

Figueroa et al. 1999; Kubo et al. 2000; Handa et al. 2001; Delage et al. 2007). The *in silico* analysis of these MRPs showed that the proteins are significantly larger than their bacterial ancestral sequences. About 75 % of the MRPs have

◀ The first column indicates the names of the ribosomal proteins according to bacterial nomenclature. Ribosomal protein names on grey bottom are those who are universally present in eubacteria, archea and eukaryotic cytosol ribosomes (Lecompte et al. 2002)

The letters *a, b, c, d, e, f, g* and *h* indicate the MRPs that have been studied in plants (*a*- Adams et al. 2002a, *b*-Grohmann et al. 1992, *c*-Kubo et al. 2000, *d*-Figueroa et al. 1999, *e*-Kadowaki et al. 1996, *f*-Handa et al. 2001, *g*-Wischmann et al. 1995, *h*-Delage et al. 2007)

(mt) indicates genes that are encoded in the mitochondrial genome

(ψ mt) indicates pseudogenes found in the mitochondrial genome

In *Arabidopsis*, the *rpl2* gene is split into two segments, with one encoded in the nucleus and the other in the mitochondrion. Thus (5') indicates the half *rpl2* gene present in *Arabidopsis* mitochondrial genome. (Adams et al. 2001)

(+) indicates the presence of nuclear candidate genes identified by bioinformatics analysis for plant MRPs derived from the ancestral bacteria (Bonen and Calixte 2006; Smits et al. 2007).

(MS/MS) indicates the proteins that have been found in the mitochondrial proteomic analysis of *A. thaliana* (Heazlewood and Millar 2005) and *C. reinhardtii* (Atteia et al. 2009).

amino-terminal extensions and some are much longer than typical amino terminal targeting signals (Bonen and Calixte 2006). This raises the possibility that such domains (that are also found in internal and carboxy-terminal sequences) perform extra functions. Indeed in *A. thaliana*, a RNA recognition motif (RRM) was found in the amino-terminal region of the RPS19 (Sanchez et al. 1996). Studies on the four RPL12 paralogues found in potato mitochondria suggest that the plant mitoribosomes are attached to the mitochondrial inner membrane, a situation already found in the case of yeast mitochondria (Delage et al. 2007).

Recruited MRPs

Most of the 54 proteins found on the present-day bacterial ribosomes have counterparts in plant mitoribosomes (Table 8.1). Indeed, bioinformatics analysis allowed the identification of 46 bacterial-like *mrp* genes in *A. thaliana* (39 nucleus-encoded and 7 mitochondria-encoded), 48 in *Oryza sativa* (37 nucleus-encoded and 11 mitochondria-encoded) and 39 in *C. reinhardtii* (all nucleus-encoded) (Bonen and Calixte 2006; Smits et al. 2007). Most of the bacterial ribosomal proteins have a homologous counterpart in plants and for this reason mitoribosomes from plants appear more bacterial-like than mammalian ones (O'Brien et al. 2005). For the few remaining missing bacterial-like MRPs, we cannot exclude that they have not been identified because of sequence divergence or because they have been replaced by a new gene. For instance, for two MRPs it has been shown that the genuine *mrp* gene was lost and a new *mrp* gene was recruited to replace it. In *A. thaliana*, a nucleus gene replaces the *rps8* mitochondrial-type gene and a duplicated copy of a nucleus-encoded chloroplastic gene replaces the *rps13* mitochondrial-type gene (Adams et al. 2002a). In addition, as in mammals and yeast, mitoribosomes in higher plants have expanded their protein content by acquiring numerous extra "supernumerary" MRPs. Nothing is known about these supernumerary proteins and a complete identification remains to be established in plants. In a comparative genomic analysis of MRPs in 18 eukaryotic species, orthologous genes for the supernumerary MRPs reported in yeast and mammals were identified in *A. thaliana* and *C. reinhardtii* (Smits et al. 2007). In *Arabidopsis* from the 16 genes identified almost all the proteins are predicted to be addressed to mitochondria and at least 3 have been found in proteomic analysis of mitochondria (Ito et al. 2009; Klodmann et al. 2011). In *Chlamydomonas* from the eight genes identified, two have been found in frame of proteomic analysis (Atteia et al. 2009).

8.2.1.3 Proteins Associated with Mitoribosomes

In *Arabidopsis*, two proteins have been characterized to be associated to mitoribosomes, namely, PNM1 and PPR336 proteins. The two proteins belong to the PPR protein family. This family is a eukaryote-specific protein family that is particularly large in higher plants. PPR proteins are composed of 35 amino acid motifs repeated in tandem and have been described as being able to bind RNA. Still, the precise nature of the association of these two proteins with mitoribosomes and their role in mitochondrial translation is unknown (Uyttewaal et al. 2008; Hammani et al. 2011).

8.2.2 *Elongation and Initiation Factors*

All the mitochondrial translation factors in plants are encoded in the nucleus. So far, only few studies on these factors have been done. Elongation Factor Tu (EF-Tu) has been characterized in maize and *Arabidopsis* (Kuhlman and Palmer 1995; Choi et al. 2000) and Elongation Factor Ts (EF-Ts) has been studied on tomato (Benichou et al. 2003). Mitochondrial proteomic studies in *Arabidopsis* have also identified an elongation factor EF-G (Heazlewood and Millar 2005). In addition, bioinformatic analysis on the nuclear genome of *Arabidopsis*, allowed the identification of putative mitochondrial translational initiation factors genes (Bonen 2004). In eubacteria, three initiation factors are necessary for protein translation, namely IF-1, IF-2, and IF-3. But in the *Arabidopsis* nuclear genome no homologue for a mitochondrial-type IF-1 gene was found and import experiments indicated that the nucleus-encoded chloroplastic IF-1 protein is not dual targeted to the mitochondrion (Millen et al. 2001).

8.2.3 *AARSs*

In higher plants, all mitochondrial aaRSs are nucleus-encoded (Figs. 8.3 and 8.4). This is also the case for the aaRSs present in the chloroplast, the second endosymbiotic compartment. Therefore, both mitochondrial and plastidial aaRSs are translated in the cytosol and are imported into the organelles. In a plant cell,

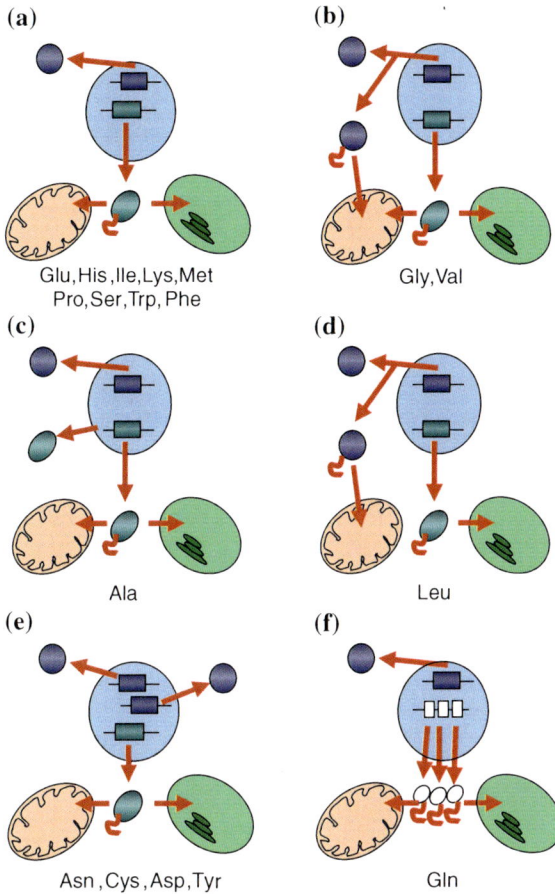

Fig. 8.3 *Arabidopsis* aaRS genes and localization of their products within the 3 cell compartments where a translation machinery is present. In (**f**), the 3 small white boxes represent the 3 genes encoding the subunits GatA, GatB, and GatC of the tRNA-dependent amidotransferase. The aaRSs belonging to each type (from **a** to **f**) are indicated by their amino acid specificity. Genes are depicted by rectangular boxes and proteins by circle or oval forms. Proteins addressed to organelles have an extension corresponding to the N-terminal targeting signals. In the case of ThrRS, a third gene coding for a cytosol specific enzyme exists (paralogous to the cyto-mito ThrRS). For ArgRS, the situation is unclear. Two genes potentially encode ArgRSs. One gene with a plastidial targeting sequence is essential for cell viability while the other one (without targeting sequence) is not (Duchêne et al. 2009)

protein synthesis occurs in three subcellular compartments, the cytosol and the two organelles. Thus, to catalyze the formation of the 20 sets of aminoacylated transfer RNAs (tRNAs) in the three compartments, a minimum total number of 60 nucleus-encoded aaRS genes is a priori required. This is unexpectedly not the case and only 45 aaRS expressed genes have been identified in the *A. thaliana* nuclear genome. This number is even less in green algae, since only 39 and 33 aaRS

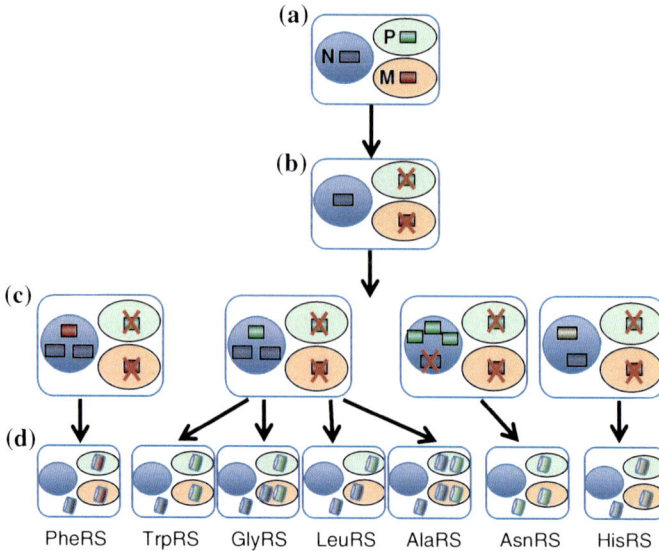

Fig. 8.4 *Arabidopsis* aaRSs have different origins and are localized in different compartments. The seven examples provided here (**d**) are representative of the complexity. In (**a**), the ancestral cell has received three sets of aaRS genes from the nucleus (*blue rectangle* (N)), the mitochondria (*red rectangle* (M)) and the chloroplast (*green rectangle* (P)). During evolution (**b**), the organellar genomes have lost their aaRS genes. In (**c**), some of them have been transferred to the nucleus. A few genes have been duplicated or triplicated. Horizontal transfer of archaeal genes may also occur (*white rectangle*). Proteins are represented by barrels with the same code color. For more details, see (Duchêne et al. 2009)

genes have been identified in *Ostreococcus tauri* and in *C. reinhardtii*, respectively (Cognat et al. 2013). This raised several intriguing questions and our present knowledge on the set, origin, and co-evolution with tRNAs is discussed below.

8.2.3.1 Distribution of aaRSs in the Cell

The reduction of the aaRS gene number (45 instead of 60 in *Arabidopsis*) implies the sharing of enzymes. Indeed, for each amino acid, one two or three genes are found and a summary of the different possibilities is presented on Fig. 8.3.

Two aaRS genes for one amino acid is the most frequent situation. This is the case for 13 amino acids. For 8 of these 13 amino acids, one gene code for a cytosol-specific enzyme while the other encodes an aaRS localized to both mitochondria and chloroplasts, thanks to the presence of N-terminal dual targeting sequences. Two main approaches were used to determine the dual role of these targeting sequences. The predicted targeting sequences were fused to a GFP reporter gene, and the localization of the fusion protein was determined *in vivo* in tobacco protoplasts and *in vitro* by import into isolated chloroplasts and mitochondria. For two other amino acids, each of the two genes encodes a dual-localized aaRS, one in the cytosol and mitochondria and the second in mitochondria and chloroplasts. For alanine, dual targeting to mitochondria and chloroplasts was also observed for one gene product while the second gene encodes an enzyme localized in the three compartments. For leucine, one aaRS is present in the cytosol and mitochondria while the second enzyme is targeted to the chloroplasts. Finally, for the 13th amino acid, arginine, the localization of the two gene products is still unclear.

A second situation found for six amino acids corresponds to the presence of three aaRS genes. In all cases, one gene product is targeted to both mitochondria and chloroplasts. In four cases, the two other genes encode cytosolic paralogues resulting from duplication events and in one case (threonine), one of the two paralogues is also addressed to the mitochondria. For phenylalanine, the two other genes encode the two subunits of the α2β2 heterotetrameric enzyme.

The last situation corresponds to GlnRS where only one gene has been identified. This gene encodes an enzyme without any potential targeting sequence and only localized in the cytosol. Thus, in the cytosol, the Gln-tRNA^{Gln} is synthesized by direct attachment of Gln to its cognate tRNA by the cytosolic GlnRS, this is the direct aminoacylation pathway. An indirect two-step transamidation pathway is used in plant organelles, as in archae and in the majority of bacteria. This second pathway doe not require a GlnRS. The mitochondrial tRNA^{Gln} is first mischarged with glutamate by a nondiscriminating dual-targeted GluRS, then, is converted into Gln-tRNA^{Gln} by a tRNA-dependent amidotransferase. This last enzyme is constituted by three subunits, called GatA, GatB and GatC, shared between mitochondria and chloroplasts (Pujol et al. 2008; Frechin et al. 2009) (Fig. 8.3).

As a whole, an extensive sharing of aaRSs between compartments is observed in *A. thaliana*. Similar observations exist in other higher plants such as rice or maize (Morgante et al. 2009; Rokov-Plavec et al. 2008). Rice paralogous proteins were found for most dual-targeted *Arabidopsis* aaRSs. Dual localization of rice GluRS and TyrRS was experimentally proven but to validate the dual-localization of the other aaRSs, experimental data are required (Morgante et al. 2009). In algae, although no extensive experimental studies have been performed, *in silico* analyses strongly suggest that most aaRSs are at least dual localized (http://plantrna.ibmp.cnrs.fr/), (Cognat et al. 2013). For example, in *Ostreococcus tauri*, from the 39 aaRS genes identified, only 2 genes were found for 16 amino acids. In *C. reinhardtii*, from the 33 aaRS genes identified, 2 genes exist for 13 amino acids, and only 1 gene is found for 7 amino acids. We cannot exclude that some aaRS genes have escaped detection, but this strongly suggests an extensive sharing of such enzymes in algae. Thus contrary to the situation found in human or yeast mitochondria where only few cases of dual-targeted aaRSs were reported, this is the rule in the plant kingdom. The dual-targeting of proteins is not restricted to aaRSs and one of the most important question to solve now is to understand how the cell can discriminate between a mitochondrial or plastidial-specific proteins and a dual-targeted one. Several nonexclusive possibilities exist such as the presence of alternative translation or transcription sites, existence of "ambiguous" targeting sequences, environmental conditions or differential mRNA sorting at the surface of the organelles (Duchêne et al. 2001; Pujol et al. 2007; Michaud et al. 2010; Ye et al. 2012).

In opposition with the reduction of aaRS gene number and the sharing of enzymes between at least two compartments, it is worth to note that two aaRSs with the same amino acid specificity can be found in the same compartment. For example, in *A. thaliana*, there are two cytosolic enzymes for five amino acids, two mitochondrial aaRSs for four amino acids, and two plastidial AlaRS. Whatever the organism, this situation is rather rare and we can wonder about the biological significance of such apparent redundancy. The *Arabidopsis* organellar duplicated aaRSs have distinct genetic origins and for two of them (GlyRS and ValRS), distinct functions have been demonstrated. The reason why aaRSs with similar specificity are present in the cytosol remains to be elucidated.

8.2.3.2 Origin of Mitochondrial aaRSs and Co-evolution Between Mitochondrial tRNAs and aaRSs

Decrypting the origin and evolution of mitochondrial aaRSs is of particular interest in the plant kingdom. First, as reported, tRNAs present in higher plant mitochondria have three different genetic origins. In addition to tRNAs coming from the ancestral α-proteobacterium at the origin of mitochondria, chloroplast-like tRNAs (deriving from the ancestral cyanobacterium at the origin of higher plant chloroplasts) are also found as well as eukaryotic nucleus-encoded tRNAs. Second, due to the two endosymbiotic events leading to mitochondria and chloroplasts, a plant cell has received three sets of aaRS genes. However, during evolution, all organellar genes encoding an aaRS have been either lost or transferred into the nucleus (Fig. 8.4). Third, to get an efficient and a faithful mitochondrial translation machinery, the disparate set of mitochondrial tRNAs has to been correctly recognized by their cognate aaRSs. Although exceptions exist, it is generally admitted that in most cases, a eukaryotic tRNA is a poor substrate for a prokaryotic aaRS and a prokaryotic tRNA is a poor substrate for a eukaryotic enzyme. Thus, the heterogeneous mix of tRNAs and the reduced number of shared aaRSs found in a present-day plant cell raises interesting questions on the origin of the aaRSs working in each compartment. Multiple alignments and distance trees analyses allowed the determination of the putative origin of *Arabidopsis* nucleus-encoded aaRSs (Duchêne et al. 2005). The most classical pattern is two nuclear aaRS genes of different origins for each amino acid but a few examples of duplication or triplication were also observed (Duchêne et al. 2009; Brandao and Silva-Filho 2011). For 17 amino acids, the genes coding for the cytosolic or for the cytosolic-mitochondrial aaRSs are very likely of eukaryotic origin. For three cytosolic aaRSs, the ancestral eukaryotic gene has been lost and replaced by genes related to archaea or to bacteria. For example, the present-day cytosolic AsnRS is of plastid origin (Peeters et al. 2000). In most cases, the plant mitochondrial-chloroplastic aaRSs or the plastidial enzymes have a bacterial origin. However, contrary to yeast or animal mitochondrial aaRSs that are expressed from ancestral mitochondrial aaRS genes transferred to the nuclear genomes, the plant organellar enzymes (with two exceptions, the SerRS and PheRS) do not originate from mitochondrial genes. For one-third of the organellar aaRSs, the plastidial (i.e., cyanobacterial) or mitochondrial (i.e., α-proteobacterial) origin cannot be determined. For the two other third, the organellar aaRSs are closely related to cyanobacterial enzymes (Duchêne et al. 2005). In the case of organellar aaRSs, accidents do also exist. Two mitochondrial-chloroplastic aaRSs are not of organellar/bacterial origins: the HisRS is related to archaea enzymes (Akashi et al. 1998a) and the ProRS is related to eukaryotic cytosolic enzymes (Duchêne et al. 2005). Such in-depth phylogenetic analysis has never been performed in other plant species and is certainly worth to achieve in other photosynthetic organisms such as algae or mosses.

As already described for *Arabidopsis* and with the exception of a few aaRSs, there is a rather good correlation between the origin of tRNAs and the origin of their cognate aminoacyl-tRNA synthetases in the chloroplast and in the cytosol. Plastidial tRNAs are of prokaryotic origin and charged by bacterial enzymes while nucleus-encoded cytosolic tRNAs are charged by eukaryotic aaRSs. In contrast, the situation appears more complex in plant mitochondria and the tRNA/aaRS co-evolution is less obvious. Most tRNAs encoded by mitochondrial genes (native and chloroplast-like) are recognized by bacterial-like aaRSs but several nucleus-encoded tRNAs imported into mitochondria are aminoacylated by bacterial-type aaRSs (e.g., GlyRS, ValRS, TrpRS, PheRS, and IleRS). To this point, it is important to note that two GlyRS and two ValRS of different origins are imported into the organelle but only the prokaryotic enzymes have retained an aminoacylation activity while the eukaryotic enzymes (also present in the cytosol) are likely inactive within the mitochondrial matrix (Duchêne et al. 2001, 2009). Why do plant mitochondria import two aaRSs with the same amino acid specificity but of different origins and with one enzyme apparently not involved directly in translation is an open question. As these enzymes are linked to imported tRNAs, it is tempted to speculate that these aaRSs are involved in the tRNA mitochondrial import process. The connection between eukaryotic tRNA mitochondrial import and eukaryotic aaRS mitochondrial import can be envisaged at different levels. The first link is at the level of the aminoacylation process. As half of the imported tRNAs are still recognized by prokaryotic-type aaRS, this suggests that the emergence of tRNA mitochondrial import preceded the acquisition of eukaryotic aaRSs. This also implies that imported tRNAs had first the possibility to be recognized by prokaryotic enzymes, as previously proposed by (Schneider 2001). Second, as already mentioned and described, eukaryotic aaRSs may play a crucial role as carrier in tRNA mitochondrial import.

What seems, however, obvious is that both the origin of the population of mitochondrial aaRSs and the origin of the population of mitochondrial tRNAs vary from one plant species to the other and very likely continue to evolve among each plant species to maintain highly functional organelles.

8.2.4 tRNAs

8.2.4.1 Mitochondrial tRNAs

The availability of several complete mitochondrial genomes of plants and their *in silico* analysis allowed the identification of mitochondrial *trn* genes (Rainaldi et al. 2003; O'Brien et al. 2009; Cognat et al. 2013). These analyses showed that mitochondrial genomes of plants encode for a variable set of tRNAs derived from the α-proteobacterial-type ancestor (called "native" tRNAs). In angiosperms, mitochondrial genomes have also acquired during evolution *trn* genes with a plastidial origin. For example, the native genes for the tRNA^{His} and tRNA^{Asn} are lost in all angiosperms studied so far and have been replaced by expressed choroplast-like genes (Fey et al. 1997). A few native or chloroplast-like tRNAs have been sequenced. At the level of both the primary sequence and the potential secondary structure, they resemble eubacterial tRNAs rather than mitochondrial tRNAs from fungi or animals. These tRNAs also appear to contain a low number of post-transcriptionally modified nucleotides, as in prokaryotes (Maréchal et al. 1985a, b, c, 1986). In addition, these analyses revealed that mitochondrial *trn* genes frequently represent an incomplete set of the tRNAs necessary for translation, i.e., to read all the sense codons of the universal genetic code used by plant mitochondria. Indeed, in angiosperms the number of *trn* genes is insufficient as *trn* genes for 5--7 amino acids are missing and in chlorophyceae (with the exception of *Scenedesmus obliquus*) only 1 to 3 *trn* genes are encoded in the mitochondrial genome. The experimental studies in a number of these organisms showed that this lack was compensated by the import of the corresponding cytosolic tRNAs (Salinas et al. 2008; Duchêne et al. 2011; Sieber et al. 2011a) (Table 8.2). These studies also showed that when the mitochondrial *trn* genes are not anymore expressed, the cytosolic tRNAs are imported, as for the "chloroplast-like" *trnW* gene in *A. thaliana* (Duchêne and Maréchal-Drouard 2001). Only very few exceptions to this rule are reported. For example, in *M. polymorpha* there is an apparent redundancy between the nucleus-encoded tRNA^{Val}(AAC) which is imported into mitochondria and the mitochondrion-encoded tRNA^{Val}(UAC) (Akashi et al. 1998b). This could be explained by a differential codon recognition profile for the two tRNAs but this remains to be demonstrated. Finally, the number of imported tRNAs is rather difficult to predict and only experimental analyses provide the most accurate data. For example, if we consider a minimal genetic code, 22 and 10 cytosolic tRNAs are predicted to be imported into *C. reinhardtii* and *T. aestivum* respectively, while 31 and 16 cytosolic tRNAs were demonstrated to be imported (Glover et al. 2001; Vinogradova et al. 2009). This is primarily due to the presence of hyper modified nucleotides at the first position of the anticodon of eukaryotic nucleus-encoded tRNAs that restrict the codon/anticodon recognition and thus increase the required number of tRNAs to be imported in order to read all sense codons. In contrast, in algae (with the exception of chlorophyceae) as well as in bryophytes the number of *trn* genes seems sufficient or nearly sufficient for mitochondrial translation.

Table 8.2 Loss of mitochondrial *trn* genes and experimental data on import of cytosolic tRNAs into mitochondria

		Other *trn* genes missing or *trn* expression lost	Import demonstrated	References
Bryophyta	*Marchantia polymorpha*	–	I(IAU), V(AAC), T(AGU)	Akashi et al. (1998b)
	Physcomitrella patens	N		
Angiosperms				
Dicots	Brassicales Loss of *trnF*	*Arabidopsis thaliana*	W, M-e	W, F	Duchêne and Maréchal-Drouard (2001), Chen et al. (1997)
	Caryophyllales	*Brassica napus*	–		
		Beta vulgaris	–		
	Solanales	*Nicotiana tabacum*	–	V, G	Delage et al. (2003b), Salinas et al. (2005)
		Solanum tuberosum	–	A, R, L, T, I, G, V	Maréchal-Drouard et al. (1990), Brubacher-Kauffmann et al. (1999), Kumar et al. (1996)
	Monocots Loss of *the second trnG (with GCC anticodon)*	*Oryza sativa*	–		Glover et al. (2001)
		Sorghum bicolor	–		
		Triticum aestivum	H	A, R, L, V, G, H, I(IAU)	
		Zea mays strain NB	–	A, R, L, T, V, G	Kumar et al. (1996)

Loss of *trnA, trnR, trnL, trnT, trnV* and one of the *trnG (with UCC anticodon)*

(continued)

Table 8.2 (continued)

		Other *trn* genes missing or *trn* expression lost	Import demonstrated	References
Gymnosperms	Coniferopsida	*Larix leptoeuropaea*	At least 11		Kumar et al. (1996)
Chlorophyta	Chlorophyceae	*Chlamydomonas reinhardtii*	All but 3	31	Vinogradova et al. (2009)
		Chlamydomonas eugametos	All but 3		
		Polytomella parva	All but 1		
		Scenedesmus obliquus	T		
	Pedinophyceae	*Pedinomonas minor*	All but 7		
	Prasinophyceae	*Nephroselmis olivacea*	–		
	Trebouxiophyceae	*Prototheca wickerhamii*	–		

8.2.4.2 Cytosolic tRNA Import Process

Mitochondrial tRNA import is not restricted to plants but is a widespread process (Rubio and Hopper 2011; Schneider 2011; Sieber et al. 2011a). In plants, the number and the identity of imported tRNAs vary from one species to another and is not always consistent with the assigned phylogenetic position. A particular feature in plants is that the import process is highly specific. Indeed as already mentioned, the mitochondrial population of nucleus-encoded tRNAs is primarily complementary to those encoded in the mitochondrial genome, meaning that only necessary cytosolic tRNAs are imported into mitochondria (Salinas et al. 2008; Duchêne et al. 2011; Sieber et al. 2011a). This implies that cell needs to discriminate between imported tRNAs and cytosol specific tRNAs. The question of tRNA import selectivity has been studied in higher plants. In these studies, tRNA mutagenesis followed by *in vitro* and *in vivo* approaches were used to find out determinants on the imported tRNAs that are necessary for mitochondrial localization. Altogether, the data obtained so far suggest that tRNA import signals are present on mature tRNAs and that they are different depending on the tRNA studied, showing the complexity of the selectivity of the import process (Small et al. 1992; Delage et al. 2003b; Laforest et al. 2005; Salinas et al. 2005). *In vivo* studies also demonstrated the requirement of aaRSs but their precise role has not been uncovered yet (Dietrich et al. 1996). Moreover, an additional layer of regulation exists in the tRNA import process. In higher plants, different studies support the existence of a differential distribution of nucleus encoded tRNAs between the cytosol and mitochondria (Glover et al. 2001; Salinas et al. 2005; Duchêne et al. 2011). In *Chlamydomonas* an in-depth study of the mitochondrial tRNA population showed that out of the 49 cytosolic tRNAs issoacceptors, 31 were present within mitochondria and that the extent of their mitochondrial steady-state levels ranged from 0.2 % to 98 % (Cognat et al. 2008, Vinogradova et al. 2009). Remarkably, the observed steady-state level of an imported tRNA is linked to the frequency of occurrence of the cognate codon in both, the mitochondrial genes and the nuclear genes. However, this fine-tuning observed in *Chlamydomonas* mitochondria between tRNA import and the codon usage appears to originate from a co-evolution process rather than from a dynamic adaptation of cytosolic tRNA import into mitochondria (Salinas et al. 2012).

Concerning the translocation machinery of tRNAs through the mitochondrial membranes little is known. The development of an *in vitro* tRNA import system (Delage et al. 2003a) together with biochemical approaches on potato mitochondria allowed the identification of some components of this translocation machinery. These investigations indicated that the Voltage Dependent Anion Channel (VDAC), known to play a role in metabolite transport, is the major component of the tRNA transport through the outer mitochondrial membrane. It showed as well, that two major components of the Translocase of the Outer mitochondrial Membrane (TOM) complex, namely TOM20 and TOM40, are likely to be important for tRNA binding at the surface of mitochondria (Salinas et al. 2006). With this *in vitro* system, tRNAs can enter in mitochondria without any added protein factors albeit at low level. Recently, a protein shuttle system was set-up in order to internalize any kind of RNA into isolated plant mitochondria. In this system, a protein with the capacity to interact with nucleic acids and that is fused to a mitochondrial targeting sequence improves the import of cytosolic tRNAs and enable the import of foreign and larger RNAs (Sieber et al. 2011b). This raises the question of the possible participation of protein carriers in the tRNA import process mitochondria and therefore aaRS would be good candidates. Understanding whether *in vivo* plant mitochondria use carrier proteins or not during the tRNA import or whether the two possibilities can coexist require further studies.

8.3 The Translation Process

8.3.1 Higher Plants

Our understanding of the mechanisms of plant mitochondrial translation is still poorly understood. Contrary to yeast or metazoan mitochondria where the genetic code is slightly modified, no variation of the genetic code has been observed in plant mitochondria. The initiation codon is prevalently the classical AUG codon. However, in a few cases, alternate initiation codons (GGG, AAU, GUG, or ACG) are likely used (for reviews see (Giegé and Brennicke 2001; Gagliardi and Binder 2007)), a feature often encountered in prokaryotes. If these alternate initiation codons are translated as Met, the formylated initiator tRNA^{fMet} must be able to read these codons under certain sequence environment. It is to note that fractionation of potato mitochondrial tRNAs by two-dimensional polyacrylamide gel electrophoresis allowed to identify two spots corresponding to tRNA^{fMet} (Maréchal-Drouard et al. 1990). This may be the result of differential post-transcriptional modifications involved in codon-anticodon interaction, thus allowing the reading of alternate start codons.

To terminate mitochondrial protein synthesis in higher plants, the UAA, UAG, and UGA stop codons of the universal genetic code are used. Quite interestingly, few cases of mitochondrial mRNAs lacking stop codons have been reported in the sunflower (*nad6*) and in *Arabidopsis* (*nad6* and *ccmC*) (Raczynska et al. 2006; Placido et al. 2009). The question of how the plant mitochondrial translation system deals with nonstop mRNAs remains to be elucidated.

The Shine-Dalgarno sequence generally located eight nucleotides upstream of the AUG codon in prokaryotic mRNAs and used as a ribosomal binding site has never been found so far in mitochondrial mRNAs from higher plants (Hazle and Bonen 2007). Twenty years ago, conserved sequence blocks have been identified in a limited number of mRNAs (Pring et al. 1992) but no experimental data supporting their role in translation initiation have been reported. Factors involved in initiation and regulation of plant mitochondrial translation remain elusive. In 2012, Manavski et al. reported a specific interaction between a novel PPR protein and the 5′ UTR of the ribosomal protein *rps3* mRNA in maize mitochondria (Manavski et al. 2012).

Many more UTR-specific PPR or other proteins are presumably involved in plant mitochondrial protein synthesis and wait for further studies.

Higher plant mitochondrial mRNAs are subjected to Cytidine (C) to Uridine (U) editing, a post-transcriptional process that modifies nucleotides identities within an RNA molecule. Most of the edited nucleotides are found in the first and second codon positions causing amino acid changes in the corresponding sequences. Analyses showed that for a given gene there are site-specific variations in editing frequencies, which result in a heterogeneously edited steady-state mRNA population. The frequency of partial editing differs in precursor mRNAs, mature mRNAs and mRNAs associated with ribosomes. Mature mRNAs are more frequently edited and mRNAs associated with ribosomes are nearly all completely edited. Some examples showed that this incomplete editing results in the synthesis of polymorphic proteins within the organelle (e.g., (Lu and Hanson 1996)). Interestingly, in the case of the ribosomal protein *rps12*, the unedited mRNA was translated but the unedited *rps12* protein failed to be incorporated and to function in the ribosome (Phreaner et al. 1996).

8.3.2 Green Algae

The universal genetic code is used in all green algae mitochondria studied so far with very few exceptions as for *P. minor* and *S. obliquus* where the genetic code deviates slightly with one and two nonstandard codons found in their respective mitochondrial genome (Turmel et al. 1999; Kuck et al. 2000). The mitochondrial codon usage is biased and for instance nine and five codons are not used at all in mitochondrial genomes of *C. reinhardtii* and *P. minor* respectively (Michaelis et al. 1990; Turmel et al. 1999). Moreover, in contrast to higher plants, editing is absent from all investigated algal mitochondria.

Since *C. reinhardtii* is a model the organism in algae all the data available are essentially from this alga. The mitochondrial genome of *C. reinhardtii* encodes for eight protein-coding genes and all of them have an AUG initiation codon and a STOP codon (UAA or UAG). The sizes of the corresponding mRNAs detected by RNA gel blots are close to the lengths of the coding regions, indicating that mRNAs have short 5' and 3' noncoding sequences. This has been confirmed in some cases by using S1 nuclease protection and primer extension to map transcript termini. The *cob*, *nd5* and *cox1* mRNAs appear to begin at the AUG initiation codon while the *nd4*, *nd1*, and *rtl* mRNAs have short 5' UTRs (equal or equivalent to 13 nt). The 3'UTRs are longer and their size is comprised between 35 and 250 nt (Boer and Gray 1986, 1988a, b). The mode of initiation of translation is not known but the compilation of the genomic sequences of the eight protein-coding genes allowed the definition of a putative ribosome-binding site ATTTTATTA or ATAATTTA, upstream of the AUG codon (Colleaux et al. 1990). However, an additional mechanism must be required for the ribosome binding for the mRNAs beginning at the AUG codon. Nothing is known about the factors involved in *Chlamydomonas* mitochondria translation regulation. However, extensive genetic studies on

Chlamydomonas chloroplast revealed various factors involved in the chloroplast translation regulation and some of these factors belong to the Octatricopeptide Repeat (OPR) family (Eberhard et al. 2011). This protein family is also present in other unicellular organisms and in bacteria but is rare in higher plants. Only four proteins of this family composed of 43 identified proteins have been characterized (Rahire et al. 2012). Among the remaining 39 OPR proteins of unknown function some seem to be addressed to mitochondria and thus could also be involved in mitochondrial translation but this is a hypothesis that remains to be confirmed.

8.4 Why Do We Know So Little? (Concluding Remarks)

Over the last decade, our knowledge on the plant mitochondrial translation process has slightly increased. The molecular processes underlying the import of aaRSs and tRNAs are better understood, and novel plant-specific and essential protein factors (e.g., members of the PPR family) either involved in translation regulation or as part of the mitoribosomes have been identified. Nevertheless, there is still a huge gap to fill before to consider that protein synthesis in plant mitochondria revealed all its secrets. As compared to other systems (e.g., translation process in yeast mitochondria or plant chloroplasts), our detailed understanding is still rudimentary. Why is this? Several problems need to be overcome. First, recent advances in proteomics improved identification but protocols to obtain highly purified soluble ribosomes are required to be able to obtain the precise composition of plant mitoribosomes Second, detailed molecular processes, identification of *cis-* and *trans-* factors wait for two major technical advances: the ability to transform higher plant mitochondrial genomes and the development of *in vitro* or *in organello* translation systems. *C. reinhardtii* mitochondrial genome can be transformed (Remacle et al. 2006) but transformation can be limited by the selection process. Although the *in vitro* protein synthesis is possible with chloroplast extract (Yukawa et al. 2007), can it be used as a model to set up a similar mitochondrial system? These are the major problems to overcome in the near future.

Acknowledgments The authors wish to thank the Centre National de la Recherche Scientifique, the French Agence Nationale de la Recherche (ANR)[ANR-09-BLAN-0240-01] and the French National Program "Investissement d'Avenir" (Labex MitoCross) for financial support.

References

Adams KL, Daley DO, Whelan J, Palmer JD (2002a) Genes for two mitochondrial ribosomal proteins in flowering plants are derived from their chloroplast or cytosolic counterparts. Plant Cell 14:931–943

Adams KL, Ong HC, Palmer JD (2001) Mitochondrial gene transfer in pieces: fission of the ribosomal protein gene *rpl2* and partial or complete gene transfer to the nucleus. Mol Biol Evol 18:2289–2297

Adams KL, Palmer JD (2003) Evolution of mitochondrial gene content: gene loss and transfer to the nucleus. Mol Phyl Evol 29:380–395

Adams KL, Qiu YL, Stoutemyer M, Palmer JD (2002b) Punctuated evolution of mitochondrial gene content: high and variable rates of mitochondrial gene loss and transfer to the nucleus during angiosperm evolution. Proc Natl Acad, USA 99:9905–9912

Agrawal RK, Sharma MR (2012) Structural aspects of mitochondrial translational apparatus. Curr Opinion Struct Biol 22:797–803

Akashi K, Grandjean O, Small I (1998a) Potential dual targeting of an *Arabidopsis* archaebacterial-like histidyl-tRNA synthetase to mitochondria and chloroplasts. FEBS Lett 431:39–44

Akashi K, Takenaka M, Yamaoka S, Suyama Y, Fukuzawa H, Ohyama K (1998b) Coexistence of nuclear DNA-encoded tRNAVal(AAC) and mitochondrial DNA-encoded tRNAVal(UAC) in mitochondria of a liverwort *Marchantia polymorpha*. Nucleic Acids Res 26:2168–2172

Atteia A, Adrait A, Brugière S, Tardif M, van Lis R, Deusch O et al (2009) A proteomic survey of *Chlamydomonas reinhardtii* mitochondria sheds new light on the metabolic plasticity of the organelle and on the nature of the alpha-proteobacterial mitochondrial ancestor. Mol Biol Evol 26:1533–1548

Benichou M, Li Z, Tournier B, Chaves A, Zegzouti H, Jauneau A et al (2003) Tomato EF-Ts(mt), a functional mitochondrial translation elongation factor from higher plants. Plant Mol Biol 53:411–422

Boer PH, Gray MW (1988a) Genes encoding a subunit of respiratory NADH dehydrogenase (ND1) and a reverse transcriptase-like protein (RTL) are linked to ribosomal RNA gene pieces in *Chlamydomonas reinhardtii* mitochondrial DNA. EMBOJ 7:3501–3508

Boer PH, Gray MW (1988b) Scrambled ribosomal RNA gene pieces in *Chlamydomonas reinhardtii* mitochondrial DNA. Cell 55:399–411

Boer PH, Gray MW (1986) The URF 5 gene of *Chlamydomonas reinhardtii* mitochondria: DNA sequence and mode of transcription. EMBO J 5:21–28

Bonen L (2004) Translational machinery in plant organelles. In: Daniell H, Chase C (eds.) Molecular biology and biotechnology of plant organelles. Springer, The Netherlands, pp. 323–345

Bonen L, Calixte S (2006) Comparative analysis of bacterial-origin genes for plant mitochondrial ribosomal proteins. Mol Biol Evol 23:701–712

Bonen L, Gray MW (1980) Organization and expression of the mitochondrial genome of plants I. The genes for wheat mitochondrial ribosomal and transfer RNA: evidence for an unusual arrangement. Nucleic Acids Res 8:319–335

Brandao MM, Silva-Filho MC (2011) Evolutionary history of *Arabidopsis thaliana* aminoacyl-tRNA synthetase dual-targeted proteins. Mol Biol Evol 28:79–85

Brubacher-Kauffmann S, Marechal-Drouard L, Cosset A, Dietrich A, Duchêne AM (1999) Differential import of nuclear-encoded tRNAGly isoacceptors into *Solanum tuberosum* mitochondria. Nucleic Acids Res 27:2037–2042

Chen HC, Viry-Moussaid M, Dietrich A, Wintz H (1997) Evolution of a mitochondrial tRNA PHE gene in *A. thaliana*: import of cytosolic tRNA PHE into mitochondria. Biochem Biophys Res Commun 237:432–437

Choi KR, Roh K, Kim J, Sim W (2000) Genomic cloning and characterization of mitochondrial elongation factor Tu (EF-Tu) gene (*tufM*) from maize (*Zea mays L.*). Gene 257:233–242

Cognat V, Deragon JM, Vinogradova E, Salinas T, Remacle C, Maréchal-Drouard L (2008) On the evolution and expression of *Chlamydomonas reinhardtii* nucleus-encoded transfer RNA genes. Genetics 179:113–123

Cognat V, Pawlak G, Duchêne AM, Daujat M, Gigant A, Salinas T et al (2013) PlantRNA, a database for tRNAs of photosynthetic eukaryotes. Nucleic Acids Res 41:273–279

Colleaux L, Michel-Wolwertz MR, Matagne RF, Dujon B (1990) The apocytochrome *b* gene of *Chlamydomonas smithii* contains a mobile intron related to both *Saccharomyces* and *Neurospora* introns. Mol Gen Genet 223:288–296

Delage L, Dietrich A, Cosset A, Maréchal-Drouard L (2003a) In vitro import of a nuclearly encoded tRNA into mitochondria of *Solanum tuberosum*. Mol Cell Biol 23:4000–4012

Delage L, Duchêne AM, Zaepfel M, Maréchal-Drouard L (2003b) The anticodon and the D-domain sequences are essential determinants for plant cytosolic tRNA(Val) import into mitochondria. Plant J 34:623–633

Delage L, Giegé P, Sakamoto M, Maréchal-Drouard L (2007) Four paralogues of RPL12 are differentially associated to ribosome in plant mitochondria. Biochimie 89:658–668

Denovan-Wright EM, Lee RW (1995) Evidence that the fragmented ribosomal RNAs of *Chlamydomonas* mitochondria are associated with ribosomes. FEBS Lett 370:222–226

Dietrich A, Maréchal-Drouard L, Carneiro V, Cosset A, Small I (1996) A single base change prevents import of cytosolic tRNA(Ala) into mitochondria in transgenic plants. Plant J 10:913–918

Duchêne AM, Farouk-Ameqrane S, Sieber F, and Maréchal-Drouard L (2011) Import of RNAs into plant mitochondria. In: Kempken F (ed.) Plant mitochondria, vol 1. Springer, New York, pp 241–260

Duchêne AM, Giritch A, Hoffmann B, Cognat V, Lancelin D, Peeters NM et al (2005) Dual targeting is the rule for organellar aminoacyl-tRNA synthetases in Arabidopsis thaliana. Proc Natl Acad Sci, U S A 102:16484–16489

Duchêne AM, Maréchal-Drouard L (2001) The chloroplast-derived *trnW* and *trnM-e* genes are not expressed in *Arabidopsis* mitochondria. Biochem Biophys Res Comm 285:1213–1216

Duchêne AM, Peeters N, Dietrich A, Cosset A, Small ID, Wintz H (2001) Overlapping destinations for two dual targeted glycyl-tRNA synthetases in *Arabidopsis thaliana* and *Phaseolus vulgaris*. J Biol Chem 276:15275–15283

Duchêne AM, Pujol C, Maréchal-Drouard L (2009) Import of tRNAs and aminoacyl-tRNA synthetases into mitochondria. Curr Genet 55:1–18

Eberhard S, Loiselay C, Drapier D, Bujaldon S, Girard-Bascou J, Kuras R et al (2011) Dual functions of the nucleus-encoded factor TDA1 in trapping and translation activation of *atpA* transcripts in *Chlamydomonas reinhardtii* chloroplasts. Plant J 67:1055–1066

Fan J, Schnare MN, Lee RW (2003) Characterization of fragmented mitochondrial ribosomal RNAs of the colorless green alga *Polytomella parva*. Nucleic Acids Res 31:769–778

Fey J, Dietrich A, Desprez T, Maréchal-Drouard L (1997) Evolutionary aspects of "chloroplast-like" *trnN* and *trnH* expression in higher-plant mitochondria. Curr Genet 32:358–360

Figueroa P, Gomez I, Carmona R, Holuigue L, Araya A, Jordana X (1999) The gene for mitochondrial ribosomal protein S14 has been transferred to the nucleus in *Arabidopsis thaliana*. Mol Gen Genet 262:139–144

Frechin M, Duchêne AM, Becker HD (2009) Translating organellar glutamine codons: a case by case scenario? RNA Biol 6:31–34

Gagliardi D and Binder S (2007) Expression if the plant mitochondrial genome. In: Logan DC (ed.) Plant mitochondria, vol 31. Blackwell publishing, Oxford pp 50–84

Giegé P, Brennicke A (2001) From gene to protein in higher plant mitochondria. C R Acad Sci III 324:209–217

Glover KE, Spencer DF, Gray MW (2001) Identification and structural characterization of nucleus-encoded transfer RNAs imported into wheat mitochondria. J Biol Chem 276:639–648

Grohmann L, Brennicke A, Schuster W (1992) The mitochondrial gene encoding ribosomal protein S12 has been translocated to the nuclear genome in *Oenothera*. Nucleic Acids Res 20:5641–5646

Hammani K, Gobert A, Small I, Giegé P (2011) A PPR protein involved in regulating nuclear genes encoding mitochondrial proteins? Plant signaling Behav 6:748–750

Handa H, Kobayashi-Uchaia A, Murayama S (2001) Characterization of a wheat cDNA encoding mitochondrial ribosomal protein L11: qualitative and quantitative tissue-specific differences in its expression. Mol Gene Genet 265:569–575

Hazle T, Bonen L (2007) Comparative analysis of sequences preceding protein-coding mitochondrial genes in flowering plants. Mol Biol Evol 24:1101–1112

Heazlewood JL, Millar AH (2005) AMPDB: the *Arabidopsis* mitochondrial protein database. Nucleic Acids Res 33:605–610

Ito J, Taylor NL, Castleden I, Weckwerth W, Millar AH, Heazlewood JL (2009) A survey of the *Arabidopsis thaliana* mitochondrial phosphoproteome. Proteomics 9:4229–4240

Kadowaki K, Kubo N, Ozawa K, Hirai A (1996) Targeting presequence acquisition after mitochondrial gene transfer to the nucleus occurs by duplication of existing targeting signals. EMBO J 15:6652–6661

Kehrein K, Bonnefoy N, and Ott M (2012) Mitochondrial protein synthesis: efficiency and accuracy. Antioxidants & redox signaling, 57:112–125

Klodmann J, Senkler M, Rode C, Braun HP (2011) Defining the protein complex proteome of plant mitochondria. Plant Phys 157:587–598

Kubo N, Jordana X, Ozawa K, Zanlungo S, Harada K, Sasaki T et al (2000) Transfer of the mitochondrial *rps10* gene to the nucleus in rice: acquisition of the 5' untranslated region followed by gene duplication. Mol Gen Genet 263:733–739

Kuck U, Jekosch K, Holzamer P (2000) DNA sequence analysis of the complete mitochondrial genome of the green alga *Scenedesmus obliquus*: evidence for UAG being a leucine and UCA being a non-sense codon. Gene 253:13–18

Kuhlman P, Palmer JD (1995) Isolation, expression, and evolution of the gene encoding mitochondrial elongation factor Tu in *Arabidopsis thaliana*. Plant Mol Biol 29:1057–1070

Kumar R, Maréchal-Drouard L, Akama K, Small I (1996) Striking differences in mitochondrial tRNA import between different plant species. Mol Gen Genet 252:404–411

Laforest MJ, Delage L, Maréchal-Drouard L (2005) The T-domain of cytosolic tRNAVal, an essential determinant for mitochondrial import. FEBS Lett 579:1072–1078

Leaver CJ, Harmey MA (1976) Higher-plant mitochondrial ribosomes contain a 5S ribosomal ribonucleic acid component. Biochemical J 157:275–277

Lecompte O, Ripp R, Thierry JC, Moras D, Poch O (2002) Comparitive analysis of ribosomal proteins in complete genomes: an example of reductive evolution at the domain scale. Nucleic Acids Res 30:5382–5390

Lu M, Hanson M (1996) Fully edited and partially edited *nad9* transcripts differ in size and both are associated with polysomes in potato mitochondria. Nucleic Acids Res 24:1369–1374

Maffey L, Degand H, Boutry M (1997) Partial purification of mitochondrial ribosomes from broad bean and identification of proteins encoded by the mitochondrial genome. Mol Gen Genet 254:365–371

Manavski N, Guyon V, Meurer J, Wienand U, Brettschneider R (2012) An essential pentatricopeptide repeat protein facilitates 5' maturation and translation initiation of *rps3* mRNA in maize mitochondria. Plant Cell 24:3087–3105

Maréchal L, Guillemaut P, Grienenberger JM, Jeannin G, Weil JH (1985a) Sequence and codon recognition of bean mitochondria and chloroplast tRNAs[Trp]: evidence for a high degree of homology. Nucleic Acids Res 13:4411–4416

Maréchal L, Guillemaut P, Grienenberger JM, Jeannin G, Weil JH (1986) Sequences of initiator and elongator methionine tRNAs in bean mitochondria. Plant Mol Biol 7:245–253

Maréchal L, Guillemaut P, Grienenberger JM, Jeannin G, Weil JH (1985b) Structure of bean mitochondrial tRNA[Phe] and localization of the tRNA[Phe] gene on the mitochondrial genomes of maize and wheat. FEBS Lett 184:289–293

Maréchal L, Guillemaut P, Weil JH (1985c) Sequence of two bean mitochondria tRNAs[Tyr] which differ in the level of post-transcriptional modification and have a prokaryotic-like large extra-loop. Plant Mol Biol 5:347–351

Maréchal-Drouard L, Guillemaut P, Cosset A, Arbogast M, Weber F, Weil JH et al (1990) Transfer RNAs of potato (*Solanum tuberosum*) mitochondria have different genetic origins. Nucleic Acids Res 18:3689–3696

Michaelis G, Vahrenholz C, Pratje E (1990) Mitochondrial DNA of *Chlamydomonas reinhardtii*: the gene for apocytochrome *b* and the complete functional map of the 15.8 kb DNA. Mol Gen Genet 223:211–216

Michaud M, Maréchal-Drouard L, Duchêne AM (2010) RNA trafficking in plant cells: targeting of cytosolic mRNAs to the mitochondrial surface. Plant Mol Biol 73:697–704

Millen RS, Olmstead RG, Adams KL, Palmer JD, Lao NT, Heggie L et al (2001) Many parallel losses of *infA* from chloroplast DNA during angiosperm evolution with multiple independent transfers to the nucleus. Plant Cell 13:645–658

Morgante CV, Rodrigues RA, Marbach PA, Borgonovi CM, Moura DS, Silva-Filho MC (2009) Conservation of dual-targeted proteins in *Arabidopsis* and rice points to a similar pattern of gene-family evolution. Mol Genet Genomics 281:525–538

Nedelcu AM (1997) Fragmented and scrambled mitochondrial ribosomal RNA coding regions among green algae: a model for their origin and evolution. Mol Biol Evol 14:506–517

O'Brien EA, Zhang Y, Wang E, Marie V, Badejoko W, Lang BF et al (2009) GOBASE: an organelle genome database. Nucleic Acids Res 37:946–950

O'Brien TW, O'Brien BJ, Norman RA (2005) Nuclear MRP genes and mitochondrial disease. Gene 354:147–151

Peeters NM, Chapron A, Giritch A, Grandjean O, Lancelin D, Lhomme T et al (2000) Duplication and quadruplication of *Arabidopsis thaliana* cysteinyl- and asparaginyl-tRNA synthetase genes of organellar origin. J Mol Evol 50:413–423

Phreaner CG, Williams MA, Mulligan RM (1996) Incomplete editing of *rps12* transcripts results in the synthesis of polymorphic polypeptides in plant mitochondria. Plant Cell 8:107–117

Pinel C, Douce R, Mache R (1986) A study of mitochondrial ribosomes from the higher plant *Solanum tuberosum L.* Molecular Biol Rep 11:93–97

Placido A, Regina TM, Quagliariello C, Volpicella M, Gallerani R, Ceci LR (2009) Mapping of 5' and 3'-ends of sunflower mitochondrial nad6 mRNAs reveals a very complex transcription pattern which includes primary transcripts lacking 5'-UTR. Biochimie 91:924–932

Pring DR, Mullen JA, Kempken F (1992) Conserved sequence blocks 5' to start codons of plant mitochondrial genes. Plant Mol Biol 19:313–317

Pujol C, Bailly M, Kern D, Maréchal-Drouard L, Becker H, Duchêne AM (2008) Dual-targeted tRNA-dependent amidotransferase ensures both mitochondrial and chloroplastic Gln-tRNAGln synthesis in plants. Proc Natl Acad Sci, U S A 105:6481–6485

Pujol C, Maréchal-Drouard L, Duchêne AM (2007) How can organellar protein N-terminal sequences be dual targeting signals? In silico analysis and mutagenesis approach. J Mol Biol 369:356–367

Raczynska KD, Le Ret M, Rurek M, Bonnard G, Augustyniak H, Gualberto JM (2006) Plant mitochondrial genes can be expressed from mRNAs lacking stop codons. FEBS Lett 580:5641–5646

Rahire M, Laroche F, Cerutti L, Rochaix JD (2012) Identification of an OPR protein involved in the translation initiation of the PsaB subunit of photosystem I. Plant J 72:652–661

Rainaldi G, Volpicella M, Licciulli F, Liuni S, Gallerani R, Ceci LR (2003) PLMItRNA, a database on the heterogeneous genetic origin of mitochondrial tRNA genes and tRNAs in photosynthetic eukaryotes. Nucleic Acids Res 31:436–438

Remacle C, Cardol P, Coosemans N, Gaisne M, Bonnefoy N (2006) High-efficiency biolistic transformation of *Chlamydomonas* mitochondria can be used to insert mutations in complex I genes. Proc Natl Acad Sci, U S A 103:4771–4776

Rokov-Plavec J, Dulic M, Duchêne AM, Weygand-Durasevic I (2008) Dual targeting of organellar seryl-tRNA synthetase to maize mitochondria and chloroplasts. Plant Cell Rep 27:1157–1168

Rubio MA, Hopper AK (2011) Transfer RNA travels from the cytoplasm to organelles. Wiley Interdiscip Rev RNA 2802–2817

Salinas T, Duby F, Larosa V, Coosemans N, Bonnefoy N, Motte P et al (2012) Co-evolution of mitochondrial tRNA import and codon usage determines translational efficiency in the green alga *Chlamydomonas*. PLoS Genets 8:e1002946

Salinas T, Duchêne AM, Delage L, Nilsson S, Glaser E, Zaepfel M et al (2006) The voltage-dependent anion channel, a major component of the tRNA import machinery in plant mitochondria. Proc Natl Acad Sci U S A 103:18362–18367

Salinas T, Duchêne AM, Maréchal-Drouard L (2008) Recent advances in tRNA mitochondrial import. Trends Biochem Sci 33:320–329

Salinas T, Schaeffer C, Maréchal-Drouard L, Duchêne AM (2005) Sequence dependence of tRNA(Gly) import into tobacco mitochondria. Biochimie 87:863–872

Sanchez H, Fester T, Kloska S, Schroder W, Schuster W (1996) Transfer of *rps19* to the nucleus involves the gain of an RNP-binding motif which may functionally replace RPS13 in *Arabidopsis* mitochondria. EMBO J 15:2138–2149

Sandoval P, Leon G, Gomez I, Carmona R, Figueroa P, Holuigue L et al (2004) Transfer of RPS14 and RPL5 from the mitochondrion to the nucleus in grasses. Gene 324:139–147

Schneider A (2001) Does the evolutionary history of aminoacyl-tRNA synthetases explain the loss of mitochondrial tRNA genes? Trends Genet 17:557–559

Schneider A (2011) Mitochondrial tRNA import and its consequences for mitochondrial translation. Annu Rev Biochem 80:1033–1053

Sharma MR, Booth TM, Simpson L, Maslov DA, Agrawal RK (2009) Structure of a mitochondrial ribosome with minimal RNA. Proc Natl Acad Sci, USA, 106:9637–9642

Sharma MR, Koc EC, Datta PP, Booth TM, Spremulli LL, Agrawal RK (2003) Structure of the mammalian mitochondrial ribosome reveals an expanded functional role for its component proteins. Cell 115:97–108

Sieber F, Duchêne AM, Maréchal-Drouard L (2011a) Mitochondrial RNA import: from diversity of natural mechanisms to potential applications. Int Rev Cell Mol Biol 287:145–190

Sieber F, Placido A, El Farouk-Ameqrane S, Duchêne AM, Maréchal-Drouard L (2011b) A protein shuttle system to target RNA into mitochondria. Nucleic Acids Res 39:e96

Small I, Maréchal-Drouard L, Masson J, Pelletier G, Cosset A, Weil JH et al (1992) In vivo import of a normal or mutagenized heterologous transfer RNA into the mitochondria of transgenic plants: towards novel ways of influencing mitochondrial gene expression? EMBO J 11:1291–1296

Smits P, Smeitink JA, van den Heuvel LP, Huynen MA, Ettema TJ (2007) Reconstructing the evolution of the mitochondrial ribosomal proteome. Nucleic Acids Res 35:4686–4703

Takemura M, Oda K, Yamato K, Ohta E, Nakamura Y, Nozato N et al (1992) Gene clusters for ribosomal proteins in the mitochondrial genome of a liverwort, *Marchantia polymorpha*. Nucleic Acids Res 20:3199–3205

Turmel M, Lemieux C, Burger G, Lang BF, Otis C, Plante I et al (1999) The complete mitochondrial DNA sequences of *Nephroselmis olivacea* and *Pedinomonas minor*. Two radically different evolutionary patterns within green algae. Plant Cell 11:1717–1730

Uyttewaal M, Mireau H, Rurek M, Hammani K, Arnal N, Quadrado M et al (2008) PPR336 is associated with polysomes in plant mitochondria. J Mol Biol 375:626–636

Vinogradova E, Salinas T, Cognat V, Remacle C, Maréchal-Drouard L (2009) Steady-state levels of imported tRNAs in *Chlamydomonas* mitochondria are correlated with both cytosolic and mitochondrial codon usages. Nucleic Acids Res 37:1521–1528

Wischmann C, Schuster W (1995) Transfer of *rps10* from the mitochondrion to the nucleus in *Arabidopsis thaliana*: evidence for RNA-mediated transfer and exon shuffling at the integration site. FEBS Lett 374:152–156

Ye W, Spanning E, Unnerstale S, Gotthold D, Glaser E, Maler L (2012) NMR investigations of the dual targeting peptide of Thr-tRNA synthetase and its interaction with the mitochondrial Tom20 receptor in *Arabidopsis thaliana*. FEBS J 279:3738–3748

Yukawa M, Kuroda H, Sugiura M (2007) A new *in vitro* translation system for non-radioactive assay from tobacco chloroplasts: effect of pre-mRNA processing on translation *in vitro*. Plant J 49:367–376

Chapter 9
Translation in Chloroplasts of Flowering Plants

Masahiro Sugiura

Abstract The chloroplast genome in flowering plants contains about 80 protein-coding genes. The chloroplast translational machinery, which is similar to that of *Escherichia coli*, reads the corresponding mRNAs. Translation initiation is critical to produce a correct protein. There are multiple possible initiation codons, either AUG or GUG, around an initiation region. Generally, *cis*-elements residing in a 5′-untranslated region and *trans*-acting factors are responsible for selection of genuine initiation codons. Unlike eubacterial mRNAs, a limited number of chloroplast mRNAs use Shine-Dalgarno-like sequences as their *cis*-elements, and many chloroplast mRNAs require specific *trans*-acting factors. As the chloroplast genome is compact, some genes (cistrons) partially overlap; namely, the start codon of a downstream cistron is located in front of the stop codon of its upstream cistron. In such a case, the downstream cistron is translated in a special manner called "translational coupling" and by an additional mechanism to produce the necessary amount of its product.

9.1 Introduction

Chloroplasts are organelles unique to plant cells that contain the entire machinery necessary for the process of oxygenic photosynthesis. The organelles also perform the biosynthesis of starch, lipids, amino acids, nucleotides, and other metabolites. Chloroplasts possess their own genomes and gene expression systems. Chloroplast genomes can be represented genetically as circular double-stranded DNA molecules. The genome in flowering plants is around 150 kilobase pairs (kbp) and contains a large inverted repeat (IR) of 20–27 kbp (Wakasugi et al. 2001; Bock 2007;

M. Sugiura (✉)
Center for Gene Research, Nagoya University, Nagoya 464-8602, Japan
e-mail: sugiura@gene.nagoya-u.ac.jp

A.-M. Duchêne (ed.), *Translation in Mitochondria and Other Organelles*,
DOI: 10.1007/978-3-642-39426-3_9, © Springer-Verlag Berlin Heidelberg 2013

Wicke et al. 2011). Some legumes are exceptions and lack the IR. The number of annotated genes is 110–115 including hypothetical chloroplast reading frames (*ycfs*), namely, conserved open reading frames (ORFs) among related species. The genes in the IR are duplicated while all the others are single-copy (e.g., 17 genes in the IR in tobacco).

A typical chloroplast genome in flowering plants contains around 80 protein-coding genes and 34 stable RNA-coding genes. The encoded proteins include ribosomal proteins, photosystem I and II subunits, NADH dehydrogenase subunits, the Rubisco large subunit, and several others. The first non-coding RNA gene (*sprA*) was discovered in tobacco and encodes a stable RNA of 218 nucleotides (nt) (Vera and Sugiura 1994). Recently, many noncoding RNAs have been reported (Lung et al. 2006; Hotto et al. 2011). There are also long ORFs relatively unique to species (e.g., 13 ORFs of over 70 codons in tobacco). Hence, new chloroplast genes could be added in the future.

Chloroplast gene expression consists of multiple steps including transcription, RNA editing, RNA splicing, RNA cleavage and trimming, and translation. Expression is regulated in part at the transcription step but primarily during post-transcriptional processes, especially during translation. Chloroplast translation has been studied by several methods: *in planta* and *in organello* studies using radioactively labeled amino acids and antibodies, and in vitro assays (see Sect. 3.1). Chloroplast transformation techniques have been applied to *Chlamydomonas reinhardtii* and tobacco chloroplasts. Genetic approaches are useful for identification of nuclear genes involved in mRNA processing and translation.

In this chapter, we discuss mainly basic translation processes in flowering plant chloroplasts. Translation in *Chlamydomonas* chloroplasts, translational control by environmental conditions, and additional aspects of chloroplast translation have been reviewed elsewhere (Rochaix 2001; Choquet and Wollman 2002; Manuell et al. 2004; Peled-Zehavi and Danon 2007; Marín-Navarro et al. 2007; Wobbe et al. 2008; Lyska et al. 2013).

9.2 Chloroplast Ribosomes

Chloroplasts evolved from endosymbiosis of a photosynthetic prokaryote. Hence, the chloroplast translation machinery is thought to be similar to that of prokaryotes.

Chloroplast ribosomes are 70S-type and contain four rRNAs (Sugiura 1992). The genes for 16S rRNA, 23S rRNA, 4.5S rRNA, and 5S rRNA are clustered and located in the IR in most flowering plants. Hence, all rRNA genes are duplicated. The cluster is transcribed as a unit and the pre-rRNAs are processed into mature rRNAs. The small (30S-type) ribosomal subunit has 16S rRNA and the large (50S-type) ribosomal subunit has 23S, 4.5S, and 5S rRNAs. The 4.5S rRNA is about 100 nucleotides (nt) long and is found in land plants. Its sequence is highly homologous to the $3'$ sequence of *Escherichia coli* (*E. coli*) 23S rRNA. A spacer of around 100 nt, which is removed during rRNA maturation, separates the 23S and 4.5S rRNAs.

A comprehensive analysis of spinach chloroplast ribosomal proteins showed that the large subunit consists of 33 proteins, of which 31 are orthologues of *E. coli* ribosomal proteins and two are plastid-specific proteins (Yamaguchi and Subramanian 2000), and that the small subunit comprises 25 proteins, of which 21 are orthologues of all *E. coli* 30S proteins, and four are plastid-specific ribosomal proteins (Yamaguchi et al. 2000). All the plastid-specific ribosomal proteins are encoded in the nuclear genome (Yamaguchi and Subramanian 2003). It was later shown that one (PSRP-1) of these six proteins is not a genuine ribosomal protein, but a ribosome-binding factor (Sharma et al. 2010). In the chloroplast genome of tobacco (a eudicot), rice (a monocot), and many other flowering plants, 20 sequences were annotated to genes for ribosomal proteins based on similarities to *E. coli* counterparts, eight for large subunit proteins and 12 for small subunit proteins (Shinozaki et al. 1986; Ohto et al. 1988; Hiratsuka et al. 1989). Occasionally, one to three genes are missing from the chloroplast genome; for example, spinach lacks the gene for L23 (Schmitz-Linneweber et al. 2001) and pea lacks genes for L22, L23, and S16 (Magee et al. 2010). An extreme case is that the parasitic flowering plant *Epifagus virginiana* lacks six genes for ribosomal proteins (Wolfe et al. 1992). These missing genes were transferred to the nuclear genome during chloroplast evolution.

The three-dimensional structure of spinach and *Chlamydomonas* chloroplast ribosomes has been studied using cryo-electron microscopes (Sharma et al. 2007; Manuell et al. 2007). The overall structure of the chloroplast ribosome is similar to bacterial ribosomes, but it is larger in size and exhibits a number of chloroplast-specific features. These features are probably involved in chloroplast-specific regulation of translation.

9.3 The Genetic Code and tRNAs

Chloroplast protein-coding genes use the universal genetic code and contain all 64 codons. At least 32 tRNA species are required to recognize all these codons according to the standard wobble rule. However, the chloroplast genome in flowering plants lacks several tRNA genes and there is no evidence for tRNA import into the chloroplasts of photosynthetic plants (Sugiura 1992). For example, sequencing the entire tobacco chloroplastic genome has identified 30 different tRNA genes (Shinozaki et al. 1986). Seven of them are located in the IR, and therefore the total number of tRNA genes is 37. Hybridization analysis using total tobacco chloroplast tRNA preparations revealed that all tRNA genes are expressed. All the tRNA species predicted from genes can form cloverleaf structures (Sugiura 1987) and none have an abnormal form, as has been reported for some mitochondrial tRNAs. Four genes encoding tRNAs that recognize codons CUU/C (Leu), CCU/C (Pro), GCU/C (Ala), and CGC/A/G (Arg) could not be found in the genome.

If the "two-out-of-three" mechanism can operate in the chloroplast, as has been shown in an in vitro protein synthesizing system from *E. coli* (Samuelsson et al. 1980), the single tRNAs tRNA^{Pro} (UGG), tRNA^{Ala} (UGC), and tRNA^{Arg} (ACG) can, respectively, read all four Pro, Ala, and Arg codons, respectively (GC pairs in the first and second codon-anticodon interaction). There is a gene for tRNA^{Leu} (UAG), and if this tRNA has an unmodified U in the first position of the antico-don, it can read all four Leu codons (CUN) by U: N wobble (Barrell et al. 1980). The bean, spinach, and soybean tRNAs^{Leu} (UAG) have unmodified Us in their anticodons (UA^{m7}G) (Pillay et al. 1984). Therefore, these 28 tRNAs have been suggested to be sufficient to read all codons in tobacco chloroplast mRNAs using the above mechanisms (Shinozaki et al. 1986). The A of tRNA^{Arg} (ACG) is gener-ally modified to inosine (I), and the modified tRNA (ICG) reads two other codons (CGC and CGA). The tRNA adenosine deaminase (A to I change) was found in *Arabidopsis* (Delannoy et al. 2009; Karcher and Bock 2009). Recent tRNA gene knockout experiments indicated that the 25 tRNA species could read all codons in tobacco chloroplasts (Alkatib et al. 2012).

The genetic code is degenerate; that is, the common 20 amino acids, except for methionine and tryptophan, are coded for by two to six codons called synonymous codons. Synonymous codons are not used with equal frequency, and are used dif-ferently by different organisms. Codon usage is generally calculated by simple summation of collected gene sequences. However, this method cannot be applied to chloroplasts and plant mitochondria because C to U RNA editing causes codon conversion at the mRNA level (Giege and Brennicke 1999; Sugiura 2008). In tobacco, RNA editing has been detected at 38 sites and 37 of them change codons (Sasaki et al. 2003). As discussed later, mRNAs often contain multiple possible initiation codons, and generally the longest ORFs are annotated as genes. Genuine initiation codons should be determined experimentally. Including these data, codon usage of the 79 tobacco chloroplast mRNAs was calculated (Nakamura and Sugiura 2007) and is shown in Table 9.1. Unlike *E. coli* and mammals, the so-called rare codons are not present in chloroplast mRNAs. Codons with cog-nate tRNAs not present are still used at frequencies similar to others (see boxes in Table 9.1). Observations in *E. coli*, yeast, and *Bacillus subtilis* showed that codon usage is positively correlated with tRNA content and tRNA gene copy num-ber, especially for highly expressed genes (Ikemura 1985; Kanaya et al. 2001). Experiments analyzing the translation efficiency of synonymous codons in *E. coli* indicated that different codons are translated at different rates and that there is generally a correlation between translation rate and codon usage or tRNA content (e.g., Robinson et al. 1984; Sørensen and Pedersen 1991). Using an in vitro trans-lation system from tobacco chloroplasts (see below), it was shown that translation efficiencies of synonymous codons are not always correlated with codon usage in chloroplasts (Nakamura and Sugiura 2007). For example, the tyrosine codons UAU and UAC are read by the same tRNA, and the translation efficiency of UAC is more than twice higher than that of UAU though the codon usage of UAU is four times higher than that of UAC. This observation suggests that individual codons possess intrinsic efficiencies (Nakamura and Sugiura 2007, 2009, 2011).

Table 9.1 Codon usage of the 79 tobacco chloroplast mRNAs

Phe	UUU	0.667	Ser	UCU	0.299	Tyr	UAU	0.804	Cys	UGU	0.752
Phe	UUC	0.333	Ser	UCC	0.152	Tyr	UAC	0.196	Cys	UGC	0.248
Leu	UUA	0.328	Ser	UCA	0.188	stop	UAA	0.519	stop	UGA	0.228
Leu	UUG	0.200	Ser	UCG	0.059	stop	UAG	0.253	Trp	UGG	1.000
Leu	**CUU**	0.214	Pro	**CCU**	0.393	His	CAU	0.770	Arg	CGU	0.219
Leu	**CUC**	0.069	Pro	**CCC**	0.187	His	CAC	0.230	Arg	**CGC**	0.063
Leu	CUA	0.128	Pro	CCA	0.286	Gln	CAA	0.756	Arg	**CGA**	0.251
Leu	CUG	0.062	Pro	CCG	0.135	Gln	CAG	0.244	Arg	**CGG**	0.074
Ile	AUU	0.495	Thr	ACU	0.392	Asn	AAU	0.768	Ser	AGU	0.209
Ile	AUC	0.200	Thr	ACC	0.197	Asn	AAC	0.232	Ser	AGC	0.059
Ile	AUA	0.306	Thr	ACA	0.304	Lys	AAA	0.754	Arg	AGA	0.289
Met	AUG	0.853	Thr	ACG	0.107	Lys	AAG	0.246	Arg	AGG	0.104
fMet	AUG	0.141									
Val	GUU	0.372	Ala	**GCU**	0.448	Asp	GAU	0.797	Gly	GGU	0.322
Val	GUC	0.120	Ala	**GCC**	0.172	Asp	GAC	0.203	Gly	GGC	0.118
Val	GUA	0.380	Ala	GCA	0.280	Glu	GAA	0.757	Gly	GGA	0.389
Val	GUG	0.128	Ala	GCG	0.100	Glu	GAG	0.243	Gly	GGG	0.171
fMet	GUG	0.006									

Codons in boxes indicate those for which no corresponding tRNA genes are present in the tobacco chloroplast genome (from Table 1 in Sugiura 1987). Numerals show codon fractions, the ratio of each codon in the family of synonymous codons (from Tables 1 in Nakamura and Sugiura 2007). The two original tables are combined and extensively modified

9.4 Translation Initiation

9.4.1 Initiation Sites

Initiation is critical for translation to produce a correct protein. This step is a major point for on-and-off regulation of translation and of the amount of a product necessary for chloroplasts. In an mRNA, there are three possible reading frames and multiple possible initiation codons in the three frames. Hence, the essential step for translation is the selection of the correct initiation codons. In chloroplasts of flowering plants, AUG and GUG codons are used, and no UUG has been reported. The *Chlorella* chloroplast gene (*infA*) encoding initiation factor-1 has a UUG initiation codon (Wakasugi et al. 1997). The synthesized *infA* mRNA was translated from the UUG codon in a tobacco chloroplast in vitro system (Hirose et al. 1999), suggesting that UUG, if present, can be an initiation codon in flowering plant chloroplasts. Among the 79 protein-coding genes in tobacco, 75 coding regions (cistrons) have AUG initiation codons, two have created AUG codons from ACG by RNA editing (*psbL* and *ndhD*), and only two cistrons have GUG codons (*rps19* and *psbC*).

Much of our knowledge about translation was gained by studying the process in extracts from broken cells (in vitro systems or cell-free systems), and the most often-used extracts are prepared from *E. coli* (Lengyel 1974). Chloroplast S30 fractions are supernatant fractions obtained by centrifugation at 30,000 $\times$ g from lysates of isolated chloroplasts. Simple chloroplast extracts show either no translation activity or incorporate ^{35}S-methionine randomly. It is necessary to find a suitable plant species, growth conditions, and leaf size. Then, procedures for chloroplast isolation, chloroplast disruption, preparation of S30 fractions, and other conditions must be optimized. The buffers used are extremely important, and their composition and concentration should be optimized individually. Reaction conditions (volume, temperature, buffer compositions, and their concentrations) need to be optimized too. Our chloroplast in vitro translation systems so prepared are highly active and support translation from exogenously added mRNA templates linearly up to 2 h at 28 °C (Yukawa et al. 2007). This system can estimate relative rate of translation under linear progression conditions and template-limiting conditions (excess translation machinery relative to template).

Figure 9.1 shows a nucleotide sequence from the 5′ end (−85) through the first 40 codons (+120) of the tobacco chloroplast *psbA* mRNA encoding the D1 protein (or 32 kDa protein) in photosystem II. Frame 1 has three possible initiation codons, GUG at −58, AUG at +1, and AUG at +109. In addition, frames 2 and 3 include three AUG codons and two GUG codons. Translation is known to start from the genuine initiation codon AUG (+1) to produce the precursor D1 protein. The C-terminal nine amino acids are removed to form the mature D1 protein of 32 kDa (Marder et al. 1984; Takahashi et al. 1988). In the 1980s, *E. coli* in vitro transcription/translation systems were used to identify *psbA* genes. The *E. coli* system (a heterologous system) started translation from AUG (+109, the 37th codon) in frame 1, and hence this AUG was assigned as the initiation site (Cohen

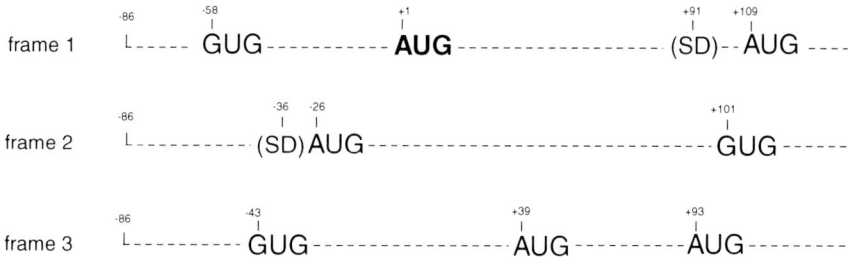

Fig. 9.1 Multiple possible initiation codons in the 5′ portion of tobacco *psbA* mRNA (Sugita and Sugiura 1984). Positions are relative to the genuine initiation codon AUG (with the A as +1). Schematic of the sequence from its transcription start site (−86) through the first 120 nt. "(SD)" indicates the SD-like sequence in the proper position. "frame 1" shows the coding frame; "frame 2" and "frame 3" are out of frame. In frame 1, *E. coli* extracts start translation from AUG at +109 (the 37 codon) but not from the genuine AUG at +1. In frame 2, *E. coli* extracts recognize AUG at −26 but translation ends at the 4th inframe stop codon, UAA

et al. 1984). Translation was probably observed because of a Shine-Dalgarno (SD)-like sequence (GGA) located 15 nt upstream of the AUG. This initiation is an artifact since *E. coli* possesses no *trans*-acting factor for *psbA* translation (see the next section). However, translation from AUG (+1) but not from AUG (+109) was observed using a reticulocyte lysate system and the *Solanum psbA* mRNA (Eyal et al. 1987). This observation may be explained by the eukaryotic ribosome entering at the 5′ end of the mRNA, scanning the 5′ UTR, and starting translation from the first AUG (+1). Well-prepared chloroplast in vitro systems start translation specifically from AUG (+1) (Hirose and Sugiura 1996; Yukawa et al. 2007). Mutation of the AUG (+1) to ACG abolishes translation completely in the in vitro translation system, indicating that the chloroplast translation machinery selects only one AUG codon (AUG at +1) among the five AUGs and three GUGs in the three frames, because the coding frame is not predetermined.

The tobacco *ndhD* mRNA has two possible initiation codons, one AUG at −27 and one GUG at −18 (Fig. 9.2). Initially, the AUG at −27 was assigned as the

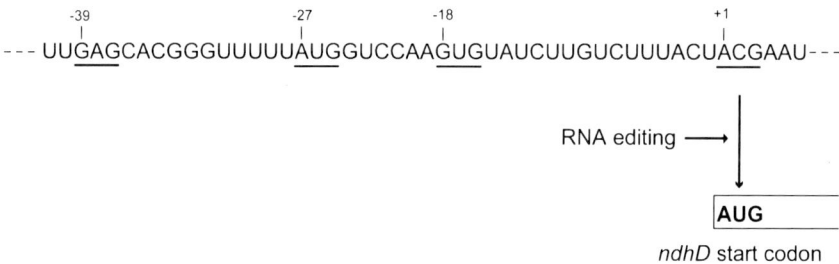

Fig. 9.2 Possible initiation codons in the coding frame of tobacco *ndhD* mRNA. There are two possible start codons, AUG at −27 and GUG at −18. An SD-like sequence (GAG) is present in the proper position from AUG at −27. The real start codon is the AUG created from ACG at +1 by RNA editing (Neckermann et al. 1994; Hirose and Sugiura 1997)

initiation codon because an SD-like sequence (GAG) is present 12 nt upstream from the AUG. Later, it was found that RNA editing converts the ACG at +1 to AUG (Neckermann et al. 1994). In vitro assays have proved that the edited AUG at +1 is the real initiation codon (Hirose and Sugiura 1997). The *ndhK* mRNA has four possible AUG codons in its coding frame in many eudicot plants. In vitro assays have also defined that the third AUG is its start codon (Yukawa and Sugiura 2008). The *psbC* mRNAs in tobacco and many other plants have two possible initiation codons, AUG and GUG, in the same frame. In vitro analysis has shown that the mRNA is translated from the GUG codon (Kuroda et al. 2007).

9.4.2 5′ Untranslated Regions and 5′ Processing

5′ Untranslated regions (5′UTRs) are known to be important for translation initiation and translational control. Chloroplast transformation analysis using tobacco has elegantly proved the involvement of 5′UTRs for translation of *psbA, rbcL,* and several other mRNAs (Staub and Maliga 1993, 1994; Eibl et al. 1999; Zou et al. 2003).

The length of 5′UTRs differs from mRNA to mRNA. In tobacco, the *psbD* mRNA has a 905 nt 5′UTR (Yao et al. 1989) and the *rpl32* mRNA has multiple 5′UTRs, the longest one up to 1,101 nt (Vera et al. 1992). Some chloroplast genes contain multiple transcription initiation sites, and hence one mRNA group consists of several mRNAs with 5′UTRs of different sizes. In addition, RNA processing occurs often in 5′UTRs. Processing of the 5′UTR is required for efficient translation of the barley *rbcL* mRNA (Reinbothe et al. 1993).

It is not easy to study effects of 5′ processing on translation using in vivo methods since a mixture of unprocessed and processed mRNA species are present in chloroplasts. In vitro analysis can evaluate each mRNA species and has found that processed *atpB* and *psbB* mRNAs are more efficiently translated than unprocessed mRNAs, while processed and unprocessed *atpH* and *rbcL* mRNAs are translated at similar rates (Yukawa et al. 2007). Processing of the 905 nt 5′UTR of tobacco *psbD* primary transcripts to 132 nt greatly enhances *psbD* translation (Adachi et al. 2012). Hence, effects of 5′ processing on translation depend on the mRNA species.

The translation mechanism in chloroplasts has been thought to be similar to that of eubacteria (Sugiura et al. 1998). Most *E. coli* mRNAs possess the SD sequence, typically GGAGG, which is located 5–7 nt upstream from the initiation codon, complementary to the 3′ terminus of 16S rRNA, and is required for selection of initiation codons together with formylmethionyl-tRNA (Chen et al. 1994). However, over half of tobacco chloroplast mRNAs lack SD-like sequences at the proper position (Sugiura et al. 1998; Hirose and Sugiura 2004a). In vitro analysis has shown that SD-like sequences located between −18 and −6 with respect to the start codon are functional for translation initiation in chloroplasts (Hirose et al. 1998; Hirose and Sugiura 2004a), whereas those located further upstream (e.g., those of *psbA* mRNA) and those located too close to the start codon (e.g., those of *petB* mRNA) are no more functional (Hirose and Sugiura 1996, 2004a). On the other hand, the SD-like sequence of *rps2* mRNA, located from −8 to −5 is not functional but is a negative regulatory element for translation (Plader and Sugiura 2003). Hence, SD-like sequences seem not to always be functional even in the proper position. Translation of *psbC* mRNA starts from GUG and requires an extended SD-like sequence, GAGGAGGU, located between −16 and −9 (Kuroda et al. 2007).

As shown in Fig. 9.3, there are several types of *cis*-elements in 5′UTRs. In tobacco, *rbcL* mRNA has a typical SD-like sequence (GGAGG), and mutation of its SD abolishes translation (Hirose and Sugiura 2004a). Tobacco *atpE* mRNA also has an SD-like sequence (GGAG) between −18 and −15, and its mutation inhibits translation. Furthermore, translation of *atpE* mRNA is inhibited by the addition of excess oligonucleotide (5′-AAAGGAGGTGAT•••) complementary to the 3′ sequence of chloroplast 16S rRNA (3′-UUUCCUCCACUA•••) (Hirose and Sugiura 2004a). This competition experiment showed that chloroplast SD-like sequences interact with the 30S ribosomal subunit via the 3′ end of chloroplast 16S rRNA. Hence, the SD sequence is a *cis*-element, as in many *E. coli* mRNAs (A). *psbA* mRNA has two separate sequences [RBS2(UGAU) and RBS2 (AAG)] that are complementary to the 3′-terminus of chloroplast 16S rRNA, and these two sequences are required for *psbA* translation (B) (Hirose and Sugiura 1996). *atpB* mRNA contains no SD-like sequence and is U-rich in its 5′UTR (between −20 and −1). This unstructured sequence encompassing the AUG initiation codon is required for translation (C) (Hirose and Sugiura 2004b). Chloroplast transformation assays showed that translation of tobacco *atpI* mRNA requires sequence element(s) upstream of the SD-like sequence (Baecker et al. 2009). Therefore,

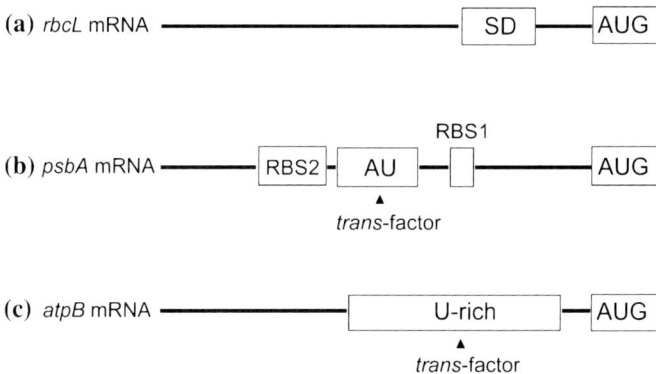

Fig. 9.3 Three types of *cis*-elements in tobacco chloroplast mRNAs. **a** *rbcL* mRNA has an SD-like sequence (SD) as its *cis*-element (Hirose and Sugiura 2004a). **b** *psbA* mRNA possesses three elements. The ribosome-binding site (RBS)1 (AAG) and RBS2 (UGAU) are possible ribosome-binding sites complementary to portions of the chloroplast 3′-16S rRNA. "AU" represents UAAAUAAA, termed the AU box, to which its *trans*-factor binds (Hirose and Sugiura 1996). **c** *atpB* mRNA contains no SD-like sequence but has a U-rich region in the 5′UTR, where a 50-kDa protein binds (Hirose and Sugiura 2004b)

SD-like sequences are not always the sole *cis*-elements involved in chloroplast translation.

9.4.3 *Trans-Acting Factors*

Since over half of chloroplast mRNAs lack SD-like sequences in the proper position, many chloroplast mRNA are likely to require specific *trans*-acting factors for translation (Marín-Navarro et al. 2007; Peled-Zehavi and Danon 2007; Wobbe et al. 2008; Lyska et al. 2013). Genetic analysis has led to the proposal of such factors. For example, the maize nuclear gene *crp1* is involved in translation of *petA-D* mRNAs (Barkan et al. 1994) and *atp-1* is required for translation of *atpB/E* mRNAs (McCormac and Barkan 1999). The *Arabidopsis* HCF173 and HCF244 proteins are involved in translation of *psbA* mRNAs (Schult et al. 2007; Link et al. 2012) and the TAB 2 product is essential for *psaB* translation (Barneche et al. 2006). Many of these proteins belong to the pentatricopeptide repeat (PPR) protein family (Schmitz-Linneweber and Small 2008). PPR proteins are RNA-binding proteins composed of tandem arrays of degenerated repeating units of about 35 amino acids. The protein family is outstanding in terrestrial plants because of the large number of its members (e.g., about 450 in *Arabidopsis*) and most PPR proteins are targeted to either chloroplasts or mitochondria. PPR domains themselves are not catalytically active but guide enzymes or ribonucleoprotein complexes to specific sequences. Hence, PPR proteins can be good candidates for specific *trans*-factors for translation of the mRNAs lacking SD-like sequences. In addition to genetic approaches, biochemical analyses (e.g., measurements of translation rates) would be powerful to elucidate the precise mechanism of translation initiation by specific factors.

For dicistronic and polycistronic mRNAs, intercistronic cleavage is often required for translation of downstream cistrons. The tobacco *psaC* and *ndhD* genes are transcribed as a dicistronic pre-mRNA that is then cleaved into monocistronic *psaC* and *ndhD* mRNAs. In vitro assays showed that the dicistronic mRNA is not functional and that the intercistronic cleavage is a prerequisite for both *psaC* and *ndhD* translation (Hirose and Sugiura 1997). Cleavage is also important for translation of other polycistronic mRNAs, and PPR proteins participate in the process (Meierhoff et al. 2003; Schmitz-Linneweber et al. 2005; Sane et al. 2005; Pfalz et al. 2009; Prikryl et al. 2011).

Biochemical approaches such as gel shift assays and UV cross-linking experiments have suggested the presence of *trans*-acting factors. For example, a 43 kDa protein interacts with the 5'UTR of spinach *psbA* mRNA (Klaff and Gruissem 1995). This protein may be ribosomal protein S1 (Alexander et al. 1998; Shteiman-Kotler and Schuster 2000). The 5'UTR of spinach *atpI* mRNAs has two regions for protein binding (Robida et al. 2002). An in vitro assay has suggested that to allow translation of tobacco *psbA* mRNA, one or more proteins bind an AU-rich sequence (the AU box) between two separate possible binding sequences at the 3′ end of 16S rRNA (Hirose and Sugiura 1996) (see Fig. 9.3b). A 50 kDa protein binds to the U-rich 5′UTR of tobacco *atpB* mRNA (Hirose and Sugiura 2004b) (see Fig. 9.3c).

9.5 Translational Coupling

In the chloroplast genome of flowering plants, some protein-coding genes overlap each other to increase the number of proteins encoded by the genome. In tobacco, eight genes partially overlap. These are *ndhC-ndhK, psbD-psbC, atpB-atpE,* and *rpl22-rps3* (Shinozaki et al. 1986). Gene overlapping implies that translation of a downstream gene (cistron) depends on that of its upstream gene (cistron), a situation known as translational coupling or termination-dependent translation, as reported for some genes from *E. coli,* its bacteriophages and some eukaryotic viruses (Jackson et al. 2007). A typical case of translational coupling is explained by the downstream cistron being translated exclusively by the ribosomes that complete translation of the upstream cistron. As the majority of the ribosomes coming from upstream cistrons are released at the stop codon, translation of downstream cistrons is very low. When the distance between the stop codon of an upstream cistron and the start codon of the downstream cistron increases, for example, by insertion of a premature termination codon in the upstream cistron, it abolishes translation of the downstream cistron.

The *ndhC* and *ndhK* genes partially overlap and are cotranscribed in most flowering plants (Matsubayashi et al. 1987). As shown in Fig. 9.4a, the AUG start codon is located 4 nt upstream from the *ndhC* stop codon UAG. Mutation of the UAG stop codon arrests translation of the downstream *ndhK* cistron (Yukawa and Sugiura 2008), indicating that *ndhK* translation depends on termination of the preceding cistron, or in other words, translational coupling (pathway 1). Hence, the amount of *ndhK* product (NdhK) is expected to be low compared to NdhC. However, in vitro assays have shown that an *ndhC/K* mRNA produces NdhC and NdhK in similar amounts. Further in vitro studies have disclosed that free ribosomes enter, with formylmethionyl-tRNA, at an internal AUG that is located in frame in the middle of the upstream *ndhC* cistron, translate the 3′ half of the *ndhC* cistron, and reach the *ndhK* start codon. Then, some of the ribosomes resume *ndhK* translation (pathway 2). We proposed that the internal initiation site AUG is not designed for synthesizing a functional isoform but for delivering additional ribosomes (Yukawa and Sugiura 2013). NdhC is synthesized via pathway 1, and pathways 1 and 2 together produce NdhK in an amount similar to NdhC (1:1), the amount required for the assembly of the NADH dehydrogenase complex.

The *psbD* and *psbC* genes, encoding the D2 and CP43 proteins, are cotranscribed in flowering plants (Yao et al. 1989; Sexton et al. 1990; Kawaguchi et al. 1992). These cistrons overlap by 14 nt, and the downstream *psbC* cistron starts with the GUG start codon (Fig. 9.4b). Translation of the downstream *psbC* cistron depends largely on that of the upstream *psbD* cistron; however, a portion is independently

(a) *ndhC/K*

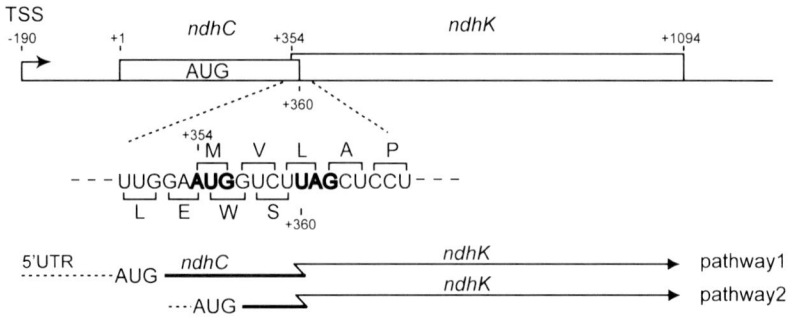

(b) *psbD/C*

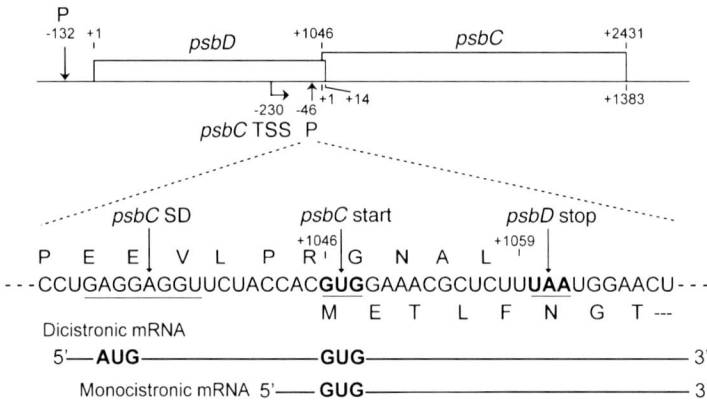

(c) *atpB/E*

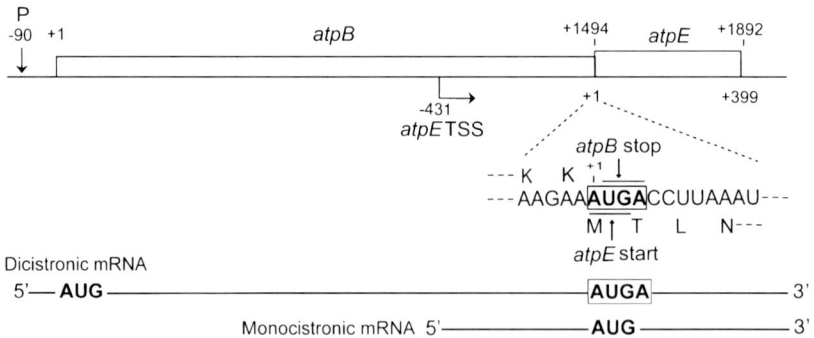

Fig. 9.4 Translation of overlapping cistrons. Schematic of three pairs of overlapping cistrons in tobacco chloroplasts. Positions are relative to the AUG start codon in an upstream cistron (with the A as +1). 5'-ends are either a transcription start site (TSS) or processing sites (P). Partial mRNA

▶ sequences around the overlapping regions are shown below. **a** In *ndhC/K* cistrons, ribosomes coming from the 5′UTR read the *ndhC* cistron and reach its stop codon UAG. Then some of the ribosomes resume translation from the *ndhK* start codon AUG (pathway 1). Free ribosomes enter at the AUG in the middle of the upstream cistron and read to its end, and again some of the ribosomes start to translate from the *ndhK* start codon AUG (pathway 2). Translation of the *ndhK* cistron depends strictly on that of its upstream cistron (Yukawa and Sugiura 2008, 2013). **b** In *psbD/C* cistrons, the *psbC* cistron is translated in part by translational coupling (Adachi et al. 2012). **c** In *atpB/E* cistrons, *atpE* translation is independent of *atpB* translation (Suzuki et al. 2011). In both cases, monocistronic mRNAs are produced from an additional promoter present within the upstream cistrons

translated (partial translational coupling) (Adachi et al. 2012). Further, an additional promoter is present within the upstream *psbD* cistron and produces the monocistronic *psbC* mRNA (Yao et al. 1989) and this mRNA is translatable (Kuroda et al. 2007). Hence, the *psbD* and *psbC* cistrons are translated by two mRNA species to produce the necessary amounts of D2 and CP34 for assembly of the photosystem II complex.

The *atpB* and *atpE* genes encode subunits β and ε of the ATP synthase complex, respectively. They are cotranscribed as dicistronic mRNAs in flowering plants. An unusual feature is an overlap (AUGA) of the *atpB* stop codon (UGA) with the *atpE* start codon (AUG). Hence, translation of these cistrons was assumed to be coupled (Zurawski et al. 1982; Krebbers et al. 1982; Shinozaki et al. 1983). Both cistrons from the tobacco dicistronic mRNA are translated in vitro, and the efficiency of *atpB* translation is higher than *atpE* (Suzuki et al. 2011). However, removal of the entire *atpB* 5′UTR arrested *atpB* translation, but *atpE* translation still proceeded. Hence, the majority of *atpE* translation does not depend on that of the upstream *atpB* cistron, and the *atpE* cistron is translated via its own *cis*-element(s) (Suzuki et al. 2011). That is, ribosomes that translate the upstream cistron are released at its stop codon and no more are available for translation of the downstream cistron. The tobacco *atpB/E* cluster produces the monocistronic *atpE* mRNA from its own promoter located within the *atpB* cistron (Kapoor et al. 1994). As described, this mRNA has an SD-like sequence essential for its translation. The chloroplast ATP synthase complex contains three β subunits, one ε subunit, and several other subunits. To meet this stoichiometry, the necessary amount of the β subunit is synthesized from the dicistronic mRNA and the ε subunit is produced from both the dicistronic and the monocistronic mRNAs.

Besides the four pairs of overlapping genes in tobacco, several genes are located close to each other. The *ndhH* and the next *ndhA* cistrons are separated by only one nt (excluding stop codon). Similarly, *rpoB* and *rpoC1* are separated by 5 nt, *psbE* and *psbF* by 9 nt, and *rpl23* and *rpl2* by 18 nt. These pairs may be translationally coupled because generally 5′UTRs of 20 nt or more are necessary for efficient translation.

9.6 Problems

The basic processes of translation elongation and termination in chloroplasts are believed to be similar to those in *E. coli*. Chloroplast elongation factors and release factors as well as initiation factors have been detected based on similarity to *E. coli* factors. Factors unique to chloroplasts may exist because translation of photosynthesis-related mRNAs is regulated by light, which does not occur in *E. coli*. *Trans*-acting factors and translation factors (except initiation factor-1) are nuclear encoded. Genetic approaches are promising to detect gene products involved in protein accumulation or translation. Biochemical studies, for example, the use of in vitro systems, are suitable to define target processes of these proteins and to measure the rate of translational initiation, elongation, or termination.

Many chloroplast genes encode subunits of large complexes such as ribosomes, photosystem II, and NADH dehydrogenase. To meet the proper stoichiometry of a complex, each subunit has to be synthesized in the needed amount. Translation is the last step of gene expression and hence this step should be precisely controlled. Therefore, quantitative analysis of translation, including the rates of translational initiation and elongation, is required. The precise mechanism of chloroplast gene translation will be a challenging field.

References

Adachi Y, Kuroda H, Yukawa Y et al (2012) Translation of partially overlapping *psbD-psbC* mRNAs in chloroplasts: the role of 5′-processing and translational coupling. Nucl Acids Res 40:3152–3158

Alexander C, Faber N, Klaff P (1998) Characterization of protein-binding to the spinach chloroplast *psbA* mRNA 5′untranslated region. Nucl Acids Res 26:2265–2272

Alkatib S, Scharff LB, Rogalski M (2012) The contributions of wobbling and superwobbling to the reading of the genetic code. PLoS Genet 8:e1003076

Baecker JJ, Sneddon JC, Hollingsworth MJ (2009) Efficient translation in chloroplasts requires element(s) upstream of the putative ribosome binding site from *atpI*. Am J Bot 96:627–636

Barkan A, Walker M, Nolasco M et al (1994) A nuclear mutation in maize blocks the processing and translation of several chloroplast mRNAs and provides evidence for the differential translation of alternative mRNA forms. EMBO J 13:3170–3181

Barneche F, Winter V, Crèvecoeur M et al (2006) *ATAB 2* is a novel factor in the signalling pathway of light-controlled synthesis of photosystem proteins. EMBO J 25:5907–5918

Barrell BG, Anderson S, Bankier AT et al (1980) Different pattern of codon recognition by mammalian mitochondrial tRNAs. Proc Natl Acad Sci USA 77:3164–3166

Bock R (2007) Structure, function, and inheritance of plastid genomes. In: Bock R (ed) Cell and molecular biology of plastids. Springer, Potsdam-Golm, pp p29–p63

Chen H, Bjerknes M, Kumar R et al (1994) Determination of the optimal aligned spacing between the Shine-Dalgarno sequence and the translation initiation codon of *Escherichia coli* mRNAs. Nucl Acids Res 22:4953–4957

Choquet Y, Wollman F-A (2002) Translational regulations as specific traits of chloroplast gene expression. FEBS Lett 529:39–42

Cohen BN, Coleman TA, Schmitt JJ et al (1984) *In vitro* expression and characterization of the translation start site of the *psbA* gene product (Q_B protein) from higher plants. Nucl Acids Res 12:6221–6230

Delannoy E, Le Ret M, Faivre-Nitschke E et al (2009) *Arabidopsis* tRNA adenosine deaminase arginine edits the wobble nucleotide of chloroplast tRNA^{Arg}(ACG) and is essential for efficient chloroplast translation. Plant Cell 21:2058–2071

Eibl C, Zou Z, Beck A et al (1999) *In vivo* analysis of plastid *psbA, rbcL* and *rpl32* UTR elements by chloroplast transformation: tobacco plastid gene expression is controlled by modulation of transcript levels and translation efficiency. Plant J 19:333–345

Eyal Y, Goloubinoff P, Edelman M (1987) The amino terminal region delimited by Met$_1$ and Met$_{37}$ is an integral part of the 32 kDa herbicide binding protein. Plant Mol Biol 8:337–343

Giege P, Brennicke A (1999) RNA editing in *Arabidopsis* mitochondria effects 441 C to U changes in ORFs. Proc Natl Acad Sci USA 96:15324–15329

Hiratsuka J, Shimada H, Whhittier R et al (1989) The complete sequence of the rice (*Oryza sativa*) chloroplast genome: Intermolecular recombination between distinct tRNA genes accounts for a major plastid DNA inversion during the evolution of the cereals. Mol Gen Genet 217:185–194

Hirose T, Sugiura M (1996) *Cis*-acting elements and *trans*-acting factors for accurate translation of chloroplast *psbA* mRNAs: development of an in vitro translation system from tobacco chloroplasts. EMBO J 15:1687–1695

Hirose T, Sugiura M (1997) Both RNA editing and RNA cleavage are required for translation of tobacco chloroplast *ndhD* mRNA: a possible regulatory mechanism for the expression of a chloroplast operon consisting of functionally unrelated genes. EMBO J 16:6804–6811

Hirose T, Sugiura M (2004a) Functional Shine-Dalgarno-like sequences for translational initiation of chloroplast mRNAs. Plant Cell Physiol 45:114–117

Hirose T, Sugiura M (2004b) Multiple elements required for translation of plastid *atpB* mRNA lacking the Shine-Dalgarno sequence. Nucl Acids Res 32:3503–3510

Hirose T, Kusumegi T, Sugiura M (1998) Translation of tobacco chloroplast *rps14* mRNA depends on a Shine-Dalgarno-like sequence in the 5'-untranslated region but not on internal RNA editing in the coding region. FEBS Lett 430:257–260

Hirose T, Ideue T, Wakasugi T et al (1999) The chloroplast *infA* gene with a functional UUG initiation codon. FEBS 445:169–172

Hotto AM, Schmitz RJ, Fei Z et al (2011) Unexpected diversity of chloroplast noncoding RNAs as revealed by deep sequencing of the Arabidopsis transcriptome. G3(1):559–570

Ikemura T (1985) Codon usage and tRNA content in unicellular and multicellular organisms. Mol Biol Evol 2:13–34

Jackson RJ, Kaminski A, Pöyry TAA (2007) Coupled termination-reinitiation events in mRNA translation. In: Mathews MB, Sorenberg N, Herskey JWB (eds) Translational control in biology and medicine, Cold Spring Harbor Lab Press, Cold Spring Harbor, pp 197–223

Kanaya S, Yamada Y, Kinouchi M et al (2001) Codon usage and tRNA genes in eukaryotes: correlation of codon usage diversity with translation efficiency and with CG-dinucleotide usage as assessed by multivariate analysis. J Mol Evol 53:290–298

Kapoor S, Wakasugi T, Deno H et al (1994) An *atpE*-specific promoter within the coding region of the *atpB* gene in tobacco chloroplast DNA. Curr Genet 26:263–268

Karcher D, Bock R (2009) Identification of the chloroplast adenosine-to-inosine tRNA editing enzyme. RNA 15:1251–1257

Kawaguchi H, Fukuda I, Shiina T et al (1992) Dynamical behavior of *psb* gene transcripts in greening wheat seedlings. I. Time course of accumulation of the *psbA* through *psbN* gene transcripts during light-induced greening. Plant Mol Biol 20:695–704

Klaff P, Gruissem W (1995) A 43 kD light-regulated chloroplast RNA-binding protein interacts with the *psbA* 5' non-translated leader RNA. Photosynth Res 46:235–248

Krebbers ET, Larrinua IM, McIntosh L et al (1982) The maize chloroplast genes for the β and ε subunits of the photosynthetic coupling factor CF$_1$ are fused. Nucl Acids Res 10:4985–5002

Kuroda H, Suzuki H, Kusumegi T et al (2007) Translation of *psbC* mRNAs starts from the downstream GUG, not the upstream AUG, and requires the extended Shine-Dalgarno sequence in tobacco chloroplasts. Plant Cell Physiol 48:1374–1378

Lengyel P (1974) The process of translation: a bird's-eye view. In: Nomura M, Tissières A, Lengyel P (eds) Ribosomes. Cold Spring Harbor Lab Press, Cold Spring Harbor, pp p13–p52

Link S, Engelmann K, Meierhoff K et al (2012) The Atypical short-chain dehydrogenases HCF173 and HCF244 are jointly involved in translational initiation of the *psbA* mRNA of Arbidopsis. Plant Physiol 160:2202–2218

Lung B, Zemann A, Madej MJ et al (2006) Identification of small non-coding RNAs from mitochondria and chloroplasts. Nucl Acids Res 34:3842–3852

Lyska D, Meierhoff K, Westhoff P (2013) How to build functional thylakoid membranes: from plastid transcription to protein complex assembly. Planta 237:413–428

Magee AM, Aspinall S, Rice DW et al (2010) Localized hypermutation and associated gene losses in legume chloroplast genomes. Genome Res 20:1700–1710

Manuell A, Beligni MV, Yamaguchi K et al (2004) Regulation of chloroplast translation: interactions of RNA elements, RNA-binding proteins and the plastid ribosome. Biochem Soc Trans 32:601–605

Manuell AL, Quispe J, Mayfield S (2007) Structure of the chloroplast ribosomes: Novel domains for translation regulation. PLoS Biol 5:e209

Marder JB, Goloubinoff P, Edelman M (1984) Molecular architecture of the rapidly metabolized 32-kilodalton protein of photosystem II. J Biol Chem 259:3900–3908

Marín-Navarro J, Manuell AL, Wu J et al (2007) Chloroplast translation regulation. Photosynth Res 94:359–374

Matsubayashi T, Wakasugi T, Shinozaki K et al (1987) Six chloroplast genes (ndhA-F) homologous to human mitochondrial genes encoding components of the respiratory chain NADH dehydrogenase are actively expressed: determination of the splice sites in ndhA and ndhB pre-mRNAs. Mol Gen Genet 210:385–393

McCormac DJ, Barkan A (1999) A nuclear gene in maize required for the translation of the chloroplast atpB/E mRNA. Plant Cell 11:1709–1716

Meierhoff K, Felder S, Nakamura T et al (2003) HCF152, an Arabidopsis RNA binding pentatricopeptide repeat protein involved in the processing of chloroplast psbB-psbT-psbH-petB-petD RNAs. Plant Cell 15:1480–1495

Nakamura M, Sugiura M (2007) Translation efficiencies of synonymous codons are not always correlated with codon usage in tobacco chloroplasts. Plant J 49:128–134

Nakamura M, Sugiura M (2009) Selection of synonymous codons for better expression of recombinant proteins in tobacco chloroplasts. Plant Biotech 26:53–56

Nakamura M, Sugiura M (2011) Translation efficiencies of synonymous codons for arginine differ dramatically and are not correlated with codon usage in chloroplasts. Gene 472:50–54

Neckermann K, Zeltz P, Igloi GL et al (1994) The role of RNA editing in conservation of start codons in chloroplast genomes. Gene 146:177–182

Ohto C, Torazawa K, Tanaka M et al (1988) Transcription of ten ribosomal protein genes from tobacco chloroplasts: a compilation of ribosomal protein genes found in the tobacco chloroplast genome. Plant Mol Biol 11:589–600

Peled-Zehavi H, Danon A (2007) Translation and translational regulation in chloroplasts. In: Bock R (ed) Cell and molecular biology of plastids. Springer, Potsdam-Golm, pp p249–p281

Pfalz J, Bayraktar OA, Prikryl J et al (2009) Site-specific binding of a PPR protein defines and stabilizes 5′ and 3′ mRNA termini in chloroplasts. EMBO J 28:2042–2052

Pillay DTN, Guillemaut G, Weil JH (1984) Nucleotide sequences of three soybean chloroplast tRNAs^{Leu} and re-examination of bean chloroplast tRNA$^{Leu}_2$ sequence. Nucl Acids Res 12:2997–3001

Plader W, Sugiura M (2003) The Shine-Dalgarno-like sequence is a negative regulatory element for translation of tobacco chloroplast rps2 mRNA: an additional mechanism for translational control in chloroplasts. Plant J 34:377–382

Prikryl J, Rojas M, Schuster G et al (2011) Mechanism of RNA stabilization and translational activation by a pentatricopeptide repeat protein. Proc Natl Acad Sci USA 108:415–420

Reinbothe S, Reinbothe C, Heintzen C et al (1993) A methyl jasmonate-induced shift in the length of the 5′ untranslated region impairs translation of the plastid rbcL transcript in barley. EMBO J 12:1505–1512

Robida MD, Merhige PM, Hollingsworth MJ (2002) Proteins are shared among RNA-protein complexes that form in the 5′untranslated regions of spinach chloroplast mRNAs. Curr Genet 41:53–62

Robinson M, Lilley R, Little S et al (1984) Codon usage can affect efficiency of translation of genes in Escherichia coli. Nucl Acids Res 12:6663–6671

Rochaix J-D (2001) Posttranscriptional control of chloroplast gene expression. From RNA to photosynthetic complex. Plant Physiol 125:142–144

Samuelsson T, Elias P, Lustig F et al (1980) Aberrations of the classic codon reading scheme during protein synthesis in vitro. J Biol Chem 255:4583–4588

Sane AP, Stein B, Westhoff P (2005) The nuclear gene *HCF107* encodes a membrane-associated R-TPR (RNA-tetratricopeptide repeat)-containing protein involved in expression of the plastidial *psbH* gene in Arabidopsis. Plant J 42:720–730

Sasaki T, Yukawa Y, Miyamoto T et al (2003) Identification of RNA editing sites in chloroplast transcripts from the maternal and paternal progenitors of tobacco (*Nicotiana tabacum*): Comparative analysis shows the involvement of distinct *trans*-factors for *ndhB* editing. Mol Biol Evol 20:1028–1035

Schmitz-Linneweber C, Small I (2008) Pentatricopeptide repeat proteins: a socket set for organelle gene expression. Trends Plant Sci 13:663–670

Schmitz-Linneweber C, Maier RM, Alcaraz J-P et al (2001) The plastid chromosome of spinach (*Spinacia oleracea*): complete nucleotide sequence and gene organization. Plant Mol Biol 45:307–315

Schmitz-Linneweber C, Williams-Carrier R, Barkan A (2005) RNA immunoprecipitation and microarray analysis show a chloroplast pentatricopeptide repeat protein to be associated with the 5′ region of mRNAs whose translation it activates. Plant Cell 17:2791–2804

Schult K, Meierhoff K, Paradies S et al (2007) The nuclear-encoded factor HCF173 is involved in the initiation of translation of the *psbA* mRNA in *Arabidopsis thaliana*. Plant Cell 19:1329–1346

Sexton TB, Christopher DA, Mullet JE (1990) Light-induced switch in barley *psbD-psbC* promoter utilization: a novel mechanism regulating chloroplast gene expression. EMBO J 9:4485–4494

Sharma MR, Wilson DN, Datta PP et al (2007) Cryo-EM study of the spinach chloroplast ribosome reveals the structural and functional roles of plastid-specific ribosomal proteins. Proc Natl Acad Sci USA 104:19315–19320

Sharma MR, Dönhöfer A, Barat C et al (2010) PSRP1 is not a ribosomal protein, but a ribosome-binding factor that is recycled by the ribosome-recycling factor (RRF) and elongation factor G (EF-G). J Biol Chem 285:4006–4014

Shinozaki K, Deno H, Kato A et al (1983) Overlap and cotranscription of the genes for the beta and epsilon subunits of tobacco chloroplast ATPase. Gene 24:147–155

Shinozaki K, Ohme M, Tanaka M et al (1986) The complete nucleotide sequence of the tobacco chloroplast genome: its gene organization and expression. EMBO J 5:2043–2049

Shteiman-Kotler A, Schuster G (2000) RNA-binding characteristics of the chloroplast S1-like ribosomal protein CS1. Nucl Acids Res 28:3310–3315

Sørensen MA, Pedersen S (1991) Absolute in vivo translation rates of individual codons in *Escherichia coli*. The two glutamic acid codons GAA and GAG are translated with a threefold difference in rate. J Mol Biol 222:265–280

Staub JM, Maliga P (1993) Accumulation of D1 polypeptide in tobacco plastids is regulated via the untranslated region of the *psbA* mRNA. EMBO J 12:601–606

Staub JM, Maliga P (1994) Translation of *psbA* mRNA is regulated by light via the 5′-untranslated region in tobacco plastids. Plant J 6:547–553

Sugita M, Sugiura M (1984) Nucleotide sequence and transcription of the gene for the 32,000 dalton thylakoid membrane protein from *Nicotiana tabacum*. Mol Gen Genet 195:308–313

Sugiura M (1987) Structure and function of the tobacco chloroplast genome. Bot Mag Tokyo 100:407–436

Sugiura M (1992) The chloroplast genome. Plant Mol Biol 19:149–168

Sugiura M (2008) RNA editing in chloroplasts. In: Goringer HU (ed) RNA editing. Springer, Berlin, pp 123–142

Sugiura M, Hirose T, Sugita M (1998) Evolution and mechanism of translation in chloroplasts. Annu Rev Genet 32:437–459

Suzuki H, Kuroda H, Yukawa Y et al (2011) The downstream *atpE* cistron is efficiently translated via its own *cis*-element in partially overlapping *atpB-atpE* dicistronic mRNAs in chloroplasts. Nucl Acids Res 39:9405–9412

Takahashi M, Shiraishi T, Asada K (1988) COOH-terminal residues of D1 and the 44 kDa CPa-2 at spinach photosystem II core complex. FEBS Lett 240:6–8

Vera A, Sugiura M (1994) A novel RNA gene in the tobacco plastid genome: its possible role in the maturation of 16S rRNA. EMBO J 13:2211–2217

Vera A, Matsubayashi T, Sugiura M (1992) Active transcription from a promoter positioned within the coding region of a divergently oriented gene: the tobacco chloroplast *rpl32* gene. Mol Gen Genet 233:151–156

Wakasugi T, Nagai T, Kapoor M et al (1997) Complete nucleotide sequence of the chloroplast genome from the green alga *Chlorella vulgaris*: The existence of genes possibly involved in chloroplast division. Proc Natl Acad Sci USA 94:5967–5972

Wakasugi T, Tsudzuki T, Sugiura M (2001) The genomics of land plant chloroplasts: Gene content and alteration of genomic information by RNA editing. Photosynth Res 70:107–118

Wicke S, Schneeweiss GM, Claude W et al (2011) The evolution of the plastid chromosome in land plants: gene content, gene order, gene function. Plant Mol Biol 76:273–297

Wobbe L, Schwarz C, Nickelsen J et al (2008) Translational control of photosynthetic gene expression in phototrophic eukaryotes. Physiol Plant 133:507–515

Wolfe KH, Morden CW, Ems SC et al (1992) Rapid evolution of the plastid translational apparatus in a nonphotosynthetic plant: loss or accelerated sequence evolution of tRNA and ribosomal protein genes. J Mol Evol 35:304–317

Yamaguchi K, Subramanian AR (2000) The plastid ribosomal proteins: identification of all the proteins in the 50S subunit of an organelle ribosome (chloroplast). J Biol Chem 275:28466–28482

Yamaguchi K, Subramanian AR (2003) Proteomic identification of all plastid-specific ribosomal proteins in higher plant chloroplast 30S ribosomal subunit PSRP-2 (U1A-type domains), PSRP-3α/β (ycf65 homologue) and PSRP-4 (Thx homologue). Eur J Biochem 270:190–205

Yamaguchi K, von Knoblauch K, Subramanian AR (2000) The plastid ribosomal proteins: identification of all the proteins in the 30S subunit of an organelle ribosome (chloroplast). J Biol Chem 275:28455–28465

Yao WB, Meng BY, Tanaka M et al (1989) An additional promoter within the protein-coding region of the *psbD-psbC* gene cluster in tobacco chloroplast DNA. Nucl Acids Res 17:9583–9591

Yukawa M, Sugiura M (2008) Termination codon-dependent translation of partially overlapping *ndhC-ndhK* transcripts in chloroplasts. Proc Natl Acad Sci USA 105:19549–19553

Yukawa M, Sugiura M (2013) Additional pathway to translate the downstream *ndhK* cistron in partially overlapping *ndhC-ndhK* mRNAs in chloroplasts. Proc Nucl Acid Sci USA 110:5701–5706

Yukawa M, Kuroda H, Sugiura M (2007) A new in vitro translation system for non-radioactive assay from tobacco chloroplasts: effect of pre-mRNA processing on translation in vitro. Plant J 49:367–376

Zou Z, Eibl C, Koop HU (2003) The stem-loop region of the tobacco *psbA* 5'UTR is an important determinant of mRNA stability and translation efficiency. Mol Genet Genomics 269:340–349

Zurawski G, Bottomley W, Whitfeld PR (1982) Structures of the genes for the β and ε subunits of spinach chloroplast ATPase indicate a dicistronic mRNA and an overlapping translation stop/start signal. Proc Natl Acad Sci USA 79:6260–6264

Chapter 10
The Chloroplasts as Platform for Recombinant Proteins Production

Nunzia Scotti, Michele Bellucci and Teodoro Cardi

Abstract Chloroplasts are a useful platform for the expression of recombinant proteins in higher plants. Transgenes can be introduced into the plastid genome (plastome) either by PEG transformation of plant protoplasts, or, more commonly, by the biolistic method, using leaves or suspension cells. Transgenes are integrated by double recombination events between flanking sequences in the vector and homologous sequences in the plastome. The genetic engineering of the plastome allows high-level foreign protein expression, site-specific gene integration, expression of multiple genes as operons, marker gene excision, and transgene containment. Since the first example of stable plastid transformation in higher plants, methods for DNA introduction, marker genes and selection strategies, vector types, and methods for marker excision have been improved. Although the plastids of some species remain difficult to transform, positive results have been shown for about 20 species. In this chapter, we summarize the basic structural and expression features of the plastid genome of higher plants, and discuss the development of a number of innovative enabling technologies for plastome transformation, the most recent and significant biotechnological applications, and the future perspectives of this technology.

N. Scotti · T. Cardi
CNR-IGV, National Research Council of Italy, Institute of Plant Genetics,
Res. Div. Portici, via Università 133, 80055 Portici, NA, Italy

M. Bellucci
CNR-IGV, National Research Council of Italy, Institute of Plant Genetics,
Res. Div. Perugia, Via Madonna Alta 130, 06128 Perugia, PG, Italy

T. Cardi (✉)
Consiglio per la Ricerca e la Sperimentazione in Agricoltura Centro di Ricerca
per l'Orticoltura, (CRA-ORT), via Cavalleggeri 25, 84098 Pontecagnano, SA, Italy
e-mail: teodoro.cardi@entecra.it

A.-M. Duchêne (ed.), *Translation in Mitochondria and Other Organelles*,
DOI: 10.1007/978-3-642-39426-3_10, © Springer-Verlag Berlin Heidelberg 2013

List of Abbreviations

aadA	aminoglycoside 3-adenylyltransferase
bar	phosphinothricin acetyl transferase gene
ELISA	Enzyme-linked immunosorbent assay
HPLC	High-performance liquid chromatography
NEP	Nucleus-encoded RNA polymerase
PEG	PolyEthylene glycol
PEP	Plastid-encoded RNA polymerase
PPR	Pentatricopeptide repeat
PTMs	Post-translation modifications
RPOTmp	Nucleus-encoded RNA polymerase localized in mitochondria and plastids
RPOTp	Nucleus-encoded RNA polymerase localized in plastids
UTRs	Untranslated regions
ROS	Reactive oxygen species
TSP	Total soluble protein

10.1 Introduction

Since the development of methods for the genetic transformation of higher plants in the early 1980s, a wide number of transgenes have been used in genetic engineering approaches to study basic biological processes or modify agronomic and physiological traits. First generation of commercially available genetically engineered crops is largely confined to insect and herbicide resistance (Jones 2011). Nevertheless, many other agronomic/physiological applications are in the pipeline (Collinge et al. 2010; Cominelli and Tonelli 2010; Ceasar and Ignacimuthu 2012; Reguera et al. 2012). Last generation transgenic plants have been also proposed as "green biofactories" for the accumulation of recombinant products (Rojas et al. 2010; Egelkrout et al. 2012).

In most transgenic plants obtained so far, transgenes are introduced into the nuclear genome of plant cells by either the well-established *Agrobacterium*-mediated transformation or direct delivery methods, as the biolistics or the PEG/electroporation-mediated transformation approaches (Altpeter et al. 2005; Craig et al. 2005; Meyers et al. 2010). In many cases, however, some concerns have been raised, calling for the development of innovative technologies. Such concerns include: the optimization of expression level and stability of transgenes and recombinant proteins (Singer et al. 2012), the containment of transgenes (Husken et al. 2010), the possibility to transfer or stack multiple genes in order to engineer complex metabolic pathways (Naqvi et al. 2010; Que et al. 2010), the development of methods for "clean" and precise integration of transgenes in the host genome (Husaini et al. 2011; Wang et al. 2011), the identification of alternative marker genes for the selection of transformed cells (Manimaran et al. 2011;

Rosellini 2012), the feasibility of correct posttranslation modifications (PTMs) in recombinant proteins (Webster and Thomas 2012), and the use of genes only from sexually compatible sources (Orzaez et al. 2010).

Alternatively to the transformation of the nuclear genome, transgenes can be introduced into the plastid genome (plastome). Following gene delivery by PEG transformation of plant protoplasts, or, more commonly, by the biolistic method, applied to leaf tissues or suspension cells, transgenes are integrated by double recombination events between flanking sequences in the vector and homologous sequences in the plastome (Meyers et al. 2010). After transgene integration into one plastid genome, repeated cycles of cell divisions and shoot regeneration are usually required to reach homoplasmy (Fig. 10.1). In comparison with the transformation of the nuclear genome, the genetic engineering of the plastome shows some attractive advantages, partly responding to concerns mentioned above and including high-level foreign protein expression, site-specific gene integration, the possibility to transfer and express multiple genes arranged in native or synthetic operons, and marker gene excision and transgene containment because of maternal inheritance of plastids in most crops (Cardi et al. 2010; Meyers et al. 2010; Maliga and Bock 2011). After the first demonstration in tobacco (Svab et al. 1990), the technology has now been

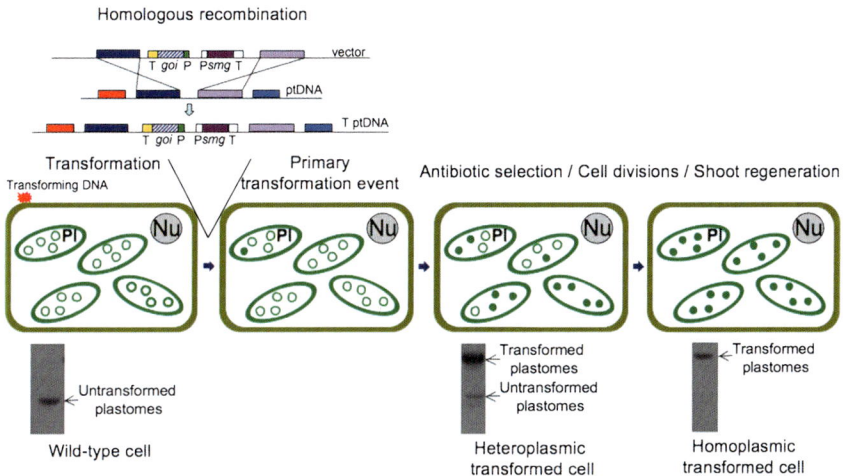

Fig. 10.1 Schematic representation of chloroplast transformation in higher plants. From *left* to *right*, a wild-type cell with untransformed plastomes (*empty circles*); the primary transformation event with a cell containing only one transformed plastome (*filled circle*); a heteroplasmic cell containing both transformed and untransformed plastomes; a homoplasmic cell with uniformly transformed plastomes. The double recombination events between vector and ptDNA sequences leading to transformed ptDNA are also shown. The Southern blots below the cells report results of restriction-digested DNA (obtained from wild type or regenerated plants) hybridized with a probe homologous to the plastome regions flanking the inserted DNA fragment. *Nu* nucleus, *Pl* plastid, *P* promoter, *T* terminator, *T ptDNA* transformed ptDNA, *goi*, gene of interest, *smg*, selectable marker gene

proved in a relatively large number of plant species belonging to several families, and for different purposes (Cardi et al. 2010; Maliga and Bock 2011; Maliga 2012). In this chapter, after reporting some basic structural and expression features of the plastid genome of higher plants, we discuss the development of a number of innovative enabling technologies for plastome transformation, aiming to overcome some inherent limitations of the procedure, and the most recent applications.

10.2 Genetic Transformation of the Plastome

The structural and expression features of higher plant plastid genomes have been recently reviewed (Bock 2007; Barkan 2011; Cardi et al. 2012; De Marchis et al. 2012; Jansen and Ruhlman 2012). Hereafter, only the principal aspects relevant to recombinant protein expression by plastid transformation are summarized.

10.2.1 Structural Features of the Plastome

The genetic system of the plant cell is organized as a network of compartments hosting the nuclear DNA (nucleus), the mitochondrial DNA or chondriome (mitochondria), and the plastid DNA or plastome (plastids). Intercompartmental exchange of genetic information among these organelles during plant evolution decisively contributed to the structure of the overall plant genome (Maier and Schmitz-Linneweber 2004). Plastids arose ~1.5 billion years ago by the endosymbiotic acquisition of a cyanobacterium and the process of transferring cyanobacterial genes to nucleus and mitochondria, which is still active both intracellularly and between species, originated the current plastomes (Jansen et al. 2011). Plastome displays structural plasticity because the genome molecules can be arranged in various circular or linear forms, either as monomers or multimers, which are organized as DNA-protein aggregates similar to bacteria nucleoids (Bendich 2004; Day and Madesis 2007; Majeran et al. 2012). A plant cell can harbor tens of plastids, each with several nucleoids, and many identical copies of the plastome can be packed in one nucleoid. Therefore, plastids have a high degree of polyploidy and plastome copy number varies with plant development and tissue type (Bock 2007). Plastid polyploidy, together with a gene-conversion mechanism, seems to be responsible for the maintenance of the low rates of mutation which characterizes the plastome. Plastid DNA (ptDNA) is largely inherited maternally, even though paternal and biparental inheritances have been observed in some cases (Bock 2007). However, even in species with strict maternal inheritance, a small percentage of paternal transmission has been described (Thyssen et al. 2012). Up to now, about 300 records exist for Eukaryota plastid genomes (http://www.ncbi.nlm.nih.gov/genomes/GenomesGroup.cgi?taxid=2759&opt=plastid), including many crop plant species.

The plastid genome architecture of embryophytes appears to be highly conserved (Wicke et al. 2011). The plastome of land plants typically is 120–160 kb and has a tetrapartite structure, with two copies of a large inverted repeat (IR) separating a large and a small single copy region, containing 100–120 unique genes (Bock 2007). Some plants, like legumes, contain no IRs (Shaver et al. 2008; Magee et al. 2010). The plastid-encoded genes can be grouped into three main functional categories: genes encoding components of the genetic apparatus, photosynthesis-related genes, and other genes (Bock 2007). A detailed description of these genes in land plants is given in Wicke et al. (2011).

10.2.2 Expression Features of the Plastome

Variable plastid types are specialized in different metabolic pathways, which are very important for the whole plant cell physiology. It is predicted that in *Arabidopsis* plastids are present from 2,000 to 3,500 proteins, but only for about 1,200 there are clear evidences of plastid localization: less than 100 are plastid encoded, while the others are nucleus encoded (van Wijk and Baginsky 2011). Plastid gene expression is a very complex and unique system, regulated at many steps, mainly posttranscriptional ones, by both nuclear-encoded and plastid-encoded factors combining eukaryotic and prokaryotic features (Fig. 10.2, Barkan 2011; Cardi et al. 2012).

Plastid genes are organized as single genes or grouped in transcriptional units (operons), with common promoters and terminators, originating polycistronic transcripts. However, due to limited ability of 3′-UTR regions to terminate transcription (Cardi et al. 2012), in transplastomic plants, polycistronic transcripts can originate also in case of independent genes arranged in the same orientation (Fig. 10.3). Two types of RNA polymerases are responsible for plastid gene transcription: one is a bacterial type, multimeric plastid-encoded polymerase (PEP), the other is a phage type, monomeric nucleus-encoded polymerase (NEP), which in dicotyledonous plants is represented by two enzymes (RPOTmp and RPOTp). According to this classification, the promoters of plastid genes have been divided into NEP and PEP promoters, but there is not a clear distinction between NEP or PEP-transcribed genes, because many plastid genes have two or even more promoters and can be transcribed by both RNA polymerases (Liere and Börner 2007). In addition, six nucleus-encoded sigma factors (SIG1-6) regulate, according to environmental and developmental cues, transcription initiation mediated by PEP (Lerbs-Mache 2011; Malik Ghulam et al. 2012).

Plastid primary transcripts can undergo a maturation process involving RNA editing, intron splicing, intercistronic and mRNA termini processing, and mRNA stabilization and decay. Mature mRNAs are then translated on 70S ribosomes. All these steps of plastid gene expression require the participation of nuclear-encoded RNA-binding factors (Stern et al. 2010; Barkan 2011), most of which are pentatricopeptide repeat proteins (PPRs), characterized by a degenerate structural

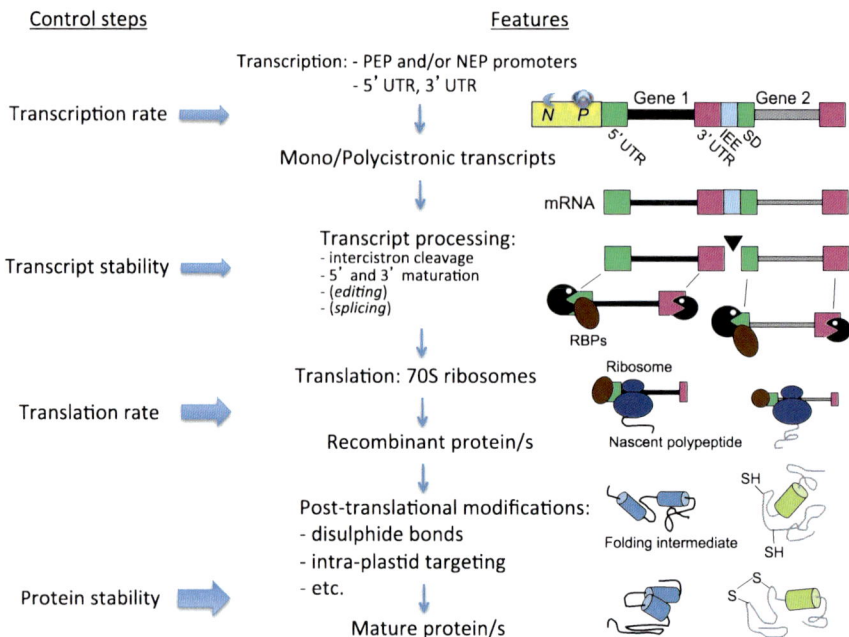

Fig. 10.2 Key control steps for the expression of recombinant proteins in transgenic plastids. The main control steps are reported on the *left* (*thicker arrows* suggest higher importance for the regulation of recombinant protein expression in plastids). The expression features reported in the figure are discussed in the text, paragraph 2.2. A dicistronic operon is represented in the figure and the *first* rectangle indicates the promoter region with NEP (N) and PEP (P) promoters recognized by nuclear-encoded monomeric and plastid-encoded multimeric RNA polymerases, respectively (number and type of promoters change in different genes). 5'-UTR and 3'-UTR are also indicated. The two genes are separated by an intercistronic expression element (IEE) which mediates intercistronic cleavage of dicistronic mRNA (Zhou et al. 2007). After the IEE, a Shine–Dalgarno (SD) sequence has been inserted to mediate translation initiation at the second cistron. However, heterologous operons inserted in the plastome have been also engineered without IEEs, because recombinant polycistrons can be efficiently translated without further processing (Quesada-Vargas et al. 2005). After intercistronic cleavage by an endonuclease (*black triangle*), the two monocistronic transcripts are processed by 5' and 3' exonucleases (pacmans) together with RNA-binding proteins (RBPs) and 3' stem-loop structures (not indicated). RNA editing and splicing are not represented in the figure because, at our knowledge, no biotechnological applications with transgene sequences requiring such transcript modifications have been reported so far

motif of 35 amino acids (Schmitz-Linneweber and Small 2008). In plastids of angiosperms, mRNA editing occurs at approximately 40 sites as a conversion of a cytidine into a uridine nucleotide (Cardi et al. 2012). Editing events are generally important for gene function, because they modify the coding sequence in order to create start codons or functional polypeptides. The plastid intron repertoire of land plants comprises approximately 21 introns (with the exception of one group I, all are group II introns) in 18 genes (Tillich and Krause 2010). Intron removal by splicing represents an additional regulatory step of plastid gene expression and

(a)

(b)

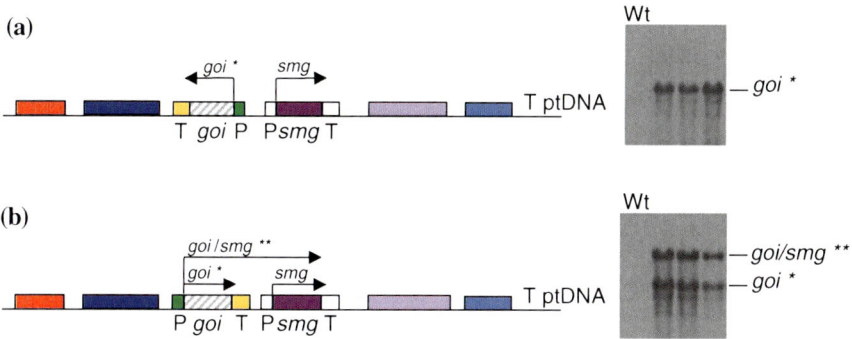

Fig. 10.3 Northern analysis showing the accumulation of mono (*goi*) and dicistronic (*goi/smg*) transcripts in plants transformed with the gene of interest (*goi*) and the selectable marker gene (*smg*) in the opposite (**a**) or same (**b**) orientation. Other boxes represent DNA flanking sequences involved in recombination-mediated transgene integration. *P* promoter, *T* terminator, *T ptDNA*, transformed ptDNA. The *goi* coding region was used as probe

a prerequisite for mRNA translation of these genes (Barkan 2011). Intercistronic processing is initiated by endonucleases, like for example RNase E (Walter et al. 2010). Mono or dicistronic translatable transcripts are then generated and maturated thanks to the concerted action of various 5′ and 3′ exonucleases with PPR-like proteins and 3′ stem-loop structures, which contribute to delimitate processed RNA termini by blocking exoribonucleases (Pfalz et al. 2009). RNA stability and decay seems to be mediated by the same ribonucleases and RNA-binding proteins described for intercistronic RNA processing (Schuster and Stern 2009; Prikryl et al. 2011), even though additional endoribonucleases and RNA-binding proteins have been characterized (Marchfelder and Binder 2004; Tillich et al. 2009).

Besides their role in RNA metabolism, PPR-like proteins also promote translation of plastid mRNAs, as recently demonstrated in maize for the ATP4 protein (Zoschke et al. 2012). To ensure efficient mRNA translation, ATP4 and other nucleus-encoded translational activators (Peled-Zehavi and Danon 2007) bind to 5′-untranslated regions (5′-UTRs), but the exact mechanism for translational activation is not completely clear yet. Most plastid mRNAs have a ribosome binding site containing a Shine–Dalgarno (SD) element, which in prokaryotes mediates ribosome recruitment. However, many plastid genes lack SD sequences; therefore, other mechanisms for translation initiation have been suggested (Marín-Navarro et al. 2007; Scharff et al. 2011). In addition, together with bacterial-type proteins, plastid ribosomes also have unique ribosomal proteins which, if down regulated, can decrease plastid translation efficiency (Tiller et al. 2012).

Even if each step in plastid gene expression is responsive to developmental and environmental conditions and seems to give a significant contribution to the overall regulation of genes located in the plastome, expression of plastid genes is primarily regulated during translation of the corresponding mRNAs (Zerges 2004; Manuell et al. 2007). Moreover, several studies conducted with transgenes inserted into the plastome suggest that posttranslational factors are extremely important for the expression of plastid genes as well (Oey et al. 2009a). Protein stability is a key determinant of protein accumulation in plastids and many aspects influencing protein stability, such as the identity of the N-terminal amino acids, protein co- or post-translational modifications and suborganelle localization, have been recently reviewed (De Marchis et al. 2012).

10.2.3 Enabling Technologies and Limiting Factors

Despite that plastid genetic engineering is today a well-established technology, the plastome of only a relatively small number of plant species and the green alga *Chlamydomonas* can be routinely engineered. Tobacco has been the first species to be transformed, using a mutant 16S rRNA gene conferring resistance to spectinomycin for selection (Svab et al. 1990). Although the use of mutated plastid genes conferring antibiotic resistance was successful in generating transplastomic plants (Dix and Kavanagh 1995; Craig et al. 2008), the real breakthrough in plastome transformation was the utilization of the selectable marker gene *aadA* (Day and Goldschmidt-Clermont 2011). Even low doses of the *aadA* enzyme detoxify spectinomycin and streptomycin; therefore, the few chloroplasts incorporating this marker gene after transformation can be selectively enriched in tissue culture until the obtainment of homoplastomic plants. Nowadays, chloroplast transformation reproducible protocols for tobacco (*Nicotiana tabacum*) and dycotyledonous species like potato, tomato, lettuce, and several others are available (Fig. 10.4), but not for monocot plants (Maliga and Bock 2011; Maliga 2012). Indeed, the main problem with cereals is that the most common antibiotics used for selection of transplastomic plants are not effective with these species. Kanamicin and streptomycin are not able to inhibit callus growth in the dark, while spectinomycin is useless due to the endogenous resistance of many cereal species to it. Therefore, a protocol based on chloramphenicol selection was developed in tobacco using the chloramphenicol acetyltransferase gene *cat* as an alternative selectable marker gene (Li et al. 2011) that could be hopefully next applied to cereal plastome transformation. Other studies tried to increase the efficacy of the existing marker genes. A recent modified version of the *aadA* gene, being both a selectable marker in tissue culture and a visual marker in transplastomic plants, was described by Tungsuchat-Huang et al. (2011). An alternative selection system has been recently described in tobacco, with the gene encoding D-amino acid oxidase used as plastid marker gene for positive/negative secondary selection once resistant plants are obtained using *aadA*-based spectinomycin selection (Gisby et al. 2012). Moreover, the anthranilate synthase α subunit has been also successfully employed as selectable marker gene (Barone et al. 2009). Marker genes can be removed after selection by: the flanking-repeat-mediated excision based on the native homologous recombination machinery, the use of site-specific foreign recombinases, the

Fig. 10.4 Plastid transformation in sugar beet (De Marchis et al. 2009). **a** Leaf petioles bombarded with gold particles coated with a plastid transformation vector expressing GFP and placed in a regeneration medium containing spectinomycin. **b** A regenerable callus (indicated with an *arrow*). **c** Selected calli placed in a medium without spectinomycin to regenerate. **d** Putative transformed regenerated shoots. **e** PCR analysis with a primer pair located outside the vector transgene sequence. From *left* to *right*: DNA molecular size marker, wild-type sugar beet DNA, DNA of three transformed plants. **f** Transplastomic sugar beet plants expressing GFP. Fluorescence (*left*) and bright-field (*right*) images from leaves of a transplastomic plant are shown

co-transformation and segregation of marker-free plastid genomes, the transient co-integration of the marker gene (reviewed in Day and Goldschmidt-Clermont 2011).

Transgenes can be targeted to any region of the plastome by providing targeting/flanking regions, typically 1–2 kb long, to obtain efficient transgene integration by recombination between plastid and vector DNA sequences (Fig. 10.1). However, the choice of the integration site may have a great importance in terms of transgene expression. The integration into a transcriptionally active site increases the level of the transcribed mRNA and that of the corresponding protein (De Cosa et al. 2001; Quesada-Vargas et al. 2005; Krichevsky et al. 2010), as well as the insertion into the plastome inverted repeat (IR) regions doubles the number of transgene copies per genome. The most used insertion sites include the regions between the *trnI-trnA*, *trnN-trnR*, *rccL-accD*, and *rrn16-rps7/12* genes.

Because of the high conservation of plastid genomes between most land plants, the first chloroplast transformation strategies utilized either homologous or heterologous flanking regions. However, the transformation efficiency using heterologous flanking regions was considerably lower in comparison to that with species-specific sequences. Indeed, even if the plastid genome is highly conserved between species with respect to protein coding regions and ribosomal RNAs, many differences can be present in the intergenic regions which are the common targets sites for homologous recombination (Saski et al. 2007). Therefore, a substantial increment in plastid transformation efficiency can be obtained with species-specific transformation vectors, as demonstrated by the decrease in the efficiency of tobacco chloroplast transformation when lettuce chloroplast targeting sequences were used to integrate transgenes into the *trnI/trnA* region of the tobacco plastome (Ruhlman et al. 2010). Similar results were obtained for potato and other species (Valkov et al. 2011, and references therein).

Several transgenes could be efficiently co-expressed in transgenic plastids arranging them in natural or synthetic operons (Quesada-Vargas et al. 2005; Krichevsky et al. 2010), and intercistronic elements facilitating the expression of monocistronic mRNAs from operons have been identified (Zhou et al. 2007). Many studies conducted with chimeric gene fusions have identified combinations of promoters, 5′-UTRs and 3′-UTRs, which can be used to achieve a high level of recombinant protein expression in chloroplasts, regulating transcript stability and translatability. Excellent results can be obtained in leaves with the promoter of the plastid rRNA operon (*Prrn*) or the *psbA* promoter, in combination with the 5′-UTR of gene 10 of the bacteriophage T7 (*T7g10*), or other 5′-UTRs from highly expressed chloroplast genes like *rbcL*, and the *E. coli rrnB* 3′ sequence as terminator (Maliga 2002; Herz et al. 2005; Tangphatsornruang et al. 2011). However, there is the need to find additional nonplastid regulatory sequences to be used in vectors designed to express multiple recombinant genes, in order to avoid unintended recombination events. Yang et al. (2013) investigated the efficacy of two additional bacteriophage 5′-UTRs (*T7g1.3* and *T4g23*) demonstrating that they can regulate *aadA* expression in chloroplasts. Conversely, transgene expression in nongreen plastids is very low, mainly due to the absence of data about gene expression in such plastid types. Recently, studies on the regulation of plastome gene expression in nongreen tissues started to fill this gap, showing a general downregulation of expression (Kahlau and Bock 2008; Valkov et al. 2009). As a consequence, expression elements that could increase transgene expression in the amyloplasts of tobacco roots and potato tubers have been identified (Valkov et al. 2011; Zhang et al. 2012). The combination of a strong plastid promoter (*psbA* promoter) with a strong *T7g10*-derived 5′-UTR that are not prone to severe developmental downregulation during fruit development has been shown to reach high-level gene expression in tomato chromoplasts, triggering GFP accumulation up to 1 % of the total protein of the fruit (Caroca et al. 2013).

A high number of studies on heterologous gene expression in plant plastids demonstrated that codon optimization of transgene sequences is not essential for efficient translation rate (Maliga and Bock 2011). This, together with the possibility to efficiently express bacterial single genes or operons, theoretically allows an easy genetic manipulation of a plethora of genes belonging to different kingdoms for expression in plant plastids. However, there are cases in which codon optimized genes of viral (Madesis et al. 2010), bacterial (Bohmert-Tatarev et al. 2011), or human (Gisby et al. 2011) origin significantly improved transgene expression in comparison with noncodon-optimized sequences.

The N-terminal sequence of recombinant proteins expressed in the chloroplasts is a key factor for both mRNA stability/translatability (Kuroda and Maliga 2001) and protein stability (Ye et al. 2001). Elghabi et al. (2011) fused N-terminal segments of highly expressed proteins in plastids to the transgene coding region, stabilizing the cyanovirin-N mRNA. Unfortunately, there are no precise rules for the best performing sequences and empiric attempts have to be made (Scotti et al. 2009; Gray et al. 2011). However, the existence of an N-end rule-like pathway with stabilizing and destabilizing N-residues has been postulated also for plastids, even though the identity of the penultimate amino acid is not the sole responsible for plastid protein stability (Apel et al. 2010). That the N-terminal part has a significant role for recombinant protein stability has been also demonstrated with the addition of extra amino acids at the 5′ end of the rotavirus VP6 protein (Inka Borchers et al. 2012). Most PTMs in plastids occur on the N-terminal part of proteins, and there are an increasing number of evidences that PTMs play a strategic role in the regulation of protein turnover (Adam et al. 2011; Bienvenut et al. 2011). In addition, many PTMs are involved in protein folding which is necessary to reach the polypeptide tertiary or quaternary stable structure. An important aspect for stability of recombinant proteins inserted into the plastome can be their subplastidial localization. Indeed, proteins can be accumulated in the stroma, the thylakoid lumen, the envelope, or the thylakoid membranes. For example, the thylakoid lumen seems to be a more adequate environment than the stroma for the accumulation of proteins that require formation of disulfide bonds (Bally et al. 2008; Lentz et al. 2012). In some cases, PTMs can be also important for protein functionality. Besides multimerization, N-terminal methionine excision and disulfide bond formation, plastids allow protein lipidation (Glenz et al. 2006), but not glycosylation, preventing the production of vaccine antigens that require glycosylation for their function and other glycoproteins.

Once accumulated in plastids, recombinant proteins must be extracted and purified for further uses but no many detailed purification methods from plastids have been published. Recently, glutathione-S-transferase and maltose-binding protein have been expressed in transplastomic plants and purified by affinity chromatography, demonstrating that they can be used as affinity tags for the rapid purification of recombinant proteins localized in the chloroplast (Ahmad et al. 2012a).

The expression of native expression elements seems to originate mutant phenotypes in some transplastomic plants, suggesting as likely cause the already reported competition for gene-specific transcription and/or translation factors (Kuroda and Maliga 2002). In other studies, the occurrence of mutant phenotypes characterized by chlorotic phenotype and growth retardation seems to be due to interference of the recombinant proteins with the plastid metabolism instead of competition for expression factors (Rigano et al. 2012). Therefore, the use of systems for inducible gene expression is desirable in such transplastomic plants (Verhounig et al. 2010; Lössl and Waheed 2011). Temporary immersion bioreactors have been also proposed as an alternative method for the production of proteins toxic for plants (Michoux et al. 2013).

10.3 Modification of Agronomic and Physiological Traits by Plastid Transformation

To date, numerous agronomic and physiological traits have been modified by plastid transformation as summarized in Table 10.1 and Fig. 10.5 and reported in recent reviews (Hasunuma et al. 2009, 2010; Rogalski and Carrer 2011; Scotti et al. 2011; Hanson et al. 2013). In this chapter, we focus on the most recent and/or significant examples.

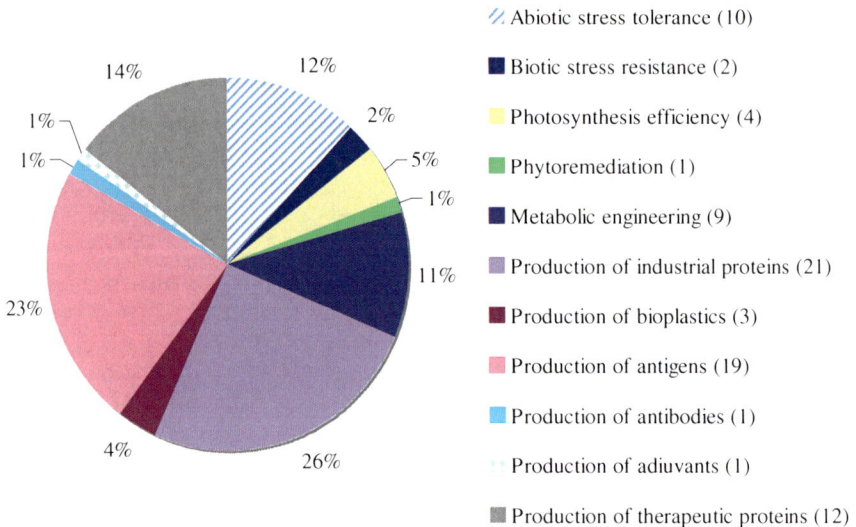

Fig. 10.5 Pie chart summarizing the transgenes engineered by plastid transformation from January 2010 to 2013

Table 10.1 Agronomic and physiological traits engineered by plastid transformation (January 2010–2013)

Trait	Enzyme/Transgene[a]	Source of transgene	Species	Yield[b]	Reference
Stress tolerance					
	Transglutaminase (*TGZ*)	Maize	Tobacco	1.1 mg g^{-1} FW	Ortigosa et al. (2010a)
	Flavodoxin (Fld)	*Anabaena* sp.	Transgenic tobacco with silenced Ferredoxin (Fd)	NR	Blanco et al. (2011)
	β-glucosidase (*Bgl-1*)	*T. reesei*	Tobacco	44.4 units g^{-1}	Jin et al. (2011)
	dehydroascorbate reductase (*DHAR*)	Rice	Tobacco	0.75 % TSP	Le Martret et al. (2011)
	glutathione-S-transferase (*GST*)	*E. coli*	Tobacco	0.75 % TSP	Le Martret et al. (2011)
	DHAR-GR	Rice-*E. coli*	Tobacco	0.75 % TSP	Le Martret et al. (2011)
	GST-GR	*E. coli*	Tobacco	0.75 % TSP	Le Martret et al. (2011)
	Mitochondrial superoxide dismutase (*MnSOD*)	*N. plumbaginifolia*	Tobacco	NR	Poage et al. (2011)
	Glutathione reductase (*GR*)	*E. coli*	Tobacco	NR	Poage et al. (2011)
	Plastid terminal oxidase 1 (*Cr-PTOX1*)	*C. reinhardtii*	Tobacco	NR	Ahmad et al. (2012b)
	Flavodoxin (*Fld*)	*Anabaena* sp.	Tobacco	11 μmol m^{-2}	Ceccoli et al. (2012)
	Agglutinin (*PTA*)	*P. ternata*	Tobacco	9.2 % TSP	Jin et al. (2012)
Photosynthesis efficiency					
	Mutated β-subunit of ATP synthase (*atpB*)	Tobacco	Tobacco	NR	Rott et al. (2011)
	rbcL–rbcS hybrid enzyme (*LLS2*)	Tobacco-Tomato	Tobacco	NR	Zhang et al. (2011)
	rbcL–rbcS hybrid enzyme (*LLS4*)	Tobacco-Tomato	Tobacco	NR	Zhang et al. (2011)
	Hypothetical chloroplast reading frame no. 4 (*ycf4*)	Tobacco	Tobacco	NR	Krech et al. (2012)
Phytoremediation					
	Metallothionein (*mt1*)	Mouse	Tobacco	NR	Ruiz et al. (2011)

(continued)

Table 10.1 (continued)

Trait	Enzyme/Transgene[a]	Source of transgene	Species	Yield[b]	Reference
Metabolic engineering					
	Operon containing six genes of the cytoplasmic mevalonate pathway (*MEV6.1*)	Synthetic	Tobacco	NR	Kumar et al. (2012)
	Tocopherol cyclase (*TC*)	Arabidopsis	Tobacco Lettuce	NR	Yabuta et al. (2013)
	γ-Tocopherol methyltransferase (γ-*TMT*)	Arabidopsis	Tobacco	NR	Yabuta et al. (2013)
	TC-γ-TMT	Arabidopsis	Tobacco	NR	Yabuta et al. (2013)

[a] *DHAR-GR* fusion between dehydroascorbate reductase and glutathione reductase (GR), *GST-GR* fusion between glutathione-S-transferase and glutathione reductase (GR), *LLS2* fused Rubisco containing tobacco small subunit (*rbcS*) and mutated (Q437R) tomato large subunit (*rbcL*), *LLS4* fused Rubisco containing tobacco small subunit (*rbcS*) and mutated (Y226F, A230T, S279T and Q437R) tomato large subunit (*rbcL*), *MEV6.1* synthetic operon containing genes encoding phosphomevalonate kinase (PMK), mevalonate kinase (MVK), mevalonate diphosphate decarboxylase (MDD), acetoacetyl CoA thiolase (AACT), C-terminal truncated 3-hydroxy-3methylglutaryl-coenzyme A reductase (HMGRt), *TC-γ-TMT* fusion between tocopherol cyclase and γ-tocopherol methyltransferase.

[b] yield according to original reference; TSP = total soluble protein; NR = not reported; FW = fresh weight

10.3.1 Stress Tolerance

Early experiments aiming to improve by plastid transformation salt/drought or cold tolerance relied on genes for the biosynthesis of the osmoprotectant glycine betaine or on Δ9 fatty acid desaturase genes, respectively (Kumar et al. 2004; Craig et al. 2008; Zhang et al. 2008). The study of Craig et al. (2008) represents the first example of transplastomic plants expressing an agronomically relevant gene produced with the "binding-type" vectors, in which chloroplast antibiotic insensitive point mutations are used to select transformants in place of heterologous marker genes.

More recently, different enzymes with anti-oxidant activities were expressed to enhance the tolerance of transplastomic tobacco plants to abiotic stresses. Genes encoding the dehydroascorbate reductase (DHAR) from rice, the glutathione-S-transferase (GST) and glutathione reductase (GR) from *E. coli* were expressed alone or as DHAR:GR or GST:GR combinations (Le Martret et al. 2011). In all transplastomic lines, the expression of the inserted genes gave an increment of specific enzyme activity and an alteration of anti-oxidant metabolism. Progeny of transplastomic plants, subjected to environmental stresses, proved to be less sensitive to low temperatures and salt stress, whereas no advantages were observed for heavy metal stress. A higher increase in tolerance to cold stress were detected in transplastomic plants expressing DHAR:GR or GST:GR gene combinations. In another study (Poage et al. 2011), the prospect for enhancing ROS scavenging and stress tolerance was pursued expressing a mitochondrial manganese superoxide dismutase (*MnSOD*) from *N. plumbaginifolia* and glutathione reductase (GR) from *E. coli* in the inverted regions of tobacco chloroplast genome. An increase of specific enzyme activity (three- and sixfold for *MnSOD* and GR, respectively) was revealed in all transplastomic plants. *MnSOD* transplastomic plants showed an enhanced tolerance to methyl-viologen and UV-B radiation, whilst GR plants an improved tolerance to the same radiation and heavy metal. No effect on photosynthetic capacity was observed in both transplastomic plants.

The expression level of proteins like ferredoxin decreases under environmental stresses in photosynthetic organisms. In cyanobacteria, this phenomenon is compensated by induction of flavodoxin, a flavoprotein with the same function not present in higher plants. In order to investigate the effect of flavodoxin on photosynthesis and stress tolerance in higher plants, Ceccoli et al. (2012) expressed a flavodoxin (Fld) of *Anabaena* sp. in the plastid genome of tobacco and compared its protective effect to tobacco nuclear transgenic lines overexpressing the same gene in the chloroplasts. Transplastomic lines expressed the flavodoxin at about 11 μmol m^{-2} of leaf tissue, fourfold more than the transgenic lines showing the highest amounts of flavodoxin. Comparative analysis between transgenic and transplastomic plants displayed a flavodoxin dose-dependent increase on photosynthetic performance and tolerance to the exposure to methyl viologen followed by a drop in high-expressing lines. The optimal photosynthetic performance and stress tolerance were observed at flavodoxin levels comparable to those of endogenous ferredoxin (about 3 μmol m^{-2}). In this case, the high increase obtained by plastid transformation resulted detrimental to plant fitness. Similarly, Ortigosa et al. (2010a) demonstrated that the overexpression of maize plastidial transglutaminase (chlTGZ) in the plastid genome of tobacco induced an oxidative stress in transplastomic plants. In particular, they evaluated several aspects such as fluorescence parameters, chloroplast ultrastructure, and oxidative and antioxidative metabolism. These analyses revealed different alterations of chloroplast ultrastructure and physiology that increased with leaf age.

As far as biotic stresses are concerned, one of the earliest biotechnological applications of plastid transformation reported the expression of the *Bacillus thuringiensis crylA(c)* gene in tobacco (McBride et al. 1995). Subsequently, De Cosa et al. (2001) produced transplastomic tobacco plants expressing the *B. thuringiensis* toxin (*cry2Aa2*) operon, at the level of about 46 % of total soluble protein (TSP), engineering two small open reading frames encoding for a chaperonin that facilitated the correct folding of Cry2Aa2 in stable crystals. Insect bioassay carried out with cotton bollworm and beet armyworm demonstrated that both insects were killed after consuming transplastomic leaves. The *B. thuringiensis* toxin (*crylAb*) was also expressed in the chloroplast genomes of soybean and cabbage (Dufourmantel et al. 2005; Liu et al. 2008), resulting in high protein yield and complete insect mortality in both cases.

Recently, the β-glucosidase gene (Bgl-1) was expressed in tobacco transplastomic plants (Jin et al. 2012). Bgl-1 transplastomic plants showed earlier flowering and increased biomass, height and leaf area compared to untransformed plants; these effects were associated with an increase of different plant hormones. Further, in the same plants, an increase in sugar esters located within the globular trichomes functioned as biopesticide against whiteflies and aphids.

10.3.2 Herbicide Resistance

The first attempt to improve herbicide tolerance by plastid transformation was pursued by Daniell et al. (1998) that introduced a mutated form of 5-enolpyruvylshikimate-3-phosphate synthase (EPSPS) from petunia in tobacco chloroplasts. The transplastomic plants obtained were tolerant to low concentration of glyphosate (5 mM) corresponding to 8 oz acre^{-1} Roundup, not enough for weed control, for which about 64 oz acre^{-1} Roundup are normally required. Better results (about 10 % TSP, corresponding to 128 oz acre^{-1} Roundup) were obtained by Ye et al. (2001) with the *Agrobacterium epsps* gene fused to the first 14 amino acids of GFP protein.

Although all forms of *bar* genes, encoding the enzyme phosphinothricin acetyl transferase (PAT) inactivating the herbicide glufosinate (phosphinothricin), were expressed at high level in transplastomic plants and these were tolerant to field-level of glufosinate, the gene did not work well as marker for direct selection of transplastomic clones (Lutz et al. 2001). Tobacco transplastomic plants expressing different mutated versions of the *als* gene from *A. thaliana* were able to grow in the presence of various herbicides (pyrimidinylcarboxylate, imidazolinone, and sulfonylurea/pyrimidinylcarboxylate), proving the possibility of using these plants on the rotation of three or more herbicide combinations (Shimizu et al. 2008).

10.3.3 Photosynthesis Efficiency

Improvement of photosynthesis efficiency would have a very high impact on agriculture, and thus many efforts are underway to engineer photosynthesis via nuclear or plastid transformation, as recently reviewed by Hanson et al. (2013).

Since the efficiency of photosynthesis is dependent on the performance of the ribulose-1,5-bisphosphate carboxylase oxygenase (Rubisco), attempts have been made to engineer this enzyme in order to enhance its catalytic turnover rate and increase its specificity for CO_2 (Whitney et al. 2009; Zhang et al. 2011). The former authors tried to manipulate Rubisco via chloroplast engineering, linking the small (S, encoded by nuclear genome) and large (L, encoded by plastid genome) subunits of tobacco by a flexible 40-amino acid tether (S40L). This strategy allowed to replace native *rbcL* in tobacco plastids with the synthetic S40L fusion gene. They found that the fusion S40L protein was able to assemble into catalytic oligomers in tobacco plastids, with higher affinities for CO_2 and O_2. Zhang et al. (2011) engineered Rubisco creating two lines of transplastomic tobacco carrying a hybrid enzyme. LLS2 plants contained tobacco small subunits (SS) and mutated large subunit (LS) with one substitution, whereas LLS4 plants displayed a tobacco SS and a tomato mutated LS bearing four substitutions. Compared to wild-type and LLS2 plants, LLS4 plants showed lower chlorophyll and Rubisco content, lower photosynthesis rates and biomass during early stages of development, but were able to reach maturity. The enzyme assay detected a carboxylase activity in both plants similar to wild type, demonstrating that hybrid enzymes were able to assemble into functional Rubisco.

Plastid transformation was also used to identify the role of plastid open reading frames in the assembly and stability of photosystem I (Krech et al. 2012), or to study the functional significance of the ATP synthase adjustment in response to the metabolic demand (Rott et al. 2011). In particular, the latter authors demonstrated that the repression of ATP synthase complex, pursued by an antisense approach on essential nuclear *atpC* gene and by the introduction of different point mutations into the translation initiation codon of the plastid *atpB* gene via plastid transformation, led to a restriction of photosynthetic electron transport due to decreased rates of plastoquinol reoxidation at the cytochrome b_6f complex, and a reduction of the quantum efficiency of CO_2 fixation due to an increased steady-state proton motive force resulting in overacidification of the thylakoid lumen.

10.3.4 Male Sterility

The cytoplasmic male sterility (CMS) phenotype is a maternally inherited trait, important to produce commercial F_1 hybrids (Schnable and Wise 1998). To date, only a manuscript described the engineering of this trait through plastid transformation by expression of the *phaA* gene encoding a β-ketothiolase (Ruiz and

Daniell 2005). Since a previous study (Lössl et al. 2003) reported that the expression of the entire polyhydroxybutyrate (PHB) operon (*phaA*, *phaB*, *phaC*) in the chloroplast genome of tobacco produced severe pleiotropic effects (stunted phenotype and male sterility) in transplastomic plants, Ruiz and Daniell (2005) investigated the specific role of *phaA* gene and evaluated its effect under a specific photoperiod or continuos illumination, using as regulatory sequences the *psbA* promoter and 5'-UTR. The transplastomic plants overexpressing β-ketothiolase, whose enzymatic activity was in the range of 14.08–14.71 units mg^{-1} plant protein during regular photoperiod, were normal except for the male sterile phenotype. In particular, electron microscopy analyses showed a collapsed morphology of pollen grains and an accelerated pattern of anther development resulting in aberrant tissue development. Further, they observed a reversibility of the male sterile phenotype (production of viable pollen and seeds) under continuous light.

10.3.5 Phytoremediation

Phytoremediation of mercury and organomercurial compounds via plastid transformation was pursued using two approaches. The first one was based on the integration of bacterial native operon containing the *merA* and *merB* genes, coding for mercuric ion reductase and organomercurial lyase, respectively, into the tobacco chloroplast genome. When grown in soil containing up to 400 μM phenylmercuric acetate (PMA), transplastomic plants showed a higher tolerance to PMA than wild-type plants (Ruiz et al. 2003). Further, they were used to investigate the uptake and translocation of different forms of mercury from roots to shoots, and their volatilization, demonstrating that they accumulated both organic and inorganic mercury forms to concentrations much higher than those usually found in the soil (Hussein et al. 2007). In transplastomic plants, the organic mercury was uptaken and translocated more efficiently than the inorganic form. The efficient translocation from roots to shoots seems to facilitate the reduction of toxic ionic mercury (Hg^{2+}) into less toxic and volatile elemental mercury (Hg^0) (Hussein et al. 2007). Although this approach demonstrated an improved phytoremediation capability in transplastomic plants, the volatile Hg^0 form can be released back into the environment where it accumulates and potentially converted again into highly toxic forms. For this reason, an alternative approach based on the expression of chelator molecules, such as metallothioneins, was developed (Ruiz et al. 2011). Transplastomic plants expressing the mouse metallothionein gene (*mt1*) were resistant up to 20 μM mercury without any phenotypic effect, and accumulated high concentrations of mercury in all tissues. In particular, the high concentration of mercury accumulated in leaves of transplastomic plants (up to 106 ng) is indicative of an active phytoremediation and translocation of mercury.

10.3.6 Metabolic Engineering

Based on the fact that several biosynthetic pathways occur in plastids, some studies demonstrated the potential of plastome engineering for the nutritional enhancement of vegetables. Recently, Yabuta et al. (2013) improved the quality and quantity of vitamin E (tocopherol), an important anti-oxidant, through the expression of different genes involved in this pathway in the chloroplast genomes of tobacco and lettuce. In particular, they observed that the overexpression of tocopherol cyclase (TC) gene in both plant species induced an increase of total tocopherol level and vitamin E activity of about 2.2- and 1.3-fold in tobacco and lettuce, respectively, compared to wild-type plants. Carotenoid biosynthesis was manipulated in tomato in order to increase the content of pro-vitamin A (β-carotene). In a first study (Wurbs et al. 2007), two microbial lycopene β-cyclase and a β-cyclase/phytoene synthase fusion gene from the fungus *Phycomyces blakesleeanus* were integrated in the chloroplast genome of tomato. HPLC analyses demonstrated that transplastomic tomato fruits accumulated a 4-fold higher provitamin A content compared to wild-type plants. A subsequent manuscript (Apel and Bock 2009) described the enhancement of provitamin A content in tomato fruits through the expression of lycopene β-cyclase from a higher plant (*Narcissus pseudonarcissus*, daffodil). In comparison with the previous study, a higher efficient conversion of lycopene to β-carotene was obtained, resulting in an accumulation of provitamin A of about 1 mg g^{-1} dry weight. In addition, the expression of lycopene β-cyclase from daffodil induced an increase of the total fruit carotenoid content and an alteration of carotenoid composition in leaves.

Another interesting example of the effectiveness of plastid transformation technology for the manipulation of metabolic pathways is the production of astaxanthin (Hasunuma et al. 2008), a carotenoid synthesized by some bacteria and fungi with wide industrial applications due to its various biological functions. This goal was pursued by the expression of plant codon optimized genes of marine bacterium *Brevundimonas* sp. encoding the ß-carotene ketolase and ß-carotene hydroxylase. Transplastomic plants expressing both genes accumulated astaxanthin at a level higher than 0.5 % (dry weight), corresponding to approximately at 70 % of total carotenoids, and synthesized also the novel carotenoid 4-ketoantheraxanthin. Further, they were characterized by a dramatic change of color in leaves and stems (reddish brown), and stigmas and corolla (pink), instead of their normal green color.

The possibility to manipulate by plastid transformation the lipid pathway has been recently discussed by Rogalski and Carrer (2011). However, besides the report of Craig et al. (2008) cited above, only another study investigated the feasibility to alter the fatty acid content by overexpression of the *accD* gene in the plastome (Madoka et al. 2002). Resultant transplastomic plants showed a significantly higher fatty acid content in leaves, extended leaf longevity and a 2-fold increase of seed yield over the controls.

10.4 Production of Recombinant Molecules in Plants by Plastid Transformation

The diffusion of the 'Molecular Farming' concept has given a significant boost to explore the suitability of chloroplast genetic engineering to produce recombinant molecules. Herein, we summarize the most successful and recent examples for each subgroup of recombinant molecules (Table 10.2 and Fig. 10.5).

10.4.1 Industrial Proteins

Most recent examples of industrial proteins produced by plastid transformation belong to enzymes involved in the hydrolysis of lignocellulosic biomass to produce fermentable sugar. Verma et al. (2010a) expressed in tobacco chloroplasts an enzyme cocktail based on genes from bacteria and fungi, and compared their yields and enzyme activities with those obtained in *E. coli*. The chloroplast-derived enzymes had higher temperature stability, wider pH range and higher enzyme activities compared to *E. coli*-derived enzymes. Further, the chloroplast-derived crude extracts of enzyme cocktails were able to release a higher amounts of glucose from filter papers, pine wood, or citrus peel than commercial enzyme cocktails. Petersen and Bock (2011) focused on the production of four enzymes from the thermophilic bacterium *Thermobifida fusca*, demonstrating also in this case that such enzymes can be successfully expressed in the chloroplast genome. By contrast with Verma et al. (2010a), the high protein yields (between 5 and 40 % of plant total soluble protein) produced by plastid transformation resulted in pigment-deficient phenotypes. Enzyme activity assay carried out on crude extracts of transplastomic plants demonstrated that all enzymes were highly active and hydrolyzed their synthetic substrates in a dose-dependent manner. On the contrary, enzyme assay carried out on natural substrates (wheat straw) showed that without a standard thermochemical treatment of plain straw it was not possible to observe a sugar release. The expression of a GH10 xylanase gene (*Xyl10B*) from *Thermotoga maritima* in the tobacco chloroplast genome gave an enzyme yield up to 15 % TSP without any phenotypic effects (Kim et al. 2011). The enzyme assay carried out with crude extracts of Xyl10B-transplastomic plants showed an exceptional catalytic activity and enabled the complete hydrolysis of natural substrates to fermentable sugars with the help of α-glucuronidase accessory enzyme. Successful results were also obtained with the expression of β-mannase gene (*man1*) from *Trichoderma reesei* (Agrawal et al. 2011).

Other important industrial polypeptides recently produced via plastid transformation include a laccase and a monellin variant. Davarpanah et al. (2012) produced transplastomic tobacco plants to develop commercial level of a fungal laccase, an enzyme with potential applications as waste detoxification and in textile industry. Although an accumulation up to 2 % of total protein was reached,

Table 10.2 Recombinant molecules produced by plastid transformation (January 2010–2013)

Sub-group	Molecule/Transgene	Source of transgene	Species	Yield[a]	Reference
Industrial proteins					
	Endoglucanase (*celD*)	*C. thermocellum*	Tobacco	4930 units g⁻¹ FW	Verma et al. (2010a)
	Exoglucanase (*celO*)	*C. thermocellum*	Tobacco	NR	Verma et al. (2010a)
	Lipase (*lipY*)	*M. tuberculosis*	Tobacco	NR	Verma et al. (2010a)
	Pectate liases (*pelA, pelB* and *pelD*)	*F. solani*	Tobacco	32 units g⁻¹ FW	Verma et al. (2010a)
	Cutinase	*F. solani*	Tobacco	NR	Verma et al. (2010a)
	Swollenin (*swo1*)	*T. reesei*	Tobacco	NR	Verma et al. (2010a)
	Xylanase (*xyn2*)	*T. reesei*	Tobacco	NR	Verma et al. (2010a)
	Acetyl xylan esterase (*axe1*)	*T. reesei*	Tobacco	NR	Verma et al. (2010a)
	β-glucosidase (*bgl1*)	*T. reesei*	Tobacco	NR	Verma et al. (2010a)
	β-mannase (*man1*)	*T. reesei*	Tobacco	25 units g⁻¹ FW	Agrawal et al. (2011)
	β-glucosidase (*bglC*)	*T. fusca*	Tobacco	12 % TSP	Gray et al. (2011)
	GH10 xylanase (*Xyl10B*)	*T. maritima*	Tobacco	15 % TSP	Kim et al. (2011)
	β-glucosidase (*bgl1C*)	*T. fusca*	Tobacco	>5 % TSP	Petersen and Bock (2011)
	Exoglucanase (*cel6B*)	*T. fusca*	Tobacco	5 % TSP	Petersen and Bock (2011)
	Endoglucanase (*cel9A*)	*T. fusca*	Tobacco	40 % TSP	Petersen and Bock (2011)
	Xyloglucanase (*xeg74*)	*T. fusca*	Tobacco	<40 % TSP	Petersen and Bock (2011)
	Laccase	*P. ostreatus*	Tobacco	2 % TP	Davarpanah et al. (2012)
	Monellin variant (*MNEI*)	synthetic	Tobacco	5 % TSP	Lee et al. (2012)
	β-glucuronidase (*GUS*)	pBI121 vector	Potato	41 % TSP	Segretin et al. (2012)
Bioplastics					
	Polyhydroxybutyrate (*phaA, phaB* and *phaC*)	*Acinetobacter* sp.	Tobacco	18.8 % DW	Bohmert-Tatarev et al. (2011)
Antigens					
	Cholera toxin-B subunit/apical membrane antigen-1 (*CTB-AMA1*)	*V. cholera-P. falciparum*	Tobacco / Lettuce	13.2 % TSP / 7.3 % TSP	Davoodi-Semiromi et al. (2010)
	Cholera toxin-B subunit/merozoite surface protein-1 (*CTB-MSP1*)	*V. cholera-P. falciparum*	Tobacco / Lettuce	10.1 % TSP / 6.1 % TSP	Davoodi-Semiromi et al. (2010)

(continued)

Table 10.2 (continued)

Sub-group	Molecule/Transgene	Source of transgene	Species	Yield[a]	Reference
	Capsid protein VP1 (*VP-β-GUS*)	Food and Mouth Disease Virus (FMDV)	Tobacco	51 % TSP	Lentz et al. (2010)
	Core polypeptide of HCV (*core^syn*)	synthetic	Tobacco	0.1 % TP	Madesis et al. (2010)
	Peptide 2L21from VP2 protein (*2L21-TD*)	Canine parvovirus (CPV)	Tobacco	6 % TSP	Ortigosa et al. (2010b)
	Protective antigen (*CTB-PA*)	*B. anthracis*	Tobacco	29.6 % TSP	Ruhlman et al. (2010)
			Lettuce	22.4 % TSP	
	Major F4ac fimbrial subunit protein (*faeG*)	*E. coli* strain C$_{83907}$	Tobacco	0.15 % TSP	Shen et al. (2010)
	Domain IV protective antigen [*PA(dIV)*]	*B. anthracis*	Tobacco	5.3 % TSP	Gorantala et al. (2011)
	Dengue-3 serotype Polyprotein (*DENV3prM/E*)	Dengue	Lettuce	NR	Kanagaraj et al. (2011)
	Viral protein 8 (*VP8*)	Bovine Rotavirus strain C486	Tobacco	600 µg g^{-1} FW	Lentz et al. (2011)
	Fragment C of tetanus toxin (*TetC*)	*C. tetani*	Tobacco	8 % TSP	Michoux et al. (2011)
	Mutated L1 (*L12xCysM*)	Human Papilloma Virus (HPV-16)	Tobacco	1.5 % TSP	Waheed et al. (2011b)
	Mutated L1 (*LTB-L12xCysM*)	Human Papilloma Virus (HPV-16)	Tobacco	2 % TSP	Waheed et al. (2011a)
	GRA4 antigen (*chlGRA4*)	*T. gondii*	Tobacco	0.2 % TP	Del L Yácono et al. (2012)
	viral protein 6 (*VP6*)	Rotavirus	Tobacco	15 % TSP	Inka Borchers et al. (2012)
	Outer surface protein A (*OspA*)	*B. burgdorferi*	Tobacco	7.6 % TSP	Michoux et al. (2013)
	Polypeptide containing V3 loop and the C4 domain from HIV gp120 (*C4V3*)	synthetic	Tobacco	25 µg g^{-1} FW	Rubio-Infante et al. (2012)
	Plastoglobulin 35-HIV/p24 (*PGL35–HIVp24*)	Arabidopsis-HIV	Tobacco	1 % TP	Shanmugabalaji et al. (2013)
	Plastoglobulin 35-HCV core protein (*PGL35–HCVcore*)	Arabidopsis-HCV	Tobacco	NR	Shanmugabalaji et al. (2013)

(continued)

Table 10.2 (continued)

Sub-group	Molecule/Transgene	Source of transgene	Species	Yield[a]	Reference
Adjuvants					
	Extra domain A from fibronectin (*EDA*)	Mouse	Tobacco	2 % TP	Farran et al. (2010)
Antibodies					
	Fragments from camelid single-chain antibodies (*VHH, GUS-E-VHH, pep-VHH*)	Camelid	Tobacco	3 % TSP	Lentz et al. (2012)
Therapeutic proteins					
	Proinsulin (*CTB-Pins*)	Human	Tobacco Lettuce	72 % TP 12.3 % TP	Ruhlman et al. (2010)
	Coagulation factor IX (*CTB-FFIX*)	Human	Tobacco	3.8 % TSP	Verma et al. (2010b)
	ß-site of amyloid precursor protein cleaving enzyme (*BACE*)	Human	Tobacco	2 % TSP	Youm et al. (2010)
	A, B, C peptides of proinsulin (*CTB-PF-x3*)	Human	Tobacco Lettuce	47 % TP 53 % TP	Boyhan and Daniell (2011)
	Cyanovirin-N (*CV-N*)	*N. ellipsosporum*	Tobacco	0.3 % TSP	Elghabi et al. (2011)
	Human transforming growth factor-ß3 (*TGFß3*)	synthetic	Tobacco	12 % TP	Gisby et al. (2011)
	Retrocyclin 101 (*RC101-GFP*)	Human	Tobacco	38 % TSP	Lee et al. (2011)
	Protegrin-1 (*PG1-GFP*)	Porcine	Tobacco	26 % TSP	Lee et al. (2011)
	Thioredoxin m-human serum albumin (*Trx m-HAS*)	Plastid-Human	Tobacco	26 % TSP	Sanz-Barrio et al. (2011)
	Thioredoxin f-human serum albumin (*Trx f-HAS*)	Plastid-Human	Tobacco	22 % TSP	Sanz-Barrio et al. (2011)
	Interferon α 5 (IFNA 5)	synthetic	Tobacco	4.4 pg g^{-1} FW	Khan and Nurjis (2012)
	Exendin-4 (*CTB-EX4*)	synthetic	Tobacco	14.3 % TP	Kwon et al. (2013)

yield according to original reference; *FW* fresh weight, *NR* not reported, *TSP* total soluble protein, *TP* total protein; *DW* dry weight no laccase enzyme activity was detected. In addition, the transplastomic plants showed a retarded growth and a pale-green phenotype. The sweet protein monellin is a natural protein derived from berries of *Dioscoreophyllum cumminsii* (tropical rainforest vine) that has potential uses as noncarbohydrate sweetener for individuals that must control their sugar intake, but it is unstable at high temperatures and acidic pH. Hence, Lee et al. (2012) evaluated plastid transformation to produce more stable monellin variants. The ELISA assay detected a yield for each variant of about 50–60 µg/mg protein comparable to those obtained from the natural source.

10.4.2 Bioplastics

Several attempts have been done to produce polyhydroxyalkanoates (PHAs), a family of biodegradable and renewable plastics, by plastid transformation. Some studies were performed with vectors including a minimal transgene expression cassettes that gave very low polymer accumulation levels (Nakashita et al. 2001; Arai et al. 2004).

The *Ralstonia eutropha phb* operon (three bacterial enzymes) was expressed in tobacco plastids using two different approaches. In a first study, Lössl et al. (2003) used the promoter and 5′-UTR of the plastid *psbA* gene. Transplastomic plants showed a PHB accumulation level very variable with the maximum yield (up to 1.7 % dry weight in leaves) at the early stages of in vitro culture. Because the highest yield achieved was associated with a growth reduction, an inducible system to regulate the transcription of the *phb* operon in tobacco plastids was subsequently developed (Lössl et al. 2005). This approach was based on a nuclear located, ethanol-inducible T7 RNA polymerase which was targeted to plastids harboring the *phb* operon under the control of T7 regulatory sequences. Double transformed plants were sprayed with a 5 % ethanol solution and 4 weeks after induction, the PHB synthesis was evaluated by gas chromatography. After induction, the highest PHB content in transformed plants was 1,383 ppm in dry weight, whereas the background level in uninduced transformed plants was 171 ppm.

In order to optimize expression and reduce the chance of unwanted rearrangements between sequences in the expression cassette and host genome, Bohmert-Tatarev et al. (2011) selected PHB biosynthetic gene sequences (*phaA*, *phaB*, and *phaC*) from two bacteria (*Acinetobacter* sp. and *Bacillus megaterium*) with GC content and codon usage similar to that of tobacco plastome, regulatory elements with limited homology to the host plastome and short spacer elements of plastidial origin upstream of each transgene. The three genes of interest plus the marker gene were cloned downstream of the *psbA* coding sequence exploiting promoter and UTRs of the same plastid gene ("operon extension strategy"). Transplastomic plants were capable of producing up to 18.8 % dry weight PHB in leaves. Furthermore, in contrast to previous results (Lössl et al. 2003), they were fertile and produced progeny with a high PHB content.

10.4.3 Antigens, Antibodies, and Adjuvants

During the past few years, exciting progress has been made with plastid-based production of pharmaceuticals and in particular with vaccine subunits, as recently summarized in several reviews (Cardi et al. 2010; Lössl and Waheed 2011; Maliga and Bock 2011; Scotti et al. 2012). Issues particularly important for this class of molecules are the achievement of high protein yields and stability, because the stimulation of the immune system, especially mucosal immunity, requires high doses of a stable antigen.

Recently, Inka Borchers et al. (2012) increased the accumulation and stability of rotavirus VP6 protein altering its $5'$-UTR and $5'$ end of the coding region. Compared to previous results (Birch-Machin et al. 2004), the inclusion of the $5'$-UTR from T7g10 and of 15 nucleotides at the $5'$ end of the *VP6* coding region increased its expression level up to 15 % of total leaf protein. Further, they observed that these sequences stabilized the protein accumulation in both young and old leaves, and that the plastid-based VP6 proteins assembled into trimeric forms similarly to rotavirus capsids (Inka Borchers et al. 2012). For other viral antigens, it has been observed that the use of a $5'$-UTR plus additional nucleotides at the $5'$ end of coding region (HPV-16 L1 or HIV-1 Pr55^{gag}) or the production of a fusion protein by adding partial or complete protein sequences at the $3'$ end of coding sequences (2L21 peptide from canine parvovirus or epitope of VP1 protein of the foot and mouth disease virus) can increase protein yields up to 51 % TSP (Lenzi et al. 2008; Scotti et al. 2009; Lentz et al. 2010; Ortigosa et al. 2010b). On the other hand, Rigano et al. (2009) demonstrated that the only use of the T7g10 $5'$-UTR was sufficient to accumulate the envelope protein (A27L) of vaccinia virus up to 18 % TSP.

In order to confer dual immunity against cholera and malaria, Davoodi-Semiromi et al. (2010) fused the cholera toxin-B subunit (CTB) to two malarial antigens, apical membrane antigen-1 (AMA1) and merozoite surface protein-1 (MSP1), and expressed them in tobacco and lettuce chloroplasts. CTB-AMA1 and CTB-MSP1 were accumulated up to 13.2 % and 10.1 % TSP in transplastomic tobacco plants and up to 7.3 % and 6.1 % TSP in transplastomic lettuce plants, respectively. The tobacco plastid-based fusion proteins were used to immunize subcutaneously or orally nine groups of mice. Significant levels of specific antibodies were detected for both diseases. In addition, the analysis of several immunological markers suggests that immunity was conferred via the Tr1 cellular and Th2 humoral immune responses. Another multicomponent vaccine that elicited both systemic and mucosal immune responses was based on fusion protein containing epitopes against diphtheria, pertussis, and tetanus (DPT) expressed in tobacco plastome (Soria-Guerra et al. 2009).

Gorantala et al. (2011) developed a vaccine against anthrax based on domain IV of protective antigen [PA(dIV)] in tobacco chloroplasts, obtaining an expression level up to 5.3 % TSP with an AT rich sequence of the gene. Further, they compared the protective response of plant- and *E. coli*-derived [PA(dIV)] in mice intraperitoneally or orally immunized with or without adjuvant. Although the highest antibody titers ($>10^5$) were detected in adjuvanted *E. coli* [PA(dIV)] groups, adjuvanted plastid-derived [PA(dIV)] also induced significant specific antibody titers ($>10^4$) in both intraperitoneal and oral immunizations. Challenge with *Bacillus anthracis* in mice intraperitoneally immunized with adjuvanted plastid-derived [PA(dIV)] conferred a lower protection (60 % vs. 100 %) than in mice immunized with adjuvanted *E. coli* [PA(dIV)].

Recently, Lentz et al. (2012) produced in tobacco chloroplasts a fragment from camelid single-chain antibodies, also known as nanobody or VHH, directed against rotavirus VP6 protein and able to neutralize rotavirus infection. They pursued three strategies: expression of the original VHH in the chloroplast stroma (VHH 3B2), a translational fusion between GUS protein and VHH (GUS-E-VHH), and the targeting of VHH to thylakoid lumen by adding a N-terminal signal peptide of the pectate lyase B of *Erwinia carotovora* (pep-VHH). Transplastomic plants expressing the nanobodies VHH 3B2 and pep-VHH were characterized by transgene instability, heteroplasmic genotype and loss of the transgene in the seeds, whereas GUS-E-VHH plants had a normal development and accumulated the nanobody in the stroma at 3 % TSP. However, the few pep-VHH homoplasmic lines obtained confirmed protein translocation into the thylakoid lumen, where the pep-VHH polypeptide was stable and its expression levels reached 2–3 % of the total soluble proteins.

To date, there is only one example of a plastid-derived adjuvant corresponding to the extra domain A (EDA) from fibronectin (Farran et al. 2010). Similarly to previous examples, the highest protein yield (2 % of total cellular protein) was obtained when the protein was translationally fused to the first 15 amino acids of the green fluorescent protein (GFP). The EDA protein was purified from tobacco leaves and used in biological assay in order to demonstrate that it retained its pro-inflammatory properties and hence that could be used as adjuvant.

10.4.4 Therapeutic Proteins

Recently, Khan and Nurjis (2012) expressed in tobacco a synthetic interferon alpha 5 gene, belonging to an important class of proteins used for the treatment of different malignancies and virologic diseases. ELISA assay carried out on total extracts of transplastomic plants demonstrated that this protein accumulated to very low yield (up to 4.4 pg/g fresh weight). Similarly, a low yield was obtained for cyanovirin-N (CV-N), a small protein (11 kDa) difficult to express in chloroplasts, able to inactivate at nanomolar concentrations all variants of HIV-1 (Elghabi et al. 2011). Various terminal fusions were tested to improve the production of CV-N in such organelle, and the highest protein yield (0.3 % TSP) was obtained with either N- or N- and C-terminal GFP fusions.

A much higher yield (12 % of total leaf protein) was obtained with a synthetic gene containing 33 % GC encoding for human transforming growth factor-β3

(TGFβ3) that was accumulated in insoluble aggregates (Gisby et al. 2011). Its insolubility facilitated initial purification and refolding in homodimeric chains linked by disulfide bonds. A biologic assay based on the ability of TGFβ3 to inhibit the proliferation of mink lung epithelial cells was carried out on refolded plastid-based protein and standard protein, and showed a similar dose-response curve between the two proteins.

Recently, various therapeutic agents, such as retrocyclin-101 (RC101) and pro-tegrin-1 (PG1), proinsulin, and coagulation factor IX, were produced in transgenic chloroplasts (Verma et al. 2010b; Boyhan and Daniell 2011; Lee et al. 2011). In order to confer protein stability, all genes were fused to GFP or CTB sequences. The protein yields of RC101 and PG1 antimicrobial peptides, promising therapeu-tic agents against bacterial and/or viral infections, especially those caused by the HIV-1 or sexually transmitted bacteria, were estimated to be approximately 35 % and 25 % TSP, respectively. The antimicrobial activity of both proteins was con-firmed by inoculation of potted plants with *E. carotovora*. Further, RC101 trans-plastomic plants were resistent to tobacco mosaic virus infections confirming also the antiviral activity (Lee et al. 2011). Hence, such peptides could have also a role against plant pathogens.

Some lysine-type antibiotics were recently developed in tobacco chloroplasts using two strategies. The highest yield (more than 70 % TSP) was obtained with the synthetic *plyGBS* gene (Oey et al. 2009a). The other two antibiotics (Cpl-1 and Pal), that proved to be unclonable into standard plastid expression cassettes, were expressed using an innovative strategy (called toxin shuttle), based on preventing lethal transgene transcription in *E. coli* by inducing premature transcription termi-nation upstream of the transgene coding region using bacterial transcription termi-nators (Oey et al. 2009b). The bacterial terminators were flanked by *loxP* sites and could therefore be excised *in planta* by site-specific recombination after chloro-plast transformation.

Since it is widely known that thioredoxins, small ubiquitous proteins, were able to enhance solubility and stability of recombinant proteins in microbial expression systems, Sanz-Barrio et al. (2011) evaluated the role of thioredoxins as modula-tors of the expression of the human serum albumin (HSA), which has been pre-viously shown to form inclusion bodies in plastids (Fernandez-San Millan et al. 2003). For such a purpose, two strategies were assayed based on the fusion of thioredoxins m and f to HSA, or on co-expression with HSA on the same vec-tor. Trx m-HSA and Trx f-HSA fusion lines accumulated, in inclusion bodies, the human serum albumin to 26 % TSP and 22 % TSP, respectively; whilst the HSA in co-expressed lines was mainly found as soluble protein with accumulation level of 1.5 and 3.1 % TSP for Trx m and Trx f, respectively. The differences observed in terms of protein accumulation were mainly due to higher HSA stability of the fused proteins.

The human proinsulin (A, B, and C peptides) was expressed in tobacco and lettuce chloroplasts and accumulated up to 47 % and 53 % of total leaf protein, respectively (Boyhan and Daniell 2011). Accumulation of this protein was stable also in senescent and dried lettuce leaves. Proinsulin was purified from tobacco leaves up to 98 % of purity. Oral and injectable delivery of plastid-based proinsulin into mice showed reduction of glucose level in blood similar to that obtained with commercial processed insulin.

Current treatment of the hemophilia disorders based on intravenous infusion of recombinant or plasma-derived coagulation factors VIII or IX can induce anaphylactic reactions. In order to prevent these reactions, Verma et al. (2010b) tested a fusion protein CTB-coagulation factor IX (CTB-FFIX) produced by plastid transformation in a murine hemophilia B model. Oral delivery of plastid-based CTB-FFIX blocked the formation of inhibitory antibodies and eliminated fatal anaphylactic reactions.

An interesting therapeutic target of Alzheimer disease, the β-site of the amyloid precursor protein cleaving enzyme (BACE), was produced at 2 % of TSP in transgenic tobacco chloroplasts (Youm et al. 2010). Mice gavaged with extracts from transplastomic plants expressing the BACE enzyme showed a specific immune response.

10.5 Conclusions and Perspectives

The first example of stable plastid transformation in higher plants (tobacco) dates back to more than 20 years ago (Svab et al. 1990). Since then, the technology has been improved in many aspects, including the development of various methods for DNA introduction, marker genes and selection strategies, vector types, and methods for marker excision. Although the plastids of some species (e.g., the monocots) remain difficult to transform, a reproducible protocol is now available for about 20 species, belonging to 8 families (Maliga 2012). Besides to biotechnology, for which some applications have been discussed in the present chapter, the transformation of the plastome is also relevant to basic studies (Maliga 2004). In the last 3 years, more than 40 original articles showing a total production of about 80 proteins for applications in different fields have been published. Important challenges for the future remain the improvement of transformation protocols in species other than tobacco and related Solanaceae, the development of inducible expression vectors, the increase of the expression level in nongreen plastids and/or for difficult-to-accumulate proteins, and the optimization of recombinant protein purification protocols. For all these aspects, however, significant improvements have been made lately, as reported in this chapter and other recent reviews (Maliga and Bock 2011; Maliga 2012). "Aspirational goals" for plastid biotechnology have been recently discussed (Clarke and Daniell 2011). In the next future, only a small number of proteins will likely remain not expressible in transgenic plastids, such as those requiring glycosylation for their functionality.

To our knowledge, no transplastomic plants have been grown so far in the field for commercialization. Field trials, however, have been conducted with transplastomic petunia, soybean and tobacco to test degree of gene containment (greenbiotech. eu/?page_id = 501), herbicide tolerance (www.faqs.org/patents/app/20120023615), and production of pharmaceuticals (Arlen et al. 2007), industrial enzymes (www .icongenetics.com/html/5935.htm) or bioplastic (www.metabolix.com/Products/ Crop-based-Technologies/Research).

References

Adam Z, Frottin F, Espagne C, Meinnel T, Giglione C (2011) Interplay between N-terminal methionine excision and FtsH protease is essential for normal chloroplast development and function in Arabidopsis. Plant Cell 23:3745–3760

Agrawal P, Verma D, Daniell H (2011) Expression of *Trichoderma reesei* β-mannanase in tobacco chloroplasts and its utilization in lignocellulosic woody biomass hydrolysis. PLoS ONE 6:e29302

Ahmad N, Michoux F, McCarthy J, Nixon PJ (2012a) Expression of the affinity tags, glu-tathione-S-transferase and maltose-binding protein, in tobacco chloroplasts. Planta 235:863–871

Ahmad N, Michoux F, Nixon PJ (2012b) Investigating the production of foreign membrane pro-teins in tobacco chloroplasts: expression of an algal plastid terminal oxidase. PLoS ONE 7:e41722

Altpeter F, Baisakh N, Beachy R, Bock R, Capell T, Christou P, Daniell H, Datta K, Datta S, Dix PJ, Fauquet C, Huang N, Kohli A, Mooibroek H, Nicholson L, Nguyen TT, Nugent G, Raemakers K, Romano A, Somers DA, Stoger E, Taylor N, Visser R (2005) Particle bombardment and the genetic enhancement of crops: myths and realities. Mol Breeding 15:305–327

Apel W, Bock R (2009) Enhancement of carotenoid biosynthesis in transplastomic tomatoes by induced lycopene-to-provitamin a conversion. Plant Physiol 151:59–66

Apel W, Schulze WX, Bock R (2010) Identification of protein stability determinants in chloro-plasts. Plant J 63:636–650

Arai Y, Shikanai T, Doi Y, Yoshida S, Yamaguchi I, Nakashita H (2004) Production of polyhy-droxybutyrate by polycistronic expression of bacterial genes in tobacco plastid. Plant Cell Physiol 45:1176–1184

Arlen PA, Falconer R, Cherukumilli S, Cole A, Cole AM, Oishi KK, Daniell H (2007) Field pro-duction and functional evaluation of chloroplast-derived interferon-α2b. Plant Biotechnol J 5:511–525

Bally J, Paget E, Droux M, Job C, Job D, Dubald M (2008) Both the stroma and thylakoid lumen of tobacco chloroplasts are competent for the formation of disulphide bonds in recombinant proteins. Plant Biotechnol J 6:46–61

Barkan A (2011) Expression of plastid genes: organelle-specific elaborations on a prokaryotic scaffold. Plant Physiol 155:1520–1532

Barone P, Zhang XH, Widholm JM (2009) Tobacco plastid transformation using the feedback-insensitive anthranilate synthase [alpha]-subunit of tobacco (ASA2) as a new selectable marker. J Exp Bot 60:3195–3202

Bendich AJ (2004) Circular chloroplast chromosomes: the grand illusion. Plant Cell 16:1661–1666

Bienvenut WV, Espagne C, Martinez A, Majeran W, Valot B, Zivy M, Vallon O, Adam Z, Meinnel T, Giglione C (2011) Dynamics of post-translational modifications and protein sta-bility in the stroma of Chlamydomonas reinhardtii chloroplasts. Proteomics 11:1734–1750

Birch-Machin I, Newell CA, Hibberd JM, Gray JC (2004) Accumulation of rotavirus VP6 protein in chloroplasts of transplastomic tobacco is limited by protein stability. Plant Biotechnol J 2:261–270

Blanco NE, Ceccoli RD, Segretin ME, Poli HO, Voss I, Melzer M, Bravo-Almonacid FF, Scheibe R, Hajirezaei MR, Carrillo N (2011) Cyanobacterial flavodoxin complements ferredoxin deficiency in knocked-down transgenic tobacco plants. Plant J 65:922–935

Bock R (2007) Structure, function, and inheritance of plastid genomes. In: Ralph B (ed) Cell and molecular biology of plastids. Topics in current genetics. Springer, Berlin, vol 19, pp 29–63

Bohmert-Tatarev K, McAvoy S, Daughtry S, Peoples OP, Snell KD (2011) High levels of bio-plastic are produced in fertile transplastomic tobacco plants engineered with a synthetic operon for the production of polyhydroxybutyrate. Plant Physiol 155:1690–1708

Boyhan D, Daniell H (2011) Low-cost production of proinsulin in tobacco and lettuce chloro-plasts for injectable or oral delivery of functional insulin and C-peptide. Plant Biotechnol J 9:585–598

Cardi T, Giegé P, Kahlau S, Scotti N (2012) Expression profiling of organellar genes. In: Bock R, Knoop V (eds) Genomics of chloroplasts and mitochondria, advances in photosynthesis and respiration. Springer, vol 35, pp 323–355

Cardi T, Lenzi P, Maliga P (2010) Chloroplasts as expression platforms for plant-produced vac-cines. Expert Rev Vaccines 9:893–911

Caroca R, Howell KA, Hasse C, Ruf S, Bock R (2013) Design of chimeric expression elements that confer high-level gene activity in chromoplasts. Plant J 73:368–379

Ceasar SA, Ignacimuthu S (2012) Genetic engineering of crop plants for fungal resistance: role of antifungal genes. Biotechnol Lett 34:995–1002

Ceccoli RD, Blanco NE, Segretin ME, Melzer M, Hanke GT, Scheibe R, Hajirezaei MR, Bravo-Almonacid FF, Carrillo N (2012) Flavodoxin displays dose-dependent effects on photosynthesis and stress tolerance when expressed in transgenic tobacco plants. Planta 236:1447–1458

Clarke J, Daniell H (2011) Plastid biotechnology for crop production: present status and future perspectives. Plant Mol Biol 76:211–220

Collinge DB, Jorgensen HJ, Lund OS, Lyngkjaer MF (2010) Engineering pathogen resistance in crop plants: current trends and future prospects. Annu Rev Phytopathol 48:269–291

Cominelli E, Tonelli C (2010) Transgenic crops coping with water scarcity. New Biotechnol 27:473–477

Craig W, Gargano D, Scotti N, Nguyen TT, Lao NT, Kavanagh TA, Dix PJ, Cardi T (2005) Direct gene transfer in potato: a comparison of particle bombardment of leaf explants and PEG-mediated transformation of protoplasts. Plant Cell Rep 24:603–611

Craig W, Lenzi P, Scotti N, De Palma M, Saggese P, Carbone V, Curran NM, Magee AM, Medgyesy P, Kavanagh TA, Dix PJ, Grillo S, Cardi T (2008) Transplastomic tobacco plants expressing a fatty acid desaturase gene exhibit altered fatty acid profiles and improved cold tolerance. Transgenic Res 17:769–782

Daniell H, Datta R, Varma S, Gray S, Lee SB (1998) Containment of herbicide resistance through genetic engineering of the chloroplast genome. Nat Biotechnol 16:345–348

Davarpanah S, Ahn J-W, Ko S, JUng S, Park Y-I, Liu J, Jeong W (2012) Stable expression of a fungal laccase protein using transplastomic tobacco. Plant Biotechnol Rep 6:305–312

Davoodi-Semiromi A, Schreiber M, Nalapalli S, Verma D, Singh ND, Banks RK, Chakrabarti D, Daniell H (2010) Chloroplast-derived vaccine antigens confer dual immunity against chol-era and malaria by oral or injectable delivery. Plant Biotechnol J 8:223–242

Day A, Goldschmidt-Clermont M (2011) The chloroplast transformation toolbox: selectable markers and marker removal. Plant Biotechnol J 9:540–553

Day A, Madesis P (2007) DNA replication, recombination, and repair in plastids. In: Bock R (ed) Cell and molecular biology of plastids, vol 19., Topics in current geneticsSpringer, Berlin, pp 65–119

De Cosa B, Moar W, Lee SB, Miller M, Daniell H (2001) Overexpression of the Bt cry2Aa2 operon in chloroplasts leads to formation of insecticidal crystals. Nat Biotechnol 19:71–74

De Marchis F, Pompa A, Bellucci M (2012) Plastid proteostasis and heterologous protein accu-mulation in transplastomic plants. Plant Physiol 160:571–581

De Marchis F, Wang Y, Stevanato P, Arcioni S, Bellucci M (2009) Genetic transformation of the sugar beet plastome. Transgenic Res 18:17–30

Del L Yácono M, Farran I, Becher ML, Sander V, Sanchez VR, Martin V, Veramendi J, Clemente M (2012) A chloroplast-derived Toxoplasma gondii GRA4 antigen used as an oral vaccine protects against toxoplasmosis in mice. Plant Biotechnol J 10:1136–1144

Dix PJ, Kavanagh TA (1995) Transforming the plastome: genetic markers and DNA delivery systems. Euphytica 85:29–34

Dufourmantel N, Tissot G, Goutorbe F, Garcon F, Muhr C, Jansens S, Pelissier B, Peltier G, Dubald M (2005) Generation and analysis of soybean plastid transformants expressing Bacillus thuringiensis Cry1Ab protoxin. Plant Mol Biol 58:659–668

Egelkrout E, Rajan V, Howard JA (2012) Overproduction of recombinant proteins in plants. Plant Sci 184:83–101

Elghabi Z, Karcher D, Zhou F, Ruf S, Bock R (2011) Optimization of the expression of the HIV fusion inhibitor cyanovirin-N from the tobacco plastid genome. Plant Biotechnol J 9:599–608

Farran I, McCarthy-Suarez I, Rio-Manterola F, Mansilla C, Lasarte JJ, Mingo-Castel AM (2010) The vaccine adjuvant extra domain a from fibronectin retains its proinflammatory properties when expressed in tobacco chloroplasts. Planta 231:977–990

Fernandez-San Millan A, Mingo-Castel A, Miller M, Daniell H (2003) A chloroplast transgenic approach to hyper-express and purify Human Serum Albumin, a protein highly susceptible to proteolytic degradation. Plant Biotechnol J 1:71–79

Gisby MF, Mellors P, Madesis P, Ellin M, Laverty H, O'Kane S, Ferguson MW, Day A (2011) A synthetic gene increases TGFβ3 accumulation by 75-fold in tobacco chloroplasts enabling rapid purification and folding into a biologically active molecule. Plant Biotechnol J 9:618–628

Gisby MF, Mudd EA, Day A (2012) Growth of transplastomic cells expressing d-amino acid oxidase in chloroplasts is tolerant to D-alanine and inhibited by D-valine. Plant Physiol 160:2219–2226

Glenz K, Bouchon B, Stehle T, Wallich R, Simon MM, Warzecha H (2006) Production of a recombinant bacterial lipoprotein in higher plant chloroplasts. Nat Biotechnol 24:76–77

Gorantala J, Grover S, Goel D, Rahi A, Jayadev Magani SK, Chandra S, Bhatnagar R (2011) A plant based protective antigen [PA(dIV)] vaccine expressed in chloroplasts demonstrates protective immunity in mice against anthrax. Vaccine 29:4521–4533

Gray BN, Yang H, Ahner BA, Hanson MR (2011) An efficient downstream box fusion allows high-level accumulation of active bacterial beta-glucosidase in tobacco chloroplasts. Plant Mol Biol 76:345–355

Hanson MR, Gray BN, Ahner BA (2013) Chloroplast transformation for engineering of photosynthesis. J Exp Bot 64:731–742

Hasunuma T, Kondo A, Miyake C (2009) Metabolic pathway engineering by plastid transformation is a powerful tool for production of compounds in higher plants. Plant Biotechnol 26:39–46

Hasunuma T, Kondo A, Miyake C (2010) Metabolic engineering by plastid transformation as a strategy to modulate isoprenoid yield in plants. In: Fett-Neto AG (ed) Plant secondary metabolism engineering. Methods in molecular biology. Humana Press, vol 643, pp 213–227

Hasunuma T, Miyazawa S, Yoshimura S, Shinzaki Y, Tomizawa K, Shindo K, Choi SK, Misawa N, Miyake C (2008) Biosynthesis of astaxanthin in tobacco leaves by transplastomic engineering. Plant J 55:857–868

Herz S, Füssl M, Steiger S, Koop HU (2005) Development of novel types of plastid transformation vectors and evaluation of factors controlling expression. Transgenic Res 14:969–982

Husaini AM, Rashid Z, Mir R, Aquil B (2011) Approaches for gene targeting and targeted gene expression in plants. GM Crops 2:150–162

Husken A, Prescher S, Schiemann J (2010) Evaluating biological containment strategies for pollen-mediated gene flow. Environ Biosafety Res 9:67–73

Hussein HS, Ruiz ON, Terry N, Daniell H (2007) Phytoremediation of mercury and organomercurials in chloroplast transgenic plants: enhanced root uptake, translocation to shoots, and volatilization. Environ Sci Technol 41:8439–8446

Inka Borchers AM, Gonzalez-Rabade N, Gray JC (2012) Increased accumulation and stability of rotavirus VP6 protein in tobacco chloroplasts following changes to the 5' untranslated region and the 5' end of the coding region. Plant Biotechnol J 10:422–434

Jansen R, Ruhlman T (2012) Plastid genomes of seed plants. In: Bock R, Knoop V (eds) Genomics of chloroplasts and mitochondria, Advances in photosynthesis and respiration. Springer, vol 35, pp 103–126

Jansen RK, Saski C, Lee SB, Hansen AK, Daniell H (2011) Complete plastid genome sequences of three Rosids (*Castanea, Prunus, Theobroma*): evidence for at least two independent transfers of rpl22 to the nucleus. Mol Biol Evol 28:835–847

Jin S, Kanagaraj A, Verma D, Lange T, Daniell H (2011) Release of hormones from conjugates: chloroplast expression of beta-glucosidase results in elevated phytohormone levels associated with significant increase in biomass and protection from aphids or whiteflies conferred by sucrose esters. Plant Physiol 155:222–235

Jin S, Zhang X, Daniell H (2012) *Pinellia ternata* agglutinin expression in chloroplasts confers broad spectrum resistance against aphid, whitefly, Lepidopteran insects, bacterial and viral pathogens. Plant Biotechnol J 10:313–327

Jones JD (2011) Why genetically modified crops? Philos Transact A Math Phys Eng Sci 369:1807–1816

Kahlau S, Bock R (2008) Plastid transcriptomics and translatomics of tomato fruit development and chloroplast-to-chromoplast differentiation: chromoplast gene expression largely serves the production of a single protein. Plant Cell 20:856–874

Kanagaraj AP, Verma D, Daniell H (2011) Expression of dengue-3 premembrane and envelope polyprotein in lettuce chloroplasts. Plant Mol Biol 76:323–333

Khan MS, Nurjis F (2012) Synthesis and expression of recombinant interferon alpha-5 gene in tobacco chloroplasts, a non-edible plant. Mol Biol Rep 39:4391–4400

Kim JY, Kavas M, Fouad WM, Nong G, Preston JF, Altpeter F (2011) Production of hyperthermostable GH10 xylanase Xyl10B from *Thermotoga maritima* in transplastomic plants enables complete hydrolysis of methylglucuronoxylan to fermentable sugars for biofuel production. Plant Mol Biol 76:357–369

Krech K, Ruf S, Masduki FF, Thiele W, Bednarczyk D, Albus CA, Tiller N, Hasse C, Schottler MA, Bock R (2012) The plastid genome-encoded Ycf4 protein functions as a nonessential assembly factor for photosystem I in higher plants. Plant Physiol 159:579–591

Krichevsky A, Meyers B, Vainstein A, Maliga P, Citovsky V (2010) Autoluminescent Plants. PLoS ONE 5:e15461

Kumar S, Dhingra A, Daniell H (2004) Plastid-expressed betaine aldehyde dehydrogenase gene in carrot cultured cells, roots, and leaves confers enhanced salt tolerance. Plant Physiol 136:2843–2854

Kumar S, Hahn FM, Baidoo E, Kahlon TS, Wood DF, McMahan CM, Cornish K, Keasling JD, Daniell H, Whalen MC (2012) Remodeling the isoprenoid pathway in tobacco by expressing the cytoplasmic mevalonate pathway in chloroplasts. Metab Eng 14:19–28

Kuroda H, Maliga P (2001) Complementarity of the 16S rRNA penultimate stem with sequences downstream of the AUG destabilizes the plastid mRNAs. Nucleic Acids Res 29:970–975

Kuroda H, Maliga P (2002) Overexpression of the clpP 5′-untranslated region in a chimeric context causes a mutant phenotype, suggesting competition for a clpP-specific RNA maturation factor in tobacco chloroplasts. Plant Physiol 129:1600–1606

Kwon KC, Nityanandam R, New JS, Daniell H (2013) Oral delivery of bioencapsulated exendin-4 expressed in chloroplasts lowers blood glucose level in mice and stimulates insulin secretion in beta-TC6 cells. Plant Biotechnol J 11:77–86

Le Martret B, Poage M, Shiel K, Nugent GD, Dix PJ (2011) Tobacco chloroplast transformants expressing genes encoding dehydroascorbate reductase, glutathione reductase, and glutathione-S-transferase, exhibit altered anti-oxidant metabolism and improved abiotic stress tolerance. Plant Biotechnol J 9:661–673

Lee SB, Kim Y, Lee J, Oh K-J, Byun M-O, Jeong M-J, Bae S-C (2012) Stable expression of the sweet protein monellin variant MNEI in tobacco chloroplasts. Plant Biotechnol Rep 6:285–295

Lee SB, Li B, Jin S, Daniell H (2011) Expression and characterization of antimicrobial peptides Retrocyclin-101 and Protegrin-1 in chloroplasts to control viral and bacterial infections. Plant Biotechnol J 9:100–115

Lentz EM, Garaicoechea L, Alfano EF, Parreño V, Wigdorovitz A, Bravo-Almonacid FF (2012) Translational fusion and redirection to thylakoid lumen as strategies to improve the accumulation of a camelid antibody fragment in transplastomic tobacco. Planta 236:703–714

Lentz EM, Mozgovoj MV, Bellido D, dus santos mj, Wigdorovitz A, Bravo-Almonacid FF (2011) VP8* antigen produced in tobacco transplastomic plants confers protection against bovine rotavirus infection in a suckling mouse model. J Biotechnol 156:100–107

Lentz EM, Segretin ME, Morgenfeld MM, Wirth SA, Dus Santos MJ, Mozgovoj MV, Wigdorovitz A, Bravo-Almonacid FF (2010) High expression level of a foot and mouth disease virus epitope in tobacco transplastomic plants. Planta 231:387–395

Lenzi P, Scotti N, Alagna F, Tornesello M, Pompa A, Vitale A, De Stradis A, Monti L, Grillo S, Buonaguro F, Maliga P, Cardi T (2008) Translational fusion of chloroplast-expressed human papillomavirus type 16 L1 capsid protein enhances antigen accumulation in transplastomic tobacco. Transgenic Res 17:1091–1102

Lerbs-Mache S (2011) Function of plastid sigma factors in higher plants: regulation of gene expression or just preservation of constitutive transcription? Plant Mol Biol 76:235–249

Li W, Ruf S, Bock R (2011) Chloramphenicol acetyltransferase as selectable marker for plastid transformation. Plant Mol Biol 76:443–451

Liere K, Börner T (2007) Transcription and transcriptional regulation in plastids. In: Ralph B (ed) Cell and molecular biology of plastids, vol 19., Topics in current geneticsSpringer, Berlin, pp 121–174

Liu CW, Lin CC, Yiu JC, Chen JJ, Tseng MJ (2008) Expression of a *Bacillus thuringiensis* toxin (cry1Ab) gene in cabbage (*Brassica oleracea* L. var. capitata L.) chloroplasts confers high insecticidal efficacy against *Plutella xylostella*. Theor Appl Genet 117:75–88

Lössl A, Bohmert K, Harloff H, Eibl C, Muhlbauer S, Koop HU (2005) Inducible trans-activation of plastid transgenes: expression of the *R. eutropha phb* operon in transplastomic tobacco. Plant Cell Physiol 46:1462–1471

Lössl A, Eibl C, Harloff HJ, Jung C, Koop HU (2003) Polyester synthesis in transplastomic tobacco (*Nicotiana tabacum* L.): significant contents of polyhydroxybutyrate are associated with growth reduction. Plant Cell Rep 21:891–899

Lössl AG, Waheed MT (2011) Chloroplast-derived vaccines against human diseases: achievements, challenges and scopes. Plant Biotechnol J 9:527–539

Lutz KA, Knapp JE, Maliga P (2001) Expression of bar in the plastid genome confers herbicide resistance. Plant Physiol 125:1585–1590

Madesis P, Osathanunkul M, Georgopoulou U, Gisby MF, Mudd EA, Nianiou I, Tsitoura P, Mavromara P, Tsaftaris A, Day A (2010) A hepatitis C virus core polypeptide expressed in chloroplasts detects anti-core antibodies in infected human sera. J Biotechnol 145:377–386

Madoka Y, Tomizawa K, Mizoi J, Nishida I, Nagano Y, Sasaki Y (2002) Chloroplast transformation with modified accD operon increases acetyl-CoA carboxylase and causes extension of leaf longevity and increase in seed yield in tobacco. Plant Cell Physiol 43:1518–1525

Magee AM, Aspinall S, Rice DW, Cusack BP, Sémon M, Perry AS, Stefanović S, Milbourne D, Barth S, Palmer JD, Gray JC, Kavanagh TA, Wolfe KH (2010) Localized hypermutation and associated gene losses in legume chloroplast genomes. Genome Res 20:1700–1710

Maier RM, Schmitz-Linneweber C (2004) Plastid genomes. In: Daniell H, Chase C (eds) Molecular biology and biotechnology of plant organelles. Springer, Dordrecht, pp 115–150

Majeran W, Friso G, Asakura Y, Qu X, Huang M, Ponnala L, Watkins KP, Barkan A, van Wijk KJ (2012) Nucleoid-enriched proteomes in developing plastids and chloroplasts from maize leaves: a new conceptual framework for nucleoid functions. Plant Physiol 158:156–189

Maliga P (2002) Engineering the plastid genome of higher plants. Curr Opin Plant Biol 5:164–172

Maliga P (2004) Plastid transformation in higher plants. Annu Rev Plant Biol 55:289–313

Maliga P (2012) Plastid transformation in flowering plants. In: Bock R, Knoop V (eds) Genomics of Chloroplasts and Mitochondria. Advances in photosynthesis and respiration. Springer, vol 35, pp 393–414

Maliga P, Bock R (2011) Plastid biotechnology: food, fuel, and medicine for the 21st century. Plant Physiol 155:1501–1510

Malik Ghulam M, Zghidi-Abouzid O, Lambert E, Lerbs-Mache S, Merendino L (2012) Transcriptional organization of the large and the small ATP synthase operons, atpI/H/F/A and atpB/E, in Arabidopsis thaliana chloroplasts. Plant Mol Biol 79:259–272

Manimaran P, Ramkumar G, Sakthivel K, Sundaram RM, Madhav MS, Balachandran SM (2011) Suitability of non-lethal marker and marker-free systems for development of transgenic crop plants: present status and future prospects. Biotechnol Adv 29:703–714

Manuell AL, Quispe J, Mayfield SP (2007) Structure of the chloroplast ribosome: novel domains for translation regulation. PLoS Biol 5:e209

Marchfelder A, Binder S (2004) Plastid and plant mitochondrial RNA processing and RNA stability. In: Daniell H, Chase C (eds) Molecular biology and biotechnology of plant organelles. Springer, Dordrecht, pp 261–294

Marín-Navarro J, Manuell AL, Wu J, Mayfield PS (2007) Chloroplast translation regulation. Photosynth Res 94:359–374

McBride KE, Svab Z, Schaaf DJ, Hogan PS, Stalker DM, Maliga P (1995) Amplification of a chimeric Bacillus gene in chloroplasts leads to an extraordinary level of an insecticidal protein in tobacco. Biotechnology (N Y) 13:362–365

Meyers B, Zaltsman A, Lacroix B, Kozlovsky SV, Krichevsky A (2010) Nuclear and plastid genetic engineering of plants: comparison of opportunities and challenges. Biotechnol Adv 28:747–756

Michoux F, Ahmad N, Hennig A, Nixon PJ, Warzecha H (2013) Production of leafy biomass using temporary immersion bioreactors: an alternative platform to express proteins in transplastomic plants with drastic phenotypes. Planta 237:903–908

Michoux F, Ahmad N, McCarthy J, Nixon PJ (2011) Contained and high-level production of recombinant protein in plant chloroplasts using a temporary immersion bioreactor. Plant Biotechnol J 9:575–584

Nakashita H, Arai Y, Shikanai T, Doi Y, Yamaguchi I (2001) Introduction of bacterial metabolism into higher plants by polycistronic transgene expression. Biosci Biotechnol Biochem 65:1688–1691

Naqvi S, Farré G, Sanahuja G, Capell T, Zhu C, Christou P (2010) When more is better: multigene engineering in plants. Trends Plant Sci 15:48–56

Oey M, Lohse M, Kreikemeyer B, Bock R (2009a) Exhaustion of the chloroplast protein synthesis capacity by massive expression of a highly stable protein antibiotic. Plant J 57:436–445

Oey M, Lohse M, Scharff LB, Kreikemeyer B, Bock R (2009b) Plastid production of protein antibiotics against pneumonia via a new strategy for high-level expression of antimicrobial proteins. Proc Natl Acad Sci USA 106:6579–6584

Ortigosa SM, Diaz-Vivancos P, Clemente-Moreno MJ, Pinto-Marijuan M, Fleck I, Veramendi J, Santos M, Hernandez JA, Torne JM (2010a) Oxidative stress induced in tobacco leaves by chloroplast over-expression of maize plastidial transglutaminase. Planta 232:593–605

Ortigosa SM, Fernández-San Millán A, Veramendi J (2010b) Stable production of peptide antigens in transgenic tobacco chloroplasts by fusion to the p53 tetramerisation domain. Transgenic Res 19:703–709

Orzaez D, Monforte AJ, Granell A (2010) Using genetic variability available in the breeder's pool to engineer fruit quality. GM Crops 1:120–127

Peled-Zehavi H, Danon A (2007) Translation and translational regulation in chloroplasts. In: Bock R (ed) Cell and molecular biology of plastids, vol 19., Topics in current genetics-sSpringer, Berlin, pp 249–281

Petersen K, Bock R (2011) High-level expression of a suite of thermostable cell wall-degrading enzymes from the chloroplast genome. Plant Mol Biol 76:311–321

Pfalz J, Bayraktar OA, Prikryl J, Barkan A (2009) Site-specific binding of a PPR protein defines and stabilizes 5′ and 3′ mRNA termini in chloroplasts. EMBO J 28:2042–2052

Poage M, Le Martret B, Jansen MA, Nugent GD, Dix PJ (2011) Modification of reactive oxygen species scavenging capacity of chloroplasts through plastid transformation. Plant Mol Biol 76:371–384

Prikryl J, Rojas M, Schuster G, Barkan A (2011) Mechanism of RNA stabilization and translational activation by a pentatricopeptide repeat protein. Proc Natl Acad Sci USA 108:415–420

Que Q, Chilton MD, de Fontes CM, He C, Nuccio M, Zhu T, Wu Y, Chen JS, Shi L (2010) Trait stacking in transgenic crops: challenges and opportunities. GM Crops 1:220–229

Quesada-Vargas T, Ruiz ON, Daniell H (2005) Characterization of heterologous multigene operons in transgenic chloroplasts: transcription, processing, and translation. Plant Physiol 138:1746–1762

Reguera M, Peleg Z, Blumwald E (2012) Targeting metabolic pathways for genetic engineering abiotic stress-tolerance in crops. Biochim Biophys Acta 1819:186–194

Rigano MM, Manna C, Giulini A, Pedrazzini E, Capobianchi M, Castilletti C, Di Caro A, Ippolito G, Beggio P, De Giuli Morghen C, Monti L, Vitale A, Cardi T (2009) Transgenic chloroplasts are efficient sites for high-yield production of the vaccinia virus envelope protein A27L in plant cells dagger. Plant Biotechnol J 7:577–591

Rigano MM, Scotti N, Cardi T (2012) Unsolved problems in plastid transformation. Bioengineered 3:329–333

Rogalski M, Carrer H (2011) Engineering plastid fatty acid biosynthesis to improve food quality and biofuel production in higher plants. Plant Biotechnol J 9:554–564

Rojas CA, Hemerly AS, Ferreira PC (2010) Genetically modified crops for biomass increase. Genes and strategies. GM Crops 1:137–142

Rosellini D (2012) Selectable markers and reporter genes: a well furnished toolbox for plant science and genetic engineering. Crit Rev Plant Sci 31:401–453

Rott M, Martins NF, Thiele W, Lein W, Bock R, Kramer DM, Schottler MA (2011) ATP synthase repression in tobacco restricts photosynthetic electron transport, CO2 assimilation, and plant growth by overacidification of the thylakoid lumen. Plant Cell 23:304–321

Rubio-Infante N, Govea-Alonso DO, Alpuche-Solis AG, Garcia-Hernandez AL, Soria-Guerra RE, Paz-Maldonado LM, Ilhuicatzi-Alvarado D, Varona-Santos JT, Verdin-Teran L, Korban SS, Moreno-Fierros L, Rosales-Mendoza S (2012) A chloroplast-derived C4V3 polypeptide from the human immunodeficiency virus (HIV) is orally immunogenic in mice. Plant Mol Biol 78:337–349

Ruhlman T, Verma D, Samson N, Daniell H (2010) The role of heterologous chloroplast sequence elements in transgene integration and expression. Plant Physiol 152:2088–2104

Ruiz ON, Alvarez D, Torres C, Roman L, Daniell H (2011) Metallothionein expression in chloroplasts enhances mercury accumulation and phytoremediation capability. Plant Biotechnol J 9:609–617

Ruiz ON, Daniell H (2005) Engineering cytoplasmic male sterility via the chloroplast genome by expression of {beta}-ketothiolase. Plant Physiol 138:1232–1246

Ruiz ON, Hussein HS, Terry N, Daniell H (2003) Phytoremediation of organomercurial compounds via chloroplast genetic engineering. Plant Physiol 132:1344–1352

Sanz-Barrio R, Millan AF, Corral-Martinez P, Segui-Simarro JM, Farran I (2011) Tobacco plastidial thioredoxins as modulators of recombinant protein production in transgenic chloroplasts. Plant Biotechnol J 9:639–650

Saski C, Lee SB, Fjellheim S, Guda C, Jansen RK, Luo H, Tomkins J, Rognli OA, Daniell H, Clarke JL (2007) Complete chloroplast genome sequences of Hordeum vulgare, Sorghum bicolor and Agrostis stolonifera, and comparative analyses with other grass genomes. Theor Appl Genet 115:571–590

Scharff LB, Childs L, Walther D, Bock R (2011) Local absence of secondary structure permits translation of mRNAs that lack ribosome-binding sites. PLoS Genet 7:e1002155

Schmitz-Linneweber C, Small I (2008) Pentatricopeptide repeat proteins: a socket set for organelle gene expression. Trends Plant Sci 13:663–670

Schnable PS, Wise RP (1998) The molecular basis of cytoplasmic male sterility and fertility restoration. Trends Plant Sci 3:175–180

Schuster G, Stern D (2009) RNA polyadenylation and decay in mitochondria and chloroplasts. In: Condon C (ed) Progress in molecular biology and translational science. Molecular biology of RNA processing and decay in prokaryotes. Elsevier Inc., vol 85, pp 393–422

Scotti N, Alagna F, Ferraiolo E, Formisano G, Sannino L, Buonaguro L, De Stradis A, Vitale A, Monti L, Grillo S, Buonaguro FM, Cardi T (2009) High-level expression of the HIV-1 Pr55(gag) polyprotein in transgenic tobacco chloroplasts. Planta 229:1109–1122

Scotti N, Gargano D, Lenzi P, Cardi T (2011) Transformation of the plastid genome in higher plants. In: Dan Y, Ow DW (eds) Historical technology developments in plant transformation. Bentham Science Publishers Ltd., pp 123–145

Scotti N, Rigano MM, Cardi T (2012) Production of foreign proteins using plastid transformation. Biotechnol Adv 30:387–397

Segretin ME, Lentz EM, Wirth SA, Morgenfeld MM, Bravo-Almonacid FF (2012) Transformation of Solanum tuberosum plastids allows high expression levels of beta-glucuronidase both in leaves and microtubers developed in vitro. Planta 235:807–818

Shanmugabalaji V, Besagni C, Piller LE, Douet V, Ruf S, Bock R, Kessler F (2013) Dual targeting of a mature plastoglobulin/fibrillin fusion protein to chloroplast plastoglobules and thylakoids in transplastomic tobacco plants. Plant Mol Biol 81:13–25

Shaver JM, Oldenburg DJ, Bendich AJ (2008) The structure of chloroplast DNA molecules and the effects of light on the amount of chloroplast DNA during development in medicago truncatula. Plant Physiol 146:1064–1074

Shen H, Qian B, Chen W, Liu Z, Yang L, Zhang D, Liang W (2010) Immunogenicity of recombinant F4 (K88) fimbrial adhesin FaeG expressed in tobacco chloroplast. Acta Biochim Biophys Sin (Shanghai) 42:558–567

Shimizu M, Goto M, Hanai M, Shimizu T, Izawa N, Kanamoto H, Tomizawa K, Yokota A, Kobayashi H (2008) Selectable tolerance to herbicides by mutated acetolactate synthase genes integrated into the chloroplast genome of tobacco. Plant Physiol 147:1976–1983

Singer SD, Liu Z, Cox KD (2012) Minimizing the unpredictability of transgene expression in plants: the role of genetic insulators. Plant Cell Rep 31:13–25

Soria-Guerra RE, Alpuche-Solis AG, Rosales-Mendoza S, Moreno-Fierros L, Bendik EM, Martinez-Gonzalez L, Korban SS (2009) Expression of a multi-epitope DPT fusion protein in transplastomic tobacco plants retains both antigenicity and immunogenicity of all three components of the functional oligomer. Planta 229:1293–1302

Stern DB, Goldschmidt-Clermont M, Hanson MR (2010) Chloroplast RNA metabolism. Annu Rev Plant Biol 61:125–155

Svab Z, Hajdukiewicz P, Maliga P (1990) Stable transformation of plastids in higher plants. Proc Natl Acad Sci USA 87:8526–8530

Tangphatsornruang S, Birch-Machin I, Newell CA, Gray JC (2011) The effect of different 3′ untranslated regions on the accumulation and stability of transcripts of a gfp transgene in chloroplasts of transplastomic tobacco. Plant Mol Biol 76:385–396

Thyssen G, Svab Z, Maliga P (2012) Exceptional inheritance of plastids via pollen in Nicotiana sylvestris with no detectable paternal mitochondrial DNA in the progeny. Plant J 72:84–88

Tiller N, Weingartner M, Thiele W, Maximova E, Schöttler MA, Bock R (2012) The plastid-specific ribosomal proteins of Arabidopsis thaliana can be divided into non-essential proteins and genuine ribosomal proteins. Plant J 69:302–316

Tillich M, Hardel SL, Kupsch C, Armbruster U, Delannoy E, Gualberto JM, Lehwark P, Leister D, Small ID, Schmitz-Linneweber C (2009) Chloroplast ribonucleoprotein CP31A is required for editing and stability of specific chloroplast mRNAs. Proc Natl Acad Sci USA 106:6002–6007

Tillich M, Krause K (2010) The ins and outs of editing and splicing of plastid RNAs: lessons from parasitic plants. New Biotechnol 27:256–266

Tungsuchat-Huang T, Slivinski KM, Sinagawa-Garcia SR, Maliga P (2011) Visual spectinomycin resistance (aadA(au)) gene for facile identification of transplastomic sectors in tobacco leaves. Plant Mol Biol 76:453–461

Valkov VT, Gargano D, Manna C, Formisano G, Dix PJ, Gray JC, Scotti N, Cardi T (2011) High efficiency plastid transformation in potato and regulation of transgene expression in leaves and tubers by alternative 5' and 3' regulatory sequences. Transgenic Res 20:137–151

Valkov VT, Scotti N, Kahlau S, MacLean D, Grillo S, Gray JC, Bock R, Cardi T (2009) Genome-wide analysis of plastid gene expression in potato leaf chloroplasts and tuber amyloplasts: transcriptional and posttranscriptional control. Plant Physiol 150:2030–2044

van Wijk KJ, Baginsky S (2011) Plastid proteomics in higher plants: current state and future goals. Plant Physiol 155:1578–1588

Verhounig A, Karcher D, Bock R (2010) Inducible gene expression from the plastid genome by a synthetic riboswitch. Proc Natl Acad Sci USA 107:6204–6209

Verma D, Kanagaraj A, Jin S, Singh ND, Kolattukudy PE, Daniell H (2010a) Chloroplast-derived enzyme cocktails hydrolyse lignocellulosic biomass and release fermentable sugars. Plant Biotechnol J 8:332–350

Verma D, Moghimi B, LoDuca PA, Singh HD, Hoffman BE, Herzog RW, Daniell H (2010b) Oral delivery of bioencapsulated coagulation factor IX prevents inhibitor formation and fatal anaphylaxis in hemophilia B mice. Proc Natl Acad Sci USA 107:7101–7106

Waheed MT, Thönes N, Müller M, Hassan SW, Gottschamel J, Lössl E, Kaul HP, Lössl AG (2011a) Plastid expression of a double-pentameric vaccine candidate containing human papillomavirus-16 L1 antigen fused with LTB as adjuvant: transplastomic plants show pleiotropic phenotypes. Plant Biotechnol J 9:651–660

Waheed MT, Thönes N, Müller M, Hassan SW, Razavi NM, Lössl E, Kaul HP, Lössl AG (2011b) Transplastomic expression of a modified human papillomavirus L1 protein leading to the assembly of capsomeres in tobacco: a step towards cost-effective second-generation vaccines. Transgenic Res 20:271–282

Walter M, Piepenburg K, Schöttler MA, Petersen K, Kahlau S, Tiller N, Drechsel O, Weingartner M, Kudla J, Bock R (2010) Knockout of the plastid RNase E leads to defective RNA processing and chloroplast ribosome deficiency. Plant J 64:851–863

Wang Y, Yau YY, Perkins-Balding D, Thomson JG (2011) Recombinase technology: applications and possibilities. Plant Cell Rep 30:267–285

Webster DE, Thomas MC (2012) Post-translational modification of plant-made foreign proteins: glycosylation and beyond. Biotechnol Adv 30:410–418

Whitney SM, Kane HJ, Houtz RL, Sharwood RE (2009) Rubisco oligomers composed of linked small and large subunits assemble in tobacco plastids and have higher affinities for CO_2 and O_2. Plant Physiol 149:1887–1895

Wicke S, Schneeweiss G, de Pamphilis C, Müller K, Quandt D (2011) The evolution of the plastid chromosome in land plants: gene content, gene order, gene function. Plant Mol Biol 76:273–297

Wurbs D, Ruf S, Bock R (2007) Contained metabolic engineering in tomatoes by expression of carotenoid biosynthesis genes from the plastid genome. Plant J 49:276–288

Yabuta Y, Tanaka H, Yoshimura S, Suzuki A, Tamoi M, Maruta T, Shigeoka S (2013) Improvement of vitamin E quality and quantity in tobacco and lettuce by chloroplast genetic engineering. Transgenic Res 22:391–402

Yang H, Gray BN, Ahner BA, Hanson MR (2013) Bacteriophage 5′ untranslated regions for control of plastid transgene expression. Planta 237:517–527

Ye GN, Hajdukiewicz PT, Broyles D, Rodriguez D, Xu CW, Nehra N, Staub JM (2001) Plastid-expressed 5-enolpyruvylshikimate-3-phosphate synthase genes provide high level glyphosate tolerance in tobacco. Plant J 25:261–270

Youm JW, Jeon JH, Kim H, Min SR, Kim MS, Joung H, Jeong WJ, Kim HS (2010) High-level expression of a human β-site APP cleaving enzyme in transgenic tobacco chloroplasts and its immunogenicity in mice. Transgenic Res 19:1099–1108

Zerges W (2004) Regulation of translation in chloroplasts. In: Daniell H, Chase C (eds) Molecular biology of plant organelles. Springer, Dordrecht, pp 443–490

Zhang J, Ruf S, Hasse C, Childs L, Scharff LB, Bock R (2012) Identification of cis-elements conferring high levels of gene expression in non-green plastids. Plant J 72:115–128

Zhang J, Tan W, Yang XH, Zhang HX (2008) Plastid-expressed choline monooxygenase gene improves salt and drought tolerance through accumulation of glycine betaine in tobacco. Plant Cell Rep 27:1113–1124

Zhang XH, Webb J, Huang YH, Lin L, Tang RS, Liu A (2011) Hybrid Rubisco of tomato large subunits and tobacco small subunits is functional in tobacco plants. Plant Sci 180:480–488

Zhou F, Karcher D, Bock R (2007) Identification of a plastid intercistronic expression element (IEE) facilitating the expression of stable translatable monocistronic mRNAs from operons. Plant J 52:961–972

Zoschke R, Kroeger T, Belcher S, Schöttler MA, Barkan A, Schmitz-Linneweber C (2012) The pentatricopeptide repeat-SMR protein ATP4 promotes translation of the chloroplast atpB/E mRNA. Plant J 72:547–558

Index